Navigating the Maze

Navigating the Maze
How Science and Technology Policies Shape America and the World

Michael S. Lubell
Mark W. Zemansky Professor of Physics, City College of the City University
of New York (CCNY), New York, NY, United States

ACADEMIC PRESS
An imprint of Elsevier

Academic Press is an imprint of Elsevier
125 London Wall, London EC2Y 5AS, United Kingdom
525 B Street, Suite 1650, San Diego, CA 92101, United States
50 Hampshire Street, 5th Floor, Cambridge, MA 02139, United States
The Boulevard, Langford Lane, Kidlington, Oxford OX5 1GB, United Kingdom

Notices
Knowledge and best practice in this field are constantly changing. As new research and experience
broaden our understanding, changes in research methods, professional practices, or medical
treatment may become necessary.

Practitioners and researchers must always rely on their own experience and knowledge in evaluating
and using any information, methods, compounds, or experiments described herein. In using such
information or methods they should be mindful of their own safety and the safety of others, including
parties for whom they have a professional responsibility.

To the fullest extent of the law, neither the Publisher nor the authors, contributors, or editors, assume
any liability for any injury and/or damage to persons or property as a matter of products liability,
negligence or otherwise, or from any use or operation of any methods, products, instructions, or
ideas contained in the material herein.

Library of Congress Cataloging-in-Publication Data
A catalog record for this book is available from the Library of Congress

British Library Cataloguing-in-Publication Data
A catalogue record for this book is available from the British Library

ISBN 978-0-12-814710-8

For information on all Academic Press publications
visit our website at https://www.elsevier.com/books-and-journals

Working together
to grow libraries in
developing countries

www.elsevier.com • www.bookaid.org

Publisher: Andre Gerhard Wolff
Acquisition Editor: Mary Preap
Editorial Project Manager: Mary Preap
Production Project Manager: Kiruthika Govindaraju
Cover Designer: Christian J Bilbow

Typeset by SPi Global, India

Dedication

**To the five people who have made my life
worth living:**

**My spouse, Laura
My daughter, Karina, and her spouse, Romain
And my grandsons, Tristan and Axel**

About the Author

Michael S. Lubell is the Mark W. Zemansky Professor of Physics at the City College of the City University of New York (CCNY) and Chair of the Aspen Institute Program on Science and Society. He has spent much of his career carrying out research in high-energy nuclear and atomic physics, as well as quantum optics and quantum chaos, and is an elected fellow of the American Association for the Advancement of Science and the American Physical Society. He is well known in public policy circles for his groundbreaking work in Washington, DC, where he served as director of public affairs of the American Physical Society for more than two decades. He has published more than 250 articles and abstracts in scientific journals and books, and has been a newspaper columnist and opinion contributor for many years. He has been active in local, state, and national politics for half a century, and has lectured widely in the United States and Europe. *Navigating the Maze* is his first full-length book.

Preface

Science, and the technologies it has spawned, have been the principal drivers of the American economy since the end of post World War II. Today, economists estimate that a whopping 85% of the U.S. gross domestic product (GDP) growth traces its origin to science and technology. The size of the impact should not be a surprise, considering the ubiquity of modern technologies.

Innovation has brought us the consumer products we take for granted: smart phones and tablets, CD and DVD players, cars that are loaded with electronics and GPS navigating tools and that rarely break down, search engines such as Google and Yahoo, the Internet, LED lights, microwave ovens, and much more. Technology has also made our military stronger and kept our nation safer. It has made food more affordable and plentiful. It has provided medical diagnostics such as MRIs, CT scanners, and genomic tests; treatments for disease, such as antibiotics, chemotherapy, immunotherapy and radiation; minimally-invasive procedures, such as laparoscopy, coronary stent insertion, and video-assisted thoracoscopy; and artificial joint and heart valve replacements.

None of those technological developments were birthed miraculously. They owe a significant part of their realization to public and private strategies and public and private investments. Collectively, the strategies and investments form the kernel of science and technology policy.

Navigating the Maze: How Science and Technology Policies Shape America and the World serves as a guide for students, scientists, engineers, technologists, entrepreneurs, business leaders, and members of the general public who want to understand the fundamentals of science and technology policymaking and how to use them to achieve beneficial outcomes. Larely using a historical narrative, the book provides unique insights into the integral nature of science, science policy, and politics, and the complex terrain on which they co-exist.

The Preamble and Introduction set the stage for the historical narrrative that follows. Part One provides a comprehensive look at the development of science and technology from 1787 to 1992, illustrating the principles of a subject that dates to the very birth of America. Part Two, which carries the narrative from 1992 until the present, focuses on concerns that are central to almost every facet of modern American life. Framed by the opportunities and challenges of globalization and the rapid pace of technological change, Part Two examines the tensions in present-day policymaking.

Acknowledgments

Two people planted the seeds for *Navigating the Maze*: Brad Fenwick of Elsevier and Christopher B. Davis of World Scientific. I owe them a debt of gratitude for suggesting the project. They provided the inspiration for the book, but there are many people who deserve recognition, two of them in particular who are no longer alive: D. Allan Bromley and Burton Richter.

Allan introduced me to the world of science and technology policy when I was a young faculty member at Yale University, and he was chairman of the Physics Department. Allan subsequently became President George H.W. Bush's science advisor, and I became director of public affairs for the American Physical Society. We became close friends, and for more than two decades, collaborated on science policy and advocacy projects in Washington until his death in 2005.

I first met Burt in the 1970s when I was a young scientist carrying out high-energy physics research at the Stanford Linear Accelerator Center, now known as SLAC National Laboratory. Burt, a Physics Nobel Laureate, was director of SLAC from 1984 to 1999, and during that time he became a fixture in Washington science and technology policy circles. I benefited greatly from his knowledge, guidance, and insights during the many years we worked together. He was eagerly looking forward to reading this book, but unfortunately he passed away in 2018, shortly before the manuscript was complete.

Two other people deserve special mention—for enticing me to Washington and supporting my efforts. Harry Lustig, who had been provost at the City College of CUNY (CCNY) and had helped recruit me there from Yale, and became the treasurer and acting executive officer of the American Physical Society (APS) after he retired from academia. In the spring of 1994, he asked me to establish a science advocacy network in Washington, close to APS's headquarters in suburban Maryland. Judy Franz, who assumed the executive officer's post later that year, became a staunch supporter of my work, and for that I remain eternally grateful.

During my 22 years in Washington, I benefited from many thoughtful discussions and collaborations with myriad scientists and policy wonks inside and outside government. A number deserve special mention: Norman R. Augustine, former under-secretary of the Army, retired CEO of Lockheed-Martin, and the author of the book's Foreword; Neal F. Lane, director of the National Science Foundation (NSF) from 1993 to 1998, and President Clinton's science advisor from 1998 to 2001; Jack Gibbons, Lane's predecessor, and previously director of the congressional Office of Technology Assessment; William E. Curry, who advised Bill Clinton on domestic policy in 1996 and 1997; Ernest Moniz, associate director for science at the Office of Science and Technology Policy (OSTP) from 1995 to 1997, and Secretary of Energy from 2013 to 2017; Arthur Bienenstock and Duncan Moore, who both served as OSTP associate directors during the Clinton Administration from 1997 to 2000, Artie for science and Duncan for technology; John H. Marburger, President George W. Bush's science advisor, with whom I had become acquainted during his years at the University of Southern California, SUNY Stony Brook and Brookhaven National Laboratory; William A. Jeffrey, director of the National Institute for Standards and Technology from 2005 to 2007; John Holdren, President Obama's science advisor and OSTP director from 2009 until 2016; Philip Rubin, Principal Assistant Director for Science at OSTP from 2012 until 2015; Thomas Kalil, who served in the National Economic Council and OSTP in the Clinton and Obama Administrations; and Jeff Smith, who filled multiple roles at NSF and OSTP, and has been a wonderful, constant friend and confidant for more than two decades.

I also benefited from many fruitful discussions with members of Congress across the political spectrum, a small number with advanced science degrees— Representatives Vernon J. Ehlers, Bill Foster, and Rush Holt, Jr., all physicists; Jerry McNerney, a mathematician; and John Olver, a chemist—and others without such professional credentials, but with a passion for science and technology, and a commitment to developing sound public policy. They include Senators Jeff Bingaman, Christopher J. Dodd, Pete V. Domenici, Bill Frist, Phil Gramm, Joseph I. Lieberman, and John D. "Jay" Rockefeller, IV, and Representatives Sherwood "Sherry" Boehlert, Judy Biggert, George E. Brown, Jr., Rosa DeLauro, Anna Eshoo, Bart Gordon, Randy Hultgren, Nita Lowey, John Porter, Louise Slaughter, and Paul Tonko. Congressional staffers who provided special insight into the workings of the legislative machinery, along with sage policy advice, over many years include William Bonvillian, Dixon Butler, Kevin Carroll, Mike Champness, Leeland Cogliani, Kevin Cook, Tom Culligan, Paul Ducette, Tony Fainberg, Elizabeth Grossman, Louis Finkel, Alex Flint, David Goldston, G. William Hoagland, Olwen Huxley, Julia Jester, Chris King, Bill McBride, Bob Palmer, Elizabeth Prostic, Peter Rooney, Adam Rosenberg, Dahlia Sokolov, Robert Simon, Jim Turner, and Harlan Watson.

In addition to Capitol Hill denizens, a number of other people merit recognition: Mitch Ambrose, Kate Bannan, Stefano Bertuzzi, Robert Birgeneau, Robert S. Boege, Marc H. Brodsky, Joanne Padron Carney, Chris Carter, Paula Collins, Douglas Comer, John C. Crowley, James Dawson, Joseph Dehmer, Patricia Dehmer, H. Frederick Dylla, Robert Eisenstein, Ray Garant, Howard Garrison, Carolyn Trupp Gil, Irwin Goodwin, William Halperin, Mike Henry, Matt Hourihan, Richard Jones, Ronald L. Kelley, Franmarie Kennedy, Kathleen "Taffy" Kingscott, Kei Koizumi, Seema Kumar, Abe Lackman, Donald Q. Lamb,

Audrey Leath, Alan J. Leshner, Kevin Marvel, William Novelli, John Palafoutas, Anthony Pignataro, Jennifer Poulakidas, Samuel Rankin, Brian Raymond, Liz Rogan, Robert Rosner, Deborah Rudolph, Glenn Ruskin, Bill Squadron, Tobin L. Smith, Albert H. Teich, Michael L. Telson, Barry Toiv, Michael Turner, Christopher Volpe, Michael S. Witherell, and Mary Woolley. I profited greatly from their insights during my two decades in Washington.

Throughout my tenure as director of public affairs of the American Physical Society, I received considerable assistance from APS staff, among them, Mark Elsesser, Don Engel, Amy Flatten, Allen Hu, Tawanda Johnson, Irving Lerch, Jodi Lieberman, Greg Mack, Brian Mosley, Steve Pierson, Jeanette Russo, Fred Schlachter, and Francis Slakey. I owe considerable thanks to several members of the APS leadership who provided guidance and support for my activities in the policy arena. They include the editors of *APS News*, Alan Chodos, and later, David Voss, who published more than one hundred of my "Inside the Beltway" columns; Martin Blume, the editor-in-chief of APS journals; and treasurer-publishers Thomas McIlrath, Joseph Serene, and Matthew Salter.

The editors of *The Hill*, *Roll Call*, the Brooks Community Newspapers (now Hearst Media), and the *Westport Minuteman* gave me the continuing opportunity to express my opinions. Walter Isaacson gave me the sage advice to tell stories rather than present facts. His prescription is largely responsible for the narrative format of the book. I thank all of them. I also thank Norm Augustine, Gerald Blazey, Neal Lane, Philip Rubin, and Jeff Smith for their careful reading of my manuscript. They caught errors, large and small, and provided invaluable commentary.

Additionally, I want to acknowledge the support of the City College of CUNY, which paid the tab while I was working on the book, and my editor at Elsevier, Mary Preap, who helped me understand the changing world of publishing and kept me focused on the project.

Finally, I want to thank my wife, Laura Appelman, who put up with my absence from daily life during the ten months it took me to write *Navigating the Maze*. Her forbearance and unwavering support were invaluable and unforgettable. I am forever in her debt.

Contents

Foreword

My mother, who was born in Colorado and lived to the age of 105, was 10 years old when the Wright brothers first flew at Kitty Hawk, a flight that could have taken place inside the Space Shuttle's fuel tank were it placed on its side. In her lifetime, she met friends of mine who had walked on the moon. She was 3 years old when Henry Ford built his first automobile. Life expectancy at birth in America was then 47 years, and telephone service had just opened between New York and Chicago. There were no X-ray machines because Röntgen had not yet discovered X-rays, and it would be 2 more years before the first radio signal was sent. More than 20 million horses provided much of America's transportation and motive power.

Of course, there were no MRIs, televisions, GPS, iPods, artificial joints, digital computers, laser eye surgery, stents, or even knowledge of such things as the Higgs particle.

Each new scientific and technological advance seemed to find its way into America's homes more rapidly than its predecessor, taking, for example, about 50 years for half the nation's homes to have electricity, 15 years for television, and 12 years for the smart phone. But such advances, fundamentally made possible by achievements in science, have created major dilemmas for the nation's policy-making enterprise—not to mention glacial legal system—as they try to keep pace with the societal issues posed.

Consider the implications for what some would characterize as a rather mundane matter: patent policy, with the first patent having been signed by George Washington for a process used in the production of fertilizer. Today, one of the driving factors in the high cost of biopharmaceuticals—some of which are indispensable to people's lives—is that years ago Congress provided the creators of certain new products protection from competition in the marketplace for a specified period of time. Critics often refer to this practice as a "legalized monopoly," in which consumer prices are generally set by whatever the market will bear. On the other hand, what motivation might a pharmaceutical company have to fund research and to develop new products if the results of its work can immediately, and legally, be copied by competitors who furthermore, did not have to bear the cost of discovery?

Policy debates over science and technology matter. Many of the most fundamental challenges facing our nation today are likely to find much of their solution in achievements in the fields of science and technology. This is the case for such matters as producing energy that does not harm the planet's environment, providing health care, rebuilding infrastructure, providing national security, and sustaining a globally competitive economy that will underpin a quality standard of living.

But establishing policies governing matters of science and technology has often proven to be agonizingly complex, in large part because scientific breakthroughs and their technological applications can be accompanied by both negative and positive consequences.

For example, how does one balance the benefits of research on clean commercial nuclear energy with concerns over proliferation of nuclear weaponry? How does one balance the enormous promise of medical developments derived from ever-expanding knowledge of the human genome with the difficult ethical issues surrounding such eventualities as designer babies? How does one address the impact of advancements in robotics that free humans from many onerous aspects of physical labor in comparison with the destructive effects these same devices can have on people's jobs? Or the substantial benefits of artificial intelligence wherein machines are increasingly able to outperform humans in some tasks, but can also destroy the jobs of people whose livelihood depends upon mental prowess? Or GPS, which is treasured by much of the public, yet can also be used by terrorists or to invade one's privacy? Or hydraulic fracking, which is freeing America of its long dependence on foreign nations to supply much of its energy, yet is outlawed in some states because of deeply held environmental concerns?

Looking to the future, issues accompanying scientific and technological breakthroughs are likely to become still more complicated. For example, widespread use of all-electric vehicles is not far down the road, based on advances in energy storage and related fields. Such research offers great promise for a much cleaner environment, but the implications for those whose jobs involve extracting, processing, and distributing traditional fuels are ominous. And what is going to replace the revenue derived from taxes on gasoline that underpin much of the nation's highway construction if no gasoline is sold?

A bit further down the road are self-driving cars and trucks. While offering the possibility of saving many lives—over 90% of fatalities in vehicular accidents in America during the past year are attributable to driver error—what might be the implications for establishing legal liability in the case of accidents, as well as replacing the livelihoods of the nation's nearly four million truck drivers? And what of the great medical breakthroughs now being achieved? Is it better to extend an elderly person's life by 6 months or to devote the requisite funds to enabling his or her grandchild to attend college? As if such conundrums were not difficult enough in their own right, the results of science and technology have a way of introducing unpredictability, as when a tree fell in Ohio and shut off electric power to much of New England and part of Canada.

One thing seems certain, and that is, the nation will need a cadre of individuals highly proficient in science and technology *as well as* public policy. A shortage of such talent is likely to lead to such outcomes as occurred a number of years ago when the legislature in one state proclaimed, for the convenience of commerce, that the figure π would forever after in that state be deemed to be precisely 3.2—so much for 3.1415926...—displaying a lack of understanding of rounding that matches a lack of understanding of mathematics.

The attitude evidenced by most scientists and some engineers is that "science is science and policy is policy, and never the twain shall meet." That principle has only one problem: it is demonstrably wrong. The argument favoring the proposition is that science is founded upon the irrevocable laws of nature, whereas "policy" is based on highly revocable laws of fallible humans. While that is true, the world does not operate in such a convenient manner. Consider the collision of science and technology with policy formulation as it affected the superconducting supercollider, stem cell research, fusion energy research, the supersonic transport, and applications of artificial intelligence and genomic research in arenas affecting individual privacy. To many scientists, the sanctity of their role would be contaminated were they to enter the politicized world of policy. The result of that position arguably has served neither science nor the nation well. We find ourselves denied the views of the very individuals who have much to contribute to debates over science and technology policy. As an example of the consequences, in one recent year, fully half the physicists in the 435-member Congress retired...leaving us with one physicist.

Further adding complexity to the promulgation of sound science and technology policy, science and technology encompass a broad spectrum of disciplines, and advances are increasingly being made at the intersection of the two. For example, one survey found that when practicing physicians were asked what recent advances had been most significant in enhancing their ability to serve their patients, three of the top five responses were from the field of engineering—not from traditional biomedical research.

The search for individuals possessing broad scientific or technological credentials accompanied by an understanding of policy formation is clearly influenced by the public's not uncommon characterization of such individuals as "nerd geeks!" (Confession: I am one of them, having spent a decade in government policy positions, several decades in industrial research and management, and a few years in academia.)

Fortunately, in this book Michael S. Lubell has provided a remarkably insightful and extraordinarily well-written treatise tying together policy and science—from the time of the founding fathers to the age of genomics, artificial intelligence, and quantum computing. It is a book that should be read by anyone with an interest in science and technology, or public policy, or both; and especially by those with an interest in neither, but whose lives are being profoundly affected by both.

In addition to a fascinating tour de force of real-world science policy issues, the author—himself a scientist with deep roots in policy-making—occasionally provides a first-hand perspective. Not everyone will, of course, agree with every word of these perspectives; but, then, not everyone will disagree either. That is the essence of science policy-making.

Norman R. Augustine
Potomac, MD, United States

Preamble

Opening doors and expanding horizons

Modern life without technology would be unthinkable, so much so that we never even consider the possibility. Rarely do we give a moment's thought to where all the technological marvels came from, how they materialized, and what policies enabled them. By the time you finish reading this book, you will be able to answer those questions. More important, you will be able to use your knowledge to help shape the future of the country and the world, improve your chances of success professionally, and achieve a comfort level with today's increasingly complex society. But first, let's take a quick tour of technology in modern life.

Let's begin with some things that almost seem like bodily appendages—smart phones, tablets, and computers. We use them for communication, entertainment, and business. They connect us to the Internet and satisfy myriad needs: sending and receiving emails and texts; making travel arrangements; getting tickets for events; buying all manner of merchandise, from the largest appliance to the smallest token; comparison shopping to get the best deals; reading the latest news (real and, unfortunately, sometimes fake); communicating with our friends and family on social media—all of it online, all from the comfort of our home or our car, or as we walk along the street.

More often than not we use plastic—credit or debit cards—instead of cash for purchases. And if we opt for cash, we get it from an ATM instead of a bank teller. At checkout counters, we rely on laser scanners to speed us on our way and to make sure we haven't been overcharged. (Stores use those scanners to keep track of their inventory and finances in real time.)

When we purchase a car, we expect it to spend little time in the repair shop and not break the bank guzzling gas. And when we drive, we rarely use paper maps any more, relying on GPS—the acronym for Global Positioning System—to get us to our destination quickly with few miscues. We value accident avoidance systems, and we assume airbags, seat harnesses, and the integrity of cabin construction will protect us in case an accident does occur. Finally, if we're really tuned in, we can see the day coming when autonomous vehicles will free us from many of the mundane tasks of driving.

On an airplane, we pay far more attention to the personal entertainment system than safety instructions, because technology has made accidents a rarity. Over a 10-year span, from 2004 to 2013, U.S. commercial airlines registered a scant 139 fatalities, flying a total of 78.7 billion miles during that period—an amazing average of only one death for every 556 million miles of travel.[1]

Navigating the Maze. https://doi.org/10.1016/B978-0-12-814710-8.09997-6

Technology has made that remarkable record possible with the use of radar guidance systems, computers, communication satellites, flight-training simulators, light-weight but strong materials, and durable jet engines.

When we shop for food, we expect the products to be safe, affordable, and plentiful. When we drink water, we assume it's free of harmful contaminants, and when we learn that it isn't, as was the case in Flint, Michigan[2] in 2016, we demand that public officials be held accountable for scientific malfeasance.

If we get sick, we know that modern medicine has extraordinary diagnostic tools at its disposal, from sophisticated blood tests and ultrasound imaging to CT scans, MRIs, and PET scans. If we need cures, antibiotics and antiviral medications for infections are available. Lasik surgery and lens replacements correct eye disorders. Radiation, proton therapy, chemotherapy, immunotherapy, and stem cell transplants fight cancers. Stents, artificial valves, and artificial hearts address cardiac disease. Artificial joints replace damaged ones. Laparoscopy, arthroscopy, and video assisted thoracoscopy minimize invasiveness in surgical procedures.

We spend a good fraction of our lives in our homes, but few of us ever pause to consider how much money energy efficient devices save us. If we did, we'd be happily surprised. Refrigerators today consume only a third of the electricity than they did 40 years ago.[3] And LED lights, which are rapidly becoming the norm, operate on a fifth of the power of comparable incandescent lamps.[4] They last more than 40 times longer, to boot.

These are a just few compelling examples of how technology beneficially affects our lives as individuals. But on a larger scale, technology and its scientific underpinnings have had perhaps an even more profound impact. The industrial and service sectors of our economy have changed dramatically in the last quarter century. The change is happening today at warp speed, with many disruptive consequences, some good and some very socially and economically challenging.

Take the case of manufacturing. Today more than 12 million Americans work in industrial manufacturing plants, according to the Bureau of Labor Statistics, but that's eight million fewer than in 1980. Manufacturing accounted for 19% of the workforce in 1980. Today, it's scarcely 8%.

The causes of the decline, according to the populist mantra that helped elect Donald Trump president in 2016, are terrible trade pacts, lack of trade enforcement, currency manipulation, and cheap Asian and Mexican labor. While all of them contributed to the workforce falloff, another instigator that is still on the cusp didn't make the list—automation.

While the data and economic analyses are still incomplete, there can be little doubt that, at the very least, automation has transformed the nature of the manufacturing workforce, altering it from less skilled to more skilled, and making greater demands on worker competencies. Automation is also very likely to be the principle driver of increased productivity—output per worker—in the future, increasing corporate profits and making companies more competitive globally.

As the United States began shedding manufacturing jobs more than a half century ago, the service sector picked up the slack. Even in the 1980s, professional, business, education, and health service jobs seemed immune to the outsourcing exodus to other countries that the manufacturing sector was experiencing. And economists spoke glowingly about a service-dominated American economy that would lead the world for years to come. But that was before the information technology revolution enabled globalized commerce.

Cheaper and more powerful computers connected to the Internet via communications satellites made it possible for service companies to outsource their work to regions of the world where labor was cheaper. Spanning many time zones, the global network also enabled companies to provide 24-7 service to their customers. But the benefits to the companies, and their customers, came at a cost to American workers—loss of jobs that were once sheltered from outsourcing to other countries.

There are two other arenas that technology has transformed remarkably. The first should come as no surprise: national security. World War II was arguably the first major conflict resolved by technology, principally the atomic bomb and radar—although cyber cryptography should be on the list, as well. After the Allied victory, the United States embarked on a major expansion of military research and development (R&D) that has carried through to the present.

The products of defense R&D have been transformative: thermonuclear weapons, ballistic missiles, missile defense systems, laser guided weaponry, stealth aircraft, GPS, drones, spy satellites, and night goggles are just a few of the many examples. Although they were developed for military applications, a number of the technologies have found their way into the consumer market place, prime among them GPS and the Internet—its defense progenitor was known as ARPANet—that are now ubiquitous.

The second arena that has undergone a revolution caused by disruptive technology is probably less obvious, except to the people who are employed by it: banking and finance. Without computers and global communication tools, Wall Street, London's The City, and financial centers throughout the world would be a shadow of the industry they have come to represent.

Trading of complex, often opaque, derivatives developed by "quants"—as quantitative analysts are commonly called—would be virtually impossible without the use of powerful computers. Once financial trading floors were filled with shouts and paper slips. No more.

Today, the trading floors of banks and hedge funds are filled with computer monitors, keyboards and bits. Stocks, bonds, commodities, and currencies are bought and sold at almost the speed of light, and the buyers and sellers might be half a world away from each other. High-speed trading has become the stock and trade of the best arbitrageurs who take advantage of the small spreads in prices that come and go faster than the blink of an eye.

Technology has generated tremendous wealth for many of the financial players, but it has also enabled some of the riskiest practices pursued by banks that are "too big to fail." Those practices produced the global financial crisis of 2007–2008, and the great recession that followed. Wall Street has recovered—with a big boost from the federal government—but Main Street and large parts of the world are still reeling from the fallout.

Arguably, the populist rebellion that began in the United States as the Tea Party movement in 2010 and ultimately propelled Donald Trump into the White House in 2017 has its roots in the great economic meltdown and the uneven recovery that followed it. But beneath the surface, the populist movement is a reflection of the impact of technology on American life and the challenges the average person faces in adjusting to the changes that are transpiring at an ever-increasing pace.

We've taken a bird's eye view of how technology and its science partner have transformed America and the world in the span of a little more than half a century. And there is no evidence that transformations will abate anytime soon.

Science and technology are crucial cogs in 21st century life. But their impacts are not accidental. They are the result of policy decisions made in the public and private arenas. If those decisions properly take into account the roles of science and technological innovation, they can have tremendously beneficial effects. If they don't, the impacts can be devastating.

The decision we made at the end of World War II, for example, to invest public money in "basic" research—which has no goal other than expanding fundamental knowledge—led eventually to the development of the laser and medical diagnostic tools such as CT scanners and MRI machines. Today, laser-enabled technologies, among them fiber optics communication, precision manufacturing devices, and DVD players—account for a staggering one-third of the American economy.[5] Without access to CT scanners and MRI machines, doctors often would be basing their diagnoses on information and observations that are far from complete, as they did decades ago.

The policy decision to make basic research a federal priority traces its lineage to Vannevar Bush, an engineer who headed the wartime Office of Scientific Research and Development (OSRD) during World War II. As the war was nearing its end, President Franklin Roosevelt asked Bush to consider what could be done to bolster American science, which had been so instrumental in the Allies' win. By the time Bush finished his report, *Science, The Endless Frontier*,[6] in July of 1945, Roosevelt had died, but President Harry Truman accepted Bush's principal recommendation: that the federal government expand its support of scientific research. It led to the creation of the National Science Foundation—although the path was tortuous—and the establishment of major research portfolios in other federal agencies.

Bush drew heavily on his wartime experience and his professional career in industry. He also recognized that the political climate was ripe for his policy proposal. He knew that Roosevelt trusted both his knowledge and his judgement, and he assumed correctly that Truman would try to follow in Roosevelt's footsteps. The result was a game changer that kept the United States at the fore-front of scientific discovery and technological innovation for more than half a century, spurring economic growth and safeguarding the nation militarily.

The financial policies advanced by President Bill Clinton and strongly supported by many members of Congress in the 1990's offer a study in contrast. In essence, the president and congressional Republicans, especially, accepted the proposition that banking had undergone a major transformation. No longer restricted to operating within individual states, banks not only were involved in financial transactions across state lines, but they had become global players. In order to compete internationally, they argued, they had to be allowed to grow, and one way to grow was to allow them to engage in both commercial (retail) and investment operations.

The Glass-Steagall Act,[7] which had been in place since 1933, prohibited them from doing so, because the combination of both activities under one roof was widely seen as leading to the risk-taking that culminated in the 1929 financial crash and the Great Depression that followed. But by 1999, most policymakers in Washington believed that market transparency and federal banking oversight were sufficient to keep risky behavior in check. They accepted Wall Street's argument that globalization of finance had altered the banking calculus. Without Glass-Steagall repeal, American banks could not compete with other securities firms at home, and especially abroad, at least that's what the banks asserted.

The Gramm-Leach-Bliley Act,[8] properly known as the Financial Services Modernization Act of 1999, passed the Senate on November 4 of that year, and President Clinton signed it into law 8 days later. The arguments that led to the passage of the bill, which tore down many of the Glass-Stegall barriers, sounded persuasive, but policymakers failed to take into account how much computers had changed the trading landscape.

Collateralized Debt Obligations (CDO) bundled mortgages into mortgage-backed securities (MBS) in such an opaque way that risk assessments became difficult. Synthetic CDOs bundled many CDOs into even more opaque securities. Credit Default Swaps (CDS), essentially insurance policies without any asset backing, were bought and sold in the darkness of private trades. Derivatives used complex mathematical algorithms—often only understood by the quants who had devised them—to combine a wide assortment of financial instruments into a single trading package.

Could all of these instruments have been created and traded without the available computer technology? Perhaps, but it would have been far more difficult. Certainly, the speed and globalization of the trading would have been nearly impossible. When the American subprime mortgage market collapsed, American banking faced the prospect of another 1929, but with technology-driven globalization, so too, did the rest of the world.

Many factors contributed to the 2007–2008 financial crisis, some of them completely unrelated to technology. But technology did play a role that policymakers seemingly never recognized when they began loosening the regulatory strictures on the finance industry.

Science and technology policy intersects modern life in countless ways. Once you understand those intersections and where they fall on the scientific, economic, business, and political landscapes, you will have a firmer grasp of the challenges facing 21st century America and the road to effective solutions. You will become more successful in your professional life and more engaged as a citizen. You will be able to open doors and expand your horizons.

Introduction

What is science and technology policy?

Science and technology policy is a dual-use phrase. It refers to the corpus of science and technology that inform public and private policies across multiple arenas. It also denotes the body of policies that affect the conduct of science, the development of technologies, the innovations that make their way into the marketplace of ideas and products and the societal impacts that ensue.

Now, conjure up an image of intersecting freeways, boulevards and streets, some with barriers and quick detours, some with high-speed lanes, each teeming with vehicles of different types and models traveling in different directions—all of it set among hills, ravines, and the occasional park. The picture fairly well describes the Los Angeles basin. It also describes the world of science and technology policy.

The roads are the policy arenas. The drivers of the vehicles are the policymakers; the legislators, and other elected officials; the bureaucrats; the science and technology practitioners; the industrialists; the bankers and venture capitalists; the influence peddlers; the advocates, and the lobbyists. The vehicles are the tools they use. And the hills, ravines, parks, barriers, and detours are all part of the political landscape on which they function. With that picture in mind, let's zoom in on some of the details.

Science (the study of the natural world) and technology (the practical application of science) are two of the freeways. Engineering (the use of science for building and achieving practical applications) and mathematics (the abstract science of numbers, equations, functions, geometry, and the like) merit the same kind of designation. All four of them are essential pathways for policy formulation.

Three of them have well-defined lanes, one less so. Let's begin with the domain of modern American technology.

At the risk of a slight oversimplification, we can divide technology into eight broad categories: aerospace, agricultural, bio-medical, communication and information, energy and environment, manufacturing, military, and transportation. Although they overlap to a small degree, for the most part, we can think of them as proceeding separately in eight individual lanes. We can do the same for engineering[1] and mathematics.[2]

It is tempting to treat the sciences in a similar fashion, using the major disciplines with which we are all familiar—astronomy, biology, chemistry, Earth science, and physics—for lane assignments. But even if we set aside other

Navigating the Maze. https://doi.org/10.1016/B978-0-12-814710-8.10000-2

highways for the social sciences, which our short list ignores, we would still be making a major error. In the 21st century, we can no longer compartmentalize the natural sciences. They have become inexorably intertwined.

Harold Varmus, Nobel Laureate, former director of the National Institutes of Health, and later, president of Memorial Sloan Kettering Cancer Center, made that point in an October 4, 2000 Washington Post Op-Ed when he wrote,

> *Medical advances may seem like wizardry. But pull back the curtain, and sitting at the lever is a high-energy physicist, a combinational chemist or an engineer. Magnetic resonance imaging is an excellent example. Perhaps the last century's greatest advance in diagnosis, MRI is the product of atomic, nuclear and high-energy physics, quantum chemistry, computer science, cryogenics, solid state physics and applied medicine.*

To use Varmus's terminology, in the modern world the sciences are "interdependent." Not only are major breakthroughs occurring at the boundaries of the individual disciplines—what policymakers call interdisciplinary advances—increasingly progress in one scientific field is dependent on progress in other fields. In some cases, scientists, themselves, move across the traditional disciplinary boundaries, bringing to bear the expertise they gained in one arena to the challenging problems in another.

In modern America, the traffic on the broad science freeway must be allowed to move unimpeded across the lane boundaries that have become increasingly blurred. Otherwise, technological progress will be stifled.

As we turn to other roads in the science and technology policy basin, we find more situations where the traffic must be able to shift lanes, as it does on a typical Los Angeles freeway. But in those cases, the movement will be more controlled and predictable. Let's begin with research and development.

Usually abbreviated as R&D, it requires a wide, multilane road to accommodate its multifaceted components. Research—the R in R&D—is defined as creative, systematic studies aimed at increasing and improving knowledge writ large. It occupies three lanes, each representing one of three major classifications: "undirected" basic research, "directed" basic research, and applied research.

Basic research is often called fundamental research, and the "undirected" label refers to activities that focus solely on scientific discovery with little or no regard for any possible application. "Directed" basic research, by contrast, is fundamental research that carries with it the goal of obtaining new knowledge in a specific arena, such as energy, agriculture, or cures for disease. Applied research is far more targeted on the endgame. It begins with the identification of some practical problem, and proceeds with work leading to a set of possible solutions.

The D part of R&D refers to the development of new products and procedures, usually building on scientific discoveries and research outcomes. For successful developments, the R&D freeway off-ramp leads to the on-ramp of an engineering highway, on which testing and evaluation take place.

On the R&D freeway, itself, traffic moves from one lane to another, as the results of one activity create new opportunities or new challenges. In some cases, a basic research outcome leads to an applied research initiative or directly to a development program. In some cases, the results of an applied research or development program produce tools that enable new, basic research endeavors. And in some cases, new, unanticipated challenges emerge during a development program, requiring additional applied research or directed basic research.

We will explore the complex R&D relationships in Chapter 12 in more detail, when we compare Pasteur's Quadrant with the linear model of innovation, attributed to Vannevar Bush's view of the scientific enterprise—incorrectly, as a careful reading of his work reveals—described in *Science, The Endless Frontier*. But for now, let's simply accept the need for R&D traffic to move across the freeway lane divides whenever progress dictates.

We turn to the education superhighway, which has its own distinctive characteristics, especially where it involves the STEM fields; science, technology, engineering, and mathematics. It starts as a very wide artery with lanes that have no identifiers and with all traffic moving in the same direction, but able to change lanes freely. Eventually, a STEM label appears on one or two of the lanes. And after some of the traffic exits, the highway bifurcates, with the well-defined STEM markings following only one branch. The narrowing of that branch continues, and by the time the highway reaches its terminus, only one of the remaining lanes carries the STEM signage.

As you have probably guessed by now, the education superhighway describes the STEM learning environment in modern America. We all begin school in very much the same way. And most of us are exposed to a common curriculum throughout our years in elementary and middle schools. In high school, we start to separate, and by the time we graduate, as 82% of us did in 2014, a STEM future will be largely foreclosed as a viable option, unless we took our science courses seriously.

If we terminate our education with only a high-school diploma, we are essentially lost to the STEM job market. And for those of us who choose to attend community college, our STEM aspirations will be significantly harder to achieve. Only if we attend 4-year colleges or universities, major in one of the STEM fields, and eventually enroll in a graduate or professional school, will the full panoply of STEM opportunities become readily apparent and accessible.

Policymakers who recognize that the health of the STEM workforce is integral to the modern American economy must travel the full length of the STEM superhighway if they want to make decisions that will be most beneficial to the nation. Focusing on only one segment of the highway—research universities, for example—is simply insufficient.

We've taken a close look at six key freeways that traverse the science and technology policy basin: science, technology, engineering, mathematics, R&D, and education. We now turn to other roads that crisscross the basin—those that represent the sectors of modern American life connected closely to technology. Healthcare, which accounted for 17.1% of the gross domestic product (GDP) of the United States in 2015 is the largest artery. And although manufacturing's portion of the GDP has been declining steadily since 1970, when it clocked in at 24.3%, it still claimed a 12.0% share in 2015. It warrants another major highway. Finance, which ballooned to 7.2% in 2015, plus energy[3] and telecommunications,[3] round out the big five highways carrying the technology traffic.

Although defense and transportation[4] represented smaller percentages of the GDP, they are both technology intensive, and will continue to contribute a substantial portion of the movement on the science and technology policy roadway system. We will give them the designations of thoroughfares. Agriculture's share of the GDP, both nationally and worldwide, has been shrinking for many years, but we will include it on our roadmap because of its historical role and its impact on other industries.[5]

Finally, although they do not appear as entries on a GDP ledger—at least not in any significant way—environmental issues, foreign affairs, global competitiveness, productivity, tax policies, trade pacts, union contracts, and wages all intersect the science and technology maze, each on its own road. Each of them informs, motivates, or constrains decision making, and their intersections with the freeways, highways, and thoroughfares occur at multiple points.

We'll conclude the Introduction with a survey of the policy traffic that traverses the science and technology basin. Every issue attracts a multiplicity of drivers: from legislators, elected officials and bureaucrats, to industrialists, financiers, and practitioners. Sometimes their interests converge, and sometimes they diverge. At any moment, they are zipping around the highways, often at high speed, but always cognizant of the landscape and the other drivers around them. They drive both offensively and defensively, depending on the circumstances, and the best of them know how to navigate unexpected barriers and take advantage of quick detours.

The cars they drive, which represent the tools they employ, depend on the policymaking sector from which they come. Legislators, elected officials, and bureaucrats, for example, traditionally promulgate policies through budget requests and appropriations bills—both of which establish spending priorities—authorization bills, tax bills, regulations, executive orders, congressional hearings, technical analyses, reports, press briefings, and in 2016, most prominently, presidential Tweets.

Individuals, corporations and advocacy groups use a different set of tools: political contributions, which open doors to lawmakers' offices; "grass-roots" lobbying, which exerts constituent pressure on elected officials; Op-Eds, which raise public awareness of a policy issue; and "grass-tops" lobbying, which employs well-known opinion makers to influence outcomes.

The STEM practitioners—the scientists, technologists, engineers, and mathematicians—have had a spotty record of policy engagement, as we will see in the succeeding chapters. Although World War II produced a historical spike in their behavior, its effect was short lived. Not until the last decade of the 20th century did the STEM community see the need for participation in the policy dialogs that shape the nature of their very work.

Whether that community remains permanently engaged is too early to forecast. But to the extent they have helped formulate, promulgate, and advance science and technology policies, they have used the same tools as any interest group, running the gamut from political contributions and grass-roots lobbying to media promotion and grass-tops lobbying.

Up to this point we have focused on policymaking in the public sector. But science and technology policymaking has a private-sector side, as well. There, however, it tends to be specific to leaders of technology companies, entrepreneurs, bankers, and venture capitalists who all make investment and strategic decisions based upon market projections, trade and tax policies, the regulatory environment, and the innovations they see on the horizon.

Regardless of the arena, public or private, policymaking must take the political landscape into account. The uphills and downhills, the barriers and quick detours, the potholes and well-paved roadways all reflect the political landscape. Ultimately, that landscape determines the incentives and disincentives that define effective science and technology policy in modern America. Healthcare, economic growth, environmental stewardship, energy production and usage, transportation, and national security are all affected by the science and technology decisions we make in the politically fraught policy arena.

Part I

Past is prologue

Courtesy of the Division of Work and Industry, National Museum of
American History, Smithsonian Institution

Chapter 1

The early years 1787–1860

It was a cool, crisp March day in Washington, but inside the White House, there was a warm glow. The year was 2000. It was President Clinton's last year in office. And with his acquittal on impeachment charges 13 months behind him, he was once again focusing exclusively on his presidential duties.

The carefully ordered rows of seats in the East Room had been filling up since three o'clock, and by the time the president strode to the podium twenty minutes later, all the chairs were filled. It was a joyful event for everyone—the president, the guests, and the honorees. It was the presentation of the National Medals of Science and Technology.

With his usual engaging smile, Bill Clinton began by recognizing several members of Congress who were attending, as well as the British Ambassador to the United States. And then he launched into the meat of his speech.[1]

Every year I look forward to this day. I always learn something from the work of the honorees. Some of you I know personally; others, I've read your books. Some of you, I'm still trying to grasp the implications of what it is I'm supposed to understand and don't quite yet ... I must say, one of the great personal joys of being President for me has been the opportunity that I've had to be involved with people who are pushing the frontiers of science and technology and to study subjects that I haven't really thought seriously about since I was in my late teens. And I thank you for that.

With that, the president reached into his pocket and pulled out a small object, hardly visible to anyone not sitting in the front rows. He continued,

When Congress minted America's first coin in 1792, one of the mottos was "Liberty, Parent of Science and Industry." Very few of those coins survived, but the Smithsonian has lent us one today. I actually have one. It's worth $300,000. Not enough to turn the head of a 25-year-old dot-com executive — but to a President, it's real money. And I thought you might like to see it because it embodies a commitment that was deep in the consciousness of Thomas Jefferson and many of our other Founders. And we could put the same inscription on your medals today.

Navigating the Maze. https://doi.org/10.1016/B978-0-12-814710-8.00001-2

It's likely the only person in the East Room who had seen the coin before was Jeffrey M. Smith, who borrowed it from the Smithsonian and had given it to Clinton to use as a prop.[2]

The story is just a vignette, but the coin[3] itself illustrates how far back in our nation's history we have to travel to capture the beginnings of American science and technology policy. The 1792 coin was a fitting capstone to the discussions the founding fathers had as they came to closure on the wording of the United States Constitution. Drafted after extensive and often contentious debate in 1787, ratified by eleven states in 1788, and finally becoming a binding document a year later, the Constitution contains the following words in Article I, Section 8.

The Congress shall have the power 1. To lay and collect Taxes, Duties, Imposts and Excises, to pay and provide for the common Defence and general Welfare of the United States... 3. To regulate Commerce with foreign Nations, and among the several States, and with the Indian Tribes ... 5. To coin Money, regulate the Value thereof, and of foreign Coin, and fix the Standard of Weights and Measures... 8. To promote the progress of Science and the useful Arts, by securing for limited Times to Authors and Inventors the exclusive Right to their respective Writings and Discoveries ...

The words of Section 8 provide the framework and justification for virtually all U.S. science and technology policy since our nation's founding. That the authors of the Constitution explicitly cited science in Clause 8 might seem surprising in a new country, recently consumed by a war of independence with scant financial resources and separated from the European Age of Enlightenment by a perilous ocean journey that could take 2 months or longer.

But by 1789, the fledgling United States already had almost two dozen institutions of higher learning[4] and two learned or scholarly societies. Moreover, among the Constitution's framers, Benjamin Franklin, who closed the 1787 Convention at age 81 and successfully urged its adoption, was a distinguished and world-renowned scientist. John Adams, another framer, and later the second elected president of the United States, had studied under John Winthrop, a noted astronomer at Harvard. And James Madison, a Princetonian and one of the principal authors of the Federalist Papers,[5] was an elected member of the American Philosophical Society,[6] the country's first scholarly organization, which Franklin founded in 1743 for the purpose of "promoting useful knowledge." Adams was also one of the first members of the American Academy of Arts and Sciences, which the Province of Massachusetts Bay had established in 1779 to "promote and encourage the knowledge of the antiquities and the natural history of America; to determine the uses which the various natural productions of country may be applied; ... and, in fine, to cultivate every art and science which may tend to advance the interest, honor, dignity and happiness of a free, independent and virtuous people."[7]

Thomas Jefferson, whose devotion and contributions to science were well known to the young nation's founders,[8] was notably absent from the convention. In 1785, he had succeeded Franklin as Minister to France and played no direct role in the Constitution's drafting. But the time Jefferson spent in France would serve him well when he assumed the presidency in 1801, and would be instrumental in America's first scientific venture.

The Constitution took effect on March 4, 1789, and less than a year later, on January 8, 1790, George Washington delivered the "First Annual Message to Congress on the State of the Union."[9] The name is longer than today's truncated title, "State of the Union Address," but the speech, extremely short by modern standards, highlighted science and technology as federal priorities.

The advancement of agriculture, commerce, and manufacturing by all proper means will not, I trust, need recommendation, but I cannot forebear intimating to you the expediency of giving effectual encouragement as well as the introduction of new and useful inventions from abroad as to the exertions of skill and genius in producing them at home.... Nor am I less persuaded that you will agree with me in opinion that there is nothing which can better deserve your patronage than the promotion of science and literature. Knowledge is in every country the surest basis of public happiness.

A little more than a dozen years had passed since the thirteen colonies declared their independence, and America's founders had already identified science and technology policy as central to the nation's future. It wouldn't be long before President Thomas Jefferson would have the opportunity to use that mandate in the first remarkably successful federal scientific venture.

Jefferson, a product of William and Mary, the South's most prestigious university at the time, returned from France in 1789—the year the French Revolution began—joining President Washington's Cabinet as the first Secretary of State shortly thereafter. But following protracted arguments with Secretary of the Treasury Alexander Hamilton, a staunch Northern Federalist with a Columbia University pedigree and a proponent of expansive national powers, Jefferson, an equally resolute anti-Federalist, submitted his resignation in 1793, much to Washington's displeasure.

Jefferson's departure from the national stage was short-lived. In November 1796, he was elected vice president of the United States as a Democratic-Republican, having finished second to John Adams, the Federalist nominee, in the general election.[10] The following March, during the same week, he was sworn into office, and he also delivered a research paper on paleontology at a meeting of the American Philosophical Society, to which he had just been elected president.

Jefferson made another run for the presidency in 1800, once more challenging Adams, this time successfully. But the outcome of an extremely ugly campaign ended with a bizarre twist. The framers of the Constitution had not anticipated that political parties would play significant roles in presidential elections, and consequently Article II, Section 1, Clause 3 required members of the Electoral College to cast ballots for two people.[11] The top vote getter would become president, as had occurred without incident in the three elections prior to 1800, and the runner up would become vice president. But in the highly partisan and geographically sectarian atmosphere that had poisoned the 1800 election, Jefferson and Aaron Burr, his Princeton-educated Democratic-Republican running mate, both received 73 votes, forcing the House of Representatives to make the final selection. Partisan rancor reached new heights in the House, and it took the chamber 36 ballots before it elected Jefferson president and Burr vice president on February 17, 1801.

Having weathered the constitutional storm, Jefferson was immediately confronted with a foreign crisis in America's backyard. In less than 3 years, though, he would be able to turn it into the nation's first significant scientific mission. It is worth a short historical digression to see how that happened, because it illustrates the way in which external events can influence science policy. What was true in the early 1800s remains true today.

The territory of Louisiana,[12] consisting of 827,000 square miles west of the Mississippi River, had been under Spanish control since the 1763 Peace Treaty of Paris that had ended the French and Indian War between Britain and France in the New World. As vice president, Jefferson had been eying it as a possible opportunity for American expansion west. But relations with Spain were good, and there was no obvious imperative for the United States to make an immediate play for it. Jefferson's calculus began to change as soon as he assumed the presidency.

Napoleon Bonaparte, who had seized power in France in 1799, made it clear he wanted to reestablish France's footprint in the New World. Within a year, he concluded a secret deal with Spain that transferred the territory back to France in exchange for a small parcel of land in Europe. In 1802, Spain officially agreed to cede control of the Territory of Louisiana and the port city of New Orleans, and Napoleon announced that after the transfer was completed, he would close the Mississippi River as well as the port of New Orleans to American shipping. His plans of reestablishing a major French presence in the New World were bold, but he would soon find them badly mistimed.

It was clear by the beginning of 1803 that the peace treaty, which Britain and France had signed in Amiens on March 25, 1802, was breaking down, and Napoleon had neither the resources nor the will to project significant French power across the Atlantic. Jefferson, who initially had hoped to arrive at a diplomatic settlement that would simply protect the nation's commercial shipping interests, sensed an opening for a more ambitious outcome. In early January 1803, he asked James Monroe, a close friend and political ally, to travel to Paris as a special envoy and assist Robert Livingston, the U.S. minister to France, in reaching an accord on the Territory of Louisiana.

A few weeks later, even before negotiations with France had started in earnest, Jefferson asked Congress to fund an expedition across the territory, ultimately reaching the Pacific. Jefferson understood that exploring the West was in the best commercial and geopolitical interests of the United States, and he wanted to move quickly following negotiations with France. But he had long harbored an intellectual interest in geographic exploration, and even if the negotiations with France had failed, there is good reason to think he would have pursued such an expedition.

By the time Monroe arrived in Paris on April 12, 1803, French Foreign Minister Charles Maurice de Tallyrand had told Livingston that France was willing to sell the entire Territory of Louisiana and the city of New Orleans. All that remained was haggling over the price. Monroe teamed up with Livingston, and on the last day of April the two of them struck a deal with French Finance Minister François, Marquis de Barbé-Marbois. The purchase price for all 827,000 square miles was $15 million. The Louisiana Purchase, which the U.S. Senate ratified on October 20, 1803, and which officially became part of the United States on December 30, doubled the territory of the young nation.

Jefferson's ambitious plan for exploring the vast acreage the nation had just purchased was well underway by the start of 1804.[13] Meriweather Lewis, a former U.S. Army captain, whom Jefferson had chosen to lead the expedition in the spring of 1803, had already completed crash courses in botany, zoology, celestial navigation, and medicine, and had overseen construction of a large keel boat suitable for navigating the Mississippi River. Lewis had also recruited William Clark to help lead the project, and together with a number of recruits, they had established a launching site on the east bank of the Mississippi near St. Louis in December 1803.

The objectives of the Corps of Discovery Expedition, as the plan was officially known when Congress appropriated $2500 for the project in February 1803, extended well beyond boosting commerce and establishing a deterrent to possible land grabs by other nations. The Lewis and Clark Expedition, as it would later be known, would also address one of Jefferson's passions: scientific discovery. The expedition would investigate plant and animal life, which Jefferson valued as a plantation owner and as a student of agronomy. It would also study geography, geology, and mineralogy. For those reasons, the American Philosophical Society endorsed its mission and helped organize science tutorials for Lewis in Philadelphia during the spring of 1803.

On May 14, 1804, Lewis, Clark, and their recruits broke camp and set off up the Mississippi, reaching what is now Bismarck, North Dakota in late November. A year later, after they had successfully crossed the Rocky Mountains, they caught sight of the Pacific Ocean and began to plan their route back. It would return them to St. Louis on September 23, 1806.

The ambitious expedition, which lasted more than 2 years, and ultimately cost the federal treasury $38,000 (more than $1 million in today's dollars), supplied Jefferson with everything he had imagined. And despite its hefty price tag, it was a bargain. It unquestionably delivered on two of its primary goals: providing a young America with a legal claim to the land and, through a far better understanding of geography and indigenous tribes, opening up new commercial opportunities for its populace. But it also provided a wealth of scientific information, and it stands as one of the best investments in research the federal government has ever made.

It is unquestionably the first use of Article I, Section 8 of the Constitution to justify federal support of science. And in that respect, it stands as a hallmark of American science and technology policy. Demonstrating Jefferson's extraordinary political savvy, it also illustrates the crucial intersection of science policy with both politics and foreign affairs. We will encounter such connections many times over.

Castles are commonplace in Britain, France, and other nations whose histories are steeped in royalty and aristocracy. But in the United States, finding a castle requires perseverance—unless you're visiting Washington, DC. Ask any tour guide or reasonably well-versed Beltway denizen, and you will be directed to an idyllic, beautifully landscaped plot of land abutting the National Mall. A striking 1855 medieval revivalist structure in red sandstone with eight crenellated towers and a slate roof occupies the site. It is the headquarters of the Smithsonian Institution, and buried in a crypt within the building are the remains of James Smithson, after whom the institution is named.

The story of the Smithsonian[14] and its eponymous founder is a curiosity of 19th century American science and technology policy and the politics of the day. The year was 1835, and Andrew Jackson was president. The first populist to occupy the White House, he had been swept into office in 1828, capturing 56% of the popular vote. Reelected in 1832 by nearly the same margin, Jackson was far better known for his military exploits, volatile temper, and identity with the common man than he was for any cultural or intellectual pursuits.

Into his lap, 3 years after being reelected, landed a strange bequest from a wealthy English gentleman who had never set foot in the United States.[15] James Smithson, a chemist who was an elected Fellow of the Royal Society, had never married and left the bulk of his fortune to his only nephew when he died in 1829. That is far from curious, but here the story takes a strange twist. His will[16] stipulated that when his nephew, Henry James Hungerford, died, if he left no children of his own, the entire estate was to be bequeathed "to the United States of America, to found at Washington, under the name of the Smithsonian Institution, an establishment for the increase and diffusion of knowledge among men."

Hungerford died childless 6 years later, and the United States government, with Jackson's support, filed an uncontested suit to secure the Smithson estate. The suit was settled on May 9, 1838, and that fall, a sum of £105,000 was deposited in the United States Treasury. By that time, Martin Van Buren had replaced Jackson, and it was under his watch that Congress began to deliberate what kind of institution Smithson had in mind. Seven years and two presidents later, after many lengthy debates, Congress finally acted. President James K. Polk signed the legislation establishing the Smithsonian Institution on August 10, 1846.

The act created a Board of Regents, which had as one of its first tasks determining the qualifications of a Secretary who would serve as a chief executive officer. The Board's December 1846 resolution is noteworthy because it set a standard for future federal science and technology agencies, foundations, institutes, and programs. It reads,[17]

> Resolved, *That it is essential for the advancement of the proper interests of the trust, that the Secretary of the Smithsonian Institution be a man possessing weight of character, and a high grade of talent; and that it is further desirable the he possess eminent scientific and general acquirements; that he be a man capable of advancing science and promoting letters by original research and effort, well qualified to act as a respected channel of communication between the Institution and scientific and literary individuals and societies in this and foreign countries; and, in a word, a man worthy to represent before the world of science and of letters the Institution over which this Board presides.*

All that remained was finding a person who met such high standards. The Board already had someone in mind, and there could be few arguments with the choice. Joseph Henry, who was born at the close of the eighteenth century, had distinguished himself as the natural successor to Benjamin Franklin as America's preeminent research scientist, devoting himself to the study of electricity and magnetism.[18]

Henry held a faculty position at the College of New Jersey, as Princeton University was then known. He was also the Secretary of National Institute for the Promotion of Science, which housed a collection of scientific items obtained by the United States Exploring Expedition.[19] With twin objectives of scientific investigation and commercial development, the Wilkes Expedition, as it is commonly called, was a global endeavor that lasted from 1838 to 1842. It was, in many respects, the natural successor to the highly-productive Lewis and Clark Expedition of the early 1800s, and at a cost of $928,000 (more than $25 million in today's dollars) it represented a 25-fold expansion in federal support of research over the span of less than 40 years.

The story of the expedition begins in 1818, 20 years before Charles Wilkes eventually set sail from Virginia on a four-year voyage that would ultimately cover 87,000 miles. It illustrates the connection between politics and policy and how pseudoscience can sometimes trump science.

John Cleves Symmes, Jr. was the nephew of John Cleves Symmes, a delegate to the Continental Congress, chief justice of New Jersey, and father-in-law of President William Henry Harrison. To say the younger Symmes was connected politically is to state the obvious. While those connections did not help him succeed in business, they did help him gain attention for his hollow earth theory and eventually congressional approval for a polar expedition to validate it.

Symmes was not the first proponent of such a theory. More than a century earlier, in 1692, Edmond Halley, the famed English astronomer, physicist, and mathematician, made a similar conjecture. Halley, most popularly known for the comet he had discovered 10 years before, and which carries his name even today, suggested that the earth was made up of concentric spherical shells.[20] He developed the hypothesis in order to account for variations of magnetic compass readings made at different locations around the globe. Despite his stature, his hollow earth proposal engendered almost universal derision.

Symmes, undeterred by the reception Halley had gotten from his scientific colleagues, upped the ante, writing in his "Circular No. 1,"[21]

I declare the earth is hollow, and habitable within; containing a number of con-centrick spheres, one within the other, and that it is open at the poles 12 or 16 degrees; I pledge my life in support of this truth, and am ready to explore the hollow, if the world will support me in the undertaking.

It took 10 years, but Symmes, with his political pull, finally succeeded in persuading Congress to approve the mission, not only to prove the veracity of the hollow earth theory, but also to find the polar holes he believed led to the earth's interior. Another decade would pass before the expedition actually began in 1838, and when it concluded in 1842, the president was none other than William Henry Harrison.

Of course, the Wilkes mission never found the holes, but, according to the Smithsonian archives, it did return with more than 4000 animal specimens, 50,000 plant specimens, and countless anthropological artifacts, minerals, gems, and fossils. They made their way into the collections of the National Institute for the Promotion of Science, and in 1855, after the building's construction was completed,[22] into the Smithsonian Institution's iconic "Castle," where many of them can still be seen.

Europe was still the unchallenged center of scientific activity, and it would remain so for most of the next 100 years. But the small cadre of advanced thinkers in the young United States was beginning to expand its horizons. On September 20, 1848, 87 of them met in Philadelphia and founded the American Association for the Advancement of Science (AAAS).[23]

Their goal was stated in the original AAAS Rules and Objectives: "By periodical and migratory meetings, to promote intercourse between those who are cultivating science in different parts of the United States, to give a stronger and more general impulse, and a more systematic direction to scientific research in our country; and to procure for the labours of scientific men, increased facilities and a wider usefulness." The AAAS expanded the purpose of a scholarly society in two ways. First, it aimed at casting as wide a net as possible across scientific disciplines. Second, it incorporated an objective of providing America's budding scientific community with better and more accessible research tools.

By the middle of the 19th century, American science had achieved significant success with two major expeditions. Its roster of scholars was growing. It could count at least three scholarly societies. And it had a new institution under construction in the nation's capital. But it was missing a mechanism to effectively develop and advance scientific objectives that would benefit the nation.

The Lewis and Clark Expedition and the Wilkes Expedition had largely been geopolitical expedients. The establishment of the Smithsonian Institution had been fortuitous. And the American Philosophical Society, the American Academy of Arts and Sciences, and the newly formed American Association for the Advancement of Science all operated outside the sphere of government.

The sole attempt at squaring the science and government circle had ended in failure when an organization called the National Institute[24] collapsed in 1846, only 4 years after Congress had incorporated it. Its brief existence carried with it lessons for future efforts.

The National Institute's life had begun with great promise in 1840, following a meeting at the home of Secretary of War Joel Poinsett, a naturalist at heart who had played a major role in organizing the Wilkes Expedition. Poinsett and the other men who attended the small gathering were motivated by a common desire to find a way of maximizing the scientific benefit of the vast collections they expected the Wilkes Expedition to be returning within the next 2 years. Poinsett, in particular, had his sights set on capturing the £105,000 Smithson had bequeathed to the U.S. government to set up a repository.

Before they left Poinsett's house, the group reached a consensus on establishing the National Institution for the Promotion of Science. In fairly short order, the Washington-based association grew to 84 resident members—among them a number of congressmen and federal officials—and 90 corresponding members from other parts of the country. But timing is everything, and the 1840 election outcome turned out to be bad news for the Institution's durability. President Martin Van Buren, a Democrat, lost his bid for reelection to the Whig standard bearer, William Henry Harrison, and as a result, most significantly, Poinsett, the Institution's guiding light, lost his Cabinet post and a great deal of his Washington influence.

Nonetheless, in 1841, the Institution convinced the new secretary of the Navy, George E. Badger, to request $5,000 from Congress to help prepare for the return of the Wilkes Expedition and its trove of scientific artifacts. Finally, in 1842, Congress took the further step of incorporating the new nongovernmental organization, changing its name to the National Institute in the process. Again, timing is everything, and by then, the Institute's promoters found themselves in competition with advocates of a different repository for the Wilkes Expedition collections: the Smithsonian Institution.

Congress was wary of both plans, but much more so of the National Institute's, which carried with it two liabilities. First, although Poinsett had in mind a Washington centric establishment—having successfully made the case for including the entire presidential Cabinet on its board of directors—a number of the Institute's members were popularizing the notion that it would be truly national in its scope. Second, even though Congress had authorized its incorporation, the Institute remained a private entity.

The die of the Institute's demise was cast shortly after the Wilkes Expedition returned. Even though Congress allocated $20,000 to the Institute, it accorded itself oversight—through what is now called the Library of Congress—over the entire collection. It did not help the Institute's cause that a clergyman, Reverend Henry King, who had been assigned the position of curator, quickly proved to be incompetent.[25] That gave Ohio Senator Benjamin Tappan ammunition to openly question the advisability, and ultimately the legality, of giving a private corporation control of government property. Within a year, Congress had assumed jurisdiction of the expedition's collections and put Wilkes, himself, in charge.

Having lost its treasure and treasury, the National Institute slowly withered away. And the first attempt to bridge the gap between government and America's burgeoning scientific enterprise came crashing down. If anything, the National Institute's history probably demonstrated that it lacked not only sufficient political savvy, but also the professionalism that science, science policy, and science management required.

Its demise and the decade-long Capitol Hill struggle the Smithsonian had encountered provided an additional cautionary note for scientists of that era: Congress was more interested in science as a practical tool than science for science's sake.[26] Legislators had made their preference clear when they provided the Franklin Institute with a grant to study boiler explosions in 1830, but repeatedly rejected funding for an astronomical observatory until 1844, when the Navy made a successful pitch for a facility that would investigate hydrography, magnetism, and meteorology, as well as astronomy. It would take many more years and a war to get scientists, policymakers, and politicians back on the same page.

Chapter 2

The Civil War era and its legacy years 1860–1870

War and science have long had a symbiotic relationship. Technological superiority, although not the sole guarantor of battlefield success, generally tips the scales of combat in favor of the side that possesses the best weaponry. And for any leader seeking to ensure the most advantageous military outcome, supporting science well ahead of looming future hostilities is an extremely shrewd strategy.

The mutual benefits to science and government have guided American policy prominently in the decades since the end of the Second World War.[1] But evidence of the reciprocity was apparent more than a century and a half ago. It underscored the establishment of the National Academy of Sciences[2] at the height of the War Between the States.

President Abraham Lincoln and America's preeminent scientists of that era well understood the value of science and technology for the conduct of war. Lincoln embraced the idea of a government institution that, according to the enabling legislation, would "whenever called upon by any Department of the Government, investigate, examine, experiment, and report upon any subject of science or art."

In the late winter of 1863, Senator Henry Wilson of Massachusetts, a consummate politician who had been shepherding the National Academy bill through the upper chamber, took advantage of the imminent adjournment of a special session of Congress and asked his colleagues "to take up a bill which will consume no time, and to which I hope there will be no opposition…. It will take but a moment, I think, and I should like to have it passed." Wishing to adjourn as quickly as possible, the members of the Senate listened to a cursory reading of the bill and passed it by voice vote without opposition. The House followed suit within hours, and later that evening Lincoln signed it into law. Science had been indelibly inscribed into the national heritage. The date was March 3, 1863.

The simple historical rendition seems very straightforward, but it belies the intrigues of the previous twelve years.[3,4] The story of the National Academy of Sciences actually began in 1851, and it had little to do with war. America's preeminent scientists simply were not satisfied with the recognition their scholarly

Navigating the Maze. https://doi.org/10.1016/B978-0-12-814710-8.00002-4

25

societies provided.[5] They craved more from the government and the public. They yearned for institutions similar to those of which their European brethren boasted.

Alexander Dallas Bache, a great-grandson of Benjamin Franklin, was about to retire as president of the AAAS in 1851. But he was still superintendent of the United States Coast Survey. And, perhaps more significantly, he held the informal title, "chief" of a group right out of a Dan Brown thriller,[6] the Scientific Lazzaroni.[7] Its members, arguably, were the best and brightest scientists in America of that era, among them Louis Agassiz, a naturalist, Benjamin Pierce, a mathematician, James Dwight Dana, a geologist, and, of course, Joseph Henry. There were, to be sure, many lesser lights.

Envious of the status science had achieved in Europe through the Royal Society of London and the French Academy, the Lazzaroni started conspiring to find a way to etch the importance of science into the American psyche. Bache and Henry led the Washington cabal, while Agassiz and Pierce were firmly entrenched in Cambridge, Massachusetts. By the time the Civil War started, all four had settled on forming a select national institution of some sort that would carry with it a government imprimatur.

Recognizing the unique prospect the war offered, Bache and Henry opportunistically enlisted the support of a military man, Charles Henry Davis, to make their case to the federal government. Davis brought with him Massachusetts bona fides, as well, having been educated at the Boston Latin School and Harvard before receiving a U.S. Navy commission in 1823. He had attained the rank of captain in 1861, and rear admiral in early 1863, just as plans for a science academy were nearing fruition.

The impetus was there, but the road to forming the National Academy was by no means smooth. For more than a decade, the Cambridge Lazzaroni had made establishing a new nationally recognized science-based university their highest priority. For them, any connection to military applications carried little, if any, weight. And in Washington, Henry, who was ever mindful of the failure of the National Institute and the decade-long congressional debate over the creation of the Smithsonian, had begun to question whether it was wise to ask Congress to charter a National Academy and fund it with government money.

The leaders of the Cambridge Lazzaroni came around first and enlisted the support of Senator Wilson to advance the enabling legislation that Lincoln ultimately signed. Fearing objections from their colleague, Bache, Agassiz, Pierce, and Davis kept Henry out of the loop in the final days before Congress surreptitiously passed Wilson's bill. The secrecy of their machinations stirred up resentment not only in Henry, but also in prominent scientists throughout the country, many of whom had been listed among the 50 founding members. And of course, many of the scientists who were not accorded membership undoubtedly viewed the academy as an elitist organization that was at odds with America's democratic principles.

The Academy's birthing pains did not subside quickly. Bache, who became the first president of the Academy in April 1863, spent his early days fending off the attacks, led most prominently by William Barton Rogers, who had established the Massachusetts Institute of Technology in 1861, just before the war began. Rogers was still leading MIT when he learned about National Academy's creation. But his opposition was probably based more on politics than philosophy. And that story is worth recounting.

For more than a year, MIT's founding president had been at odds with a number of his Massachusetts compatriots, especially two of the leaders of the Cambridge Lazzaroni, Louis Agassiz and Benjamin Pierce. Both of them had been strong advocates of legislation that still stands today as one of the hallmarks and greatest successes of American Science and Technology policy: The Land Grant College Act of 1862.[8] But at the time, Rogers viewed it as an existential threat to his legacy in establishing MIT.

The Land-Grant initiative originated in Strafford, Vermont, a village that even today retains its quintessential New England charm. The travel website *Happy Vermont*[9] calls it "The Prettiest Vermont Town You've Probably Never Visited." And every March since 1801, according to the website, its residents— today numbering a scant 1000—have held a town meeting in an iconic white clapboard building that looks more like a church than a local seat of government. Despite its low profile and out of the way location, the bucolic town was the birthplace of two prominent Americans who left oversized footprints on the fabric of the nation.

William Sloane Coffin, who was born in Strafford in 1923, became one of the most famous antiwar activists of the Vietnam War era. He was chaplain of Yale University, and later senior minister at New York City's Riverside Church, and is probably the better known of the two, at least to modern American generations. But more than a century and a half before Coffin's name had become synonymous with the Vietnam War protest movement, Strafford had been home to William Morrill, who was born there in 1810. By age 30, Morrill had achieved sufficient financial success that he could afford to retire and live the rest of his life as a gentleman farmer. But he found himself drawn to politics, and that is how he left his enduring mark.

First elected to Congress as a Whig in 1854, Morris almost immediately cast his lot with the founders of the new Republican party, and thereafter ran as a Republican, first for reelection to the House of Representatives, and then for the Senate, beginning in 1866. Morrill's interest in agriculture made him a natural advocate for the establishment of agricultural colleges, which promoters had been agitating for since the late 1830s. Morrill's first foray into the arena ended in failure, when President James Buchanan, a Democrat, vetoed his legislation in 1861, siding with southern states that generally opposed the allocation of federal lands for agricultural institutions, in part, because they had less federal land available than their northern counterparts.

Morrill returned to the drawing board and broadened the land grant college mandate to include academic areas well beyond agriculture. Once the southern states seceded, Morrill resubmitted his amended legislation, which Republican President Abraham Lincoln signed into law on July 2, 1862. The act, which has become known as the Morrill Land Grant Act of 1862, transformed the American higher education landscape. It mandated the transfer of federal lands to the states for the purpose of establishing[10]

> *"...at least one college where the leading object shall be, without excluding other scientific and classical studies, and including military tactics, to teach such branches of learning as are related to agriculture and the mechanic arts, in such manner as the legislatures of the States may respectively prescribe, in order to promote the liberal and practical education of the industrial classes in the several pursuits and professions in life."*

It was the first direct federal promotion of science and technology in higher education.

Although the 1862 statute applied only to the states remaining in the Union, it was amended after the war to include the entire Confederacy. In 1890, mindful of the plight of emancipated slaves, Morrill shepherded another bill through Congress, requiring land grant institutions to admit blacks without discrimination. The Second Morrill Act, as it is known, resulted in the creation of a separate set of Negro Land Grant Colleges, when 17 southern and border states objected to racially blind admissions policies for their existing institutions. Finally, in 1994, Congress added 31 tribal colleges to the Land Grant roster, and today the full list contains 106 institutions of higher learning, many of them with preeminent programs in science and technology.[11]

Ironically, before it opened its doors to students for the first time in 1865, MIT had taken advantage of the 1862 act and had become a land grant college. Today, of course, it is one of the preeminent private American universities, although it's a fair bet that few, if any, of its students or faculty know about their first president's early recriminations.

William Barton Rogers was not the only Cambridge denizen who was at odds with the Lazzaroni and their new academy. Asa Gray, the country's best-known botanist, wearing a Harvard professorship as a badge, was a mainstay and an officer of the Boston-based American Academy of Arts and Sciences. He feared that the National Academy of Sciences would undermine the effectiveness of the American Academy and other scholarly associations, such as the American Philosophical Society and the American Association for the Advancement of Science. In addition, as a staunch believer of evolution, he had little use for Louis Agassiz, who was one of the foremost critics of Darwin's theory.

Gray also had long been a confidant of Henry, and his vocal opposition to the National Academy left the nation's renowned physicist and Smithsonian president with a difficult choice: to abandon Gray and support the Academy, or to join Gray as one of its critics. Henry cast his lot with the Lazzaroni and presided over the 1863 organizational meeting in New York that led to Bache's election. Undeterred, Gray joined forces with Rodgers, and less than two months later, he became president of the American Academy, with Rodgers serving as recording secretary.

With Gray and Rodgers hectoring the science community, Bache knew he had to act swiftly to establish the National Academy as the government's go-to organization for science and technology advice. He took it upon himself and formed committees to address problems deemed critical to a nation at war. With Admiral Davis's assistance, Bache settled on five subjects for the Academy to study:[12] (1) the uniformity of weights, measures, and coinage; (2) the impact of salt water erosion on hulls of ships; (3) the effect of iron hulls on magnetic compasses; (4) the utility of the Saxton hydrometer; and (5) the reliability of existing wind and ocean current charts. The committees worked carefully and quickly, but their findings proved to be of limited utility. Four studies initiated in 1864 resulted in similar inconsequential impacts.

In the spring of 1864, Bache fell seriously ill, and for the remainder of the war years, the National Academy of Sciences was little more than a shell organization. By design, unlike its European counterparts, it was forbidden to receive government subsidies, and surviving on contributions from its limited membership proved challenging. Bache survived to see the Civil War end in May 1865, but his incapacity almost killed the Academy. His death on February 16, 1867 was probably all that saved it. The following stipulation in his will demonstrates how much of his heart he had put into its creation, and how committed he remained to it to the day he died:[13]

> Item.—*As to all the rest and residue of my Estate, including the sum of Five thousand dollars placed at the disposal of my wife in case she should not desire to make any disposition of the same, I direct my executors hereinafter named to apply the income thereof after the death of my wife according to and under the directions of Joseph Henry of Washington, Louis Agassiz and Benjamin Peirce of Harvard College, Massachusetts, to the prosecution of researches in Physical and Natural Science by assisting experimentalists and observers in such manner and in such sums as shall be agreed upon by the three above-named gentlemen, or any two of them, whom I constitute a Board of Direction for the application of the income of my residuary estate for the above objects, after the death of my said wife. The class of subjects to be selected by this Board, and the results of such observations and experiments, to be published at the expense of my Trust Estate under their direction out of the income thereof but without encroaching on the principal.*

Bache effectively placed the financially strapped National Academy in temporary receivership, naming three leaders of the Lazzaroni as administrators. Henry, the most prominent of them, acceded to the wishes of Bache's widow and reluctantly took over as president of the Academy. When Nancy Clark (Fowler) Bates died three years later, in accordance with her late husband's will, the Academy was the beneficiary of a $42,000 trust bequest.[14]

Joseph Henry did not sit idly by waiting for the day the Academy would become solvent. Upon assuming his presidency, he moved to fundamentally alter the structure and purpose of the institution. Instead of solely providing the government with expertise to solve practical problems, the Academy would begin to emphasize science in the abstract. Membership in the Academy, which would be enlarged well beyond the original 50, would be based principally on proven distinction in original research. Finally, reflecting Henry's Washington base, meetings of the Academy would only take place in the nation's capital once a year.[15]

The Civil War left extremely deep, hurtful scars on America's soul, and more than a century and a half later, many of them are still visible. However, for science and technology, the War era's indelible impacts have been extraordinarily positive. The epoch produced the National Academy of Sciences and the Land Grant Colleges. But, as the clouds of war cleared, it was evident that those two tangible outcomes presaged something of far greater significance. A new era was dawning in which the role of science and technology in American life would be impossible to ignore. New policies and new structures inside and outside government would be needed to address the extraordinary changes that were in store.

It is difficult to say whether the Civil War's end in 1865 was a sharp turning point, or just an inflection, but without question, the last 35 years of the 19th century produced rapid industrialization, unfettered extraction of natural resources, development of advanced agricultural methods, explosive economic growth, and extraordinary wealth disparity. Technology was the driver and enabler of the dramatic changes that swept the nation.

The 1862 Homestead Act,[16] for example, opened up millions of acres of farmland, but it was the transcontinental railroads—technological marvels at the time—that made the farms significant economic contributors. That was also true for mining and lumber.

Small electric motors played a similarly transformative role. Developed in the 1880s and 1890s, they freed factories and mills from a need to be sited on rivers or a reliance on centralized steam power. Urbanization was one result. Dramatically improved worker productivity was another, although it was often at the expense of poorly paid laborers suffering under abysmal factory conditions.

During that go-go era, industry and government generally worked hand in glove, with corruption not in short supply. Mark Twain (Samuel Langhorne Clemens) captured the essence of the period in a satirical novel,[17] *The Gilded Age*, which he co-authored with Charles Dudley Warner in 1873. The name they appended to that era of extraordinary excess, wealth creation, inequality, and rapid technological change has endured.

As the Gilded Age emerged from the ashes of war, scientific bureaus, commissions, and offices soon began to proliferate within the federal government. The stage for the growth of a science bureaucracy—modest by today's standards, but quite significant for those times—had actually been set in 1862, three years before the war ended. On May 15 of that year, President Lincoln had signed legislation creating the Department of Agriculture, although not according it Cabinet status. Four days later he had signed legislation providing subsidies and loans for constructing a transcontinental railroad,[18] and a day after that, the Homestead Act. Finally, on July 2, he put his stamp of approval on Morrill's Land Grant bill. If the South had not seceded, it is doubtful Congress would have been able to pass any of those bills, because they represented an expansion of federal authority, which the Confederate states would have found anathema. Taken together though, the four acts presaged a move toward a more expansive interpretation of Article I, Section 8 of the Constitution. The more generous reading soon would be used to justify federal sponsorship of scientific research well beyond the military. It would begin with agriculture and geology.

The enabling legislation that created the Department of Agriculture makes it clear science was intended to play a central role, and to this day, the department retains much of its early science flavor. The actual language is worth considering:[19]

Be it enacted by the Senate and House of Representatives of the United States of America in Congress assembled, *That there is hereby established at the seat of Government of the United States a Department of Agriculture, the general designs and duties of which shall be to acquire and to diffuse among the people of the United States useful information on subjects connected to agriculture in the most general and comprehensive sense of the word, and to procure, propagate, and distribute among the people new and valuable seeds and plants.*

Section 2 And be it further enacted, *That there shall be appointed by the President, by and with the advice and consent of the Senate, a "Commissioner of Agriculture," who shall be the chief executive officer of the Department of Agriculture, who shall hold his office by a tenure similar to that of other civil officers appointed by the President, and who shall receive for his compensation a salary of three thousand dollars per annum.*

Section 3. And be it further enacted, *That it shall be the duty of the Commissioner of Agriculture to acquire and preserve in his Department all information concerning agriculture which he can obtain by means of books and correspondence, and by practical and scientific experiments, (accurate records of which shall be kept in his office,) by the collection of statistics, and by any other appropriate means within his power; to collect, as he may be able, new and valuable seeds and plants; to test, by cultivation, the value of such of them as may require such tests; to propagate such as may be worthy of propagation, and to distribute them among agriculturist. He shall annually make a general report in writing of his acts to the President and to Congress, in which he may recommend the publication of papers forming parts of or accompanying his report, which report shall also contain an account of all moneys received and expended by him...to acquire and preserve...all information concerning Agriculture...by means of books and correspondence, and by practical and scientific experiments... make a general report...in which he may recommend the publication of papers...*

The words of Section 3 embody what we recognize to this day as the essence of scientific research: how we obtain new information, gain new insights, and make discoveries, and what we do with the products of our research once we have completed the work.

Although the intent of the legislation was clear, a quarter of a century would pass before Congress recognized the importance of connecting the Agriculture Department's research mandate to the core competencies of the Land Grant colleges. The Hatch Act[20] of 1872 made that connection by establishing agricultural experiment stations on the campuses. But communicating the results of the research to farmers, especially in rural areas, was still a problem. In 1914, Congress addressed the glaring deficiency, passing the Smith-Lever Act,[21] which created a cooperative extension system, managed by the Land Grant institutions, to help farmers take advantage of agricultural research carried out at those institutions. By that time, the Department of Agriculture had additional heft, having been accorded Cabinet level status in 1889.[22]

Chapter 3

The Gilded Age 1870–1900

As the Lazzaroni recognized when they successfully convinced Congress and President Lincoln, in the midst of the Civil War, to establish the National Academy, science was generally a high-value proposition for military interests within the federal government. But as the Gilded Age unfolded, the cozy relationship between science and the military slowly began to fray, at least in the minds of many members of Congress.

In fact, on Capitol Hill, those tensions were nothing new, as a flashback to 1825 reveals. That year, John Quincy Adams became the sixth president of the United States, having been chosen by the House of Representatives following an election in which no candidate received a majority of Electoral College votes. JQA, as he was known, was not just a successful politician from Massachusetts, he was also a Fellow of the American Academy of Arts and Sciences, the holder of the Boylston Professorship of Rhetoric and Oratory at Harvard, and a renowned statesman.

In his first annual address[1] to Congress on December 6, 1825, Adams made two requests related to science. First, he argued for federal support for university research on "weights and measures," a subject that had a strong Constitutional connection through Article I, Section 8. Second, he proposed an astronomical observatory—for which there was no direct Constitutional mandate—making his case with the following words:

> Connected with the establishment of an university, or separate from it, might be undertaken the erection of an astronomical observatory, with provision for the support of an astronomer, to be in constant attendance of observation upon the phenomena of the heavens, and for the periodical publication of his observances. it is with no feeling of pride as an American that the remark may be made that on the comparatively small territorial surface of Europe there are existing upward of 130 of these light-houses of the skies, while throughout the whole American hemisphere there is not one. If we reflect a moment upon the discoveries which in the last four centuries have been made in the physical constitution of the universe by the means of these buildings and of observers stationed in them, shall we doubt of their usefulness to every nation? And while scarcely a year passes over our heads without bringing some new astronomical discovery to light, which we must fain receive at second hand from Europe, are we not cutting ourselves off from the

Navigating the Maze. https://doi.org/10.1016/B978-0-12-814710-8.00003-6

means of returning light for light while we have neither observatory nor observer upon our half of the globe and the earth revolves in perpetual darkness to our unsearching eyes?

The response to both proposals was tepid, although 2 years after Adams lost his 1828 reelection bid to Andrew Jackson, Secretary of the Treasury, Samuel D. Ingraham, at the request of the Senate, authorized an activity within his department that would eventually become the Bureau of Weights and Measures.[2] That same year, 1830, Congress established the Depot of Charts and Instruments under the auspices of the Navy.[3] But it wouldn't be until 1844, with Adams by then occupying a seat in the House of Representatives,[4] that Congress finally acquiesced and expanded the Depot's responsibilities to include astronomy when it consolidated a number of maritime military research activities within a new institution, The Naval Observatory.

The Survey of the Coast,[5] sometimes called the first federal science agency, followed a similar, perhaps even more fraught, congressional trajectory. At the request of Thomas Jefferson in 1807, Congress approved the project, tasking the Treasury Department with its management. But lack of adequate equipment, and the outbreak of the War of 1812, caused it to remain in a state of suspended animation until 1816. By that time, Jefferson had left office, and James Madison was already completing his second term. More significantly, Congress had begun to examine whether the Army and the Navy were better equipped to carry out the project more efficiently, faster, and at far lower cost as part of their routine charting activities.

In 1818, a little more than a year after the Survey had finally begun to make measurements, Congress reached its decision: the superintendent of the project, Ferdinand Rudolph Hassler,[6] one of the world's leading meteorologists, an émigré from Switzerland and, by many reports, something of a character, would need to find another job. The Survey of the Coast was now in the hands of the Navy, with civilians prohibited from participating. It remained that way until 1832, when Congress decided to revisit the issue under pressure from maritime enterprises. It had become increasingly clear that the Navy had botched the task. Whatever little it had generated in the way of charts were deemed almost worthless.[7] The Survey had suffered terribly, and so, too, had Hassler, whose life, both personal as well as professional, had nearly come undone.

Reversing itself and disregarding the strenuous objections of one of the members of the House of Representatives,[8] Congress reestablished the project as the Coast Survey under the terms of the 1807 act. Turning the clock back completely, it named Hassler—who had spent the last 2 years leading the weights and measures activity in the Treasury Department—as the superintendent for the second time. Hassler was very eager to resume his work on the Survey, but not before he made two demands of Congress.

First, noting that commercial interests would be major beneficiaries of the project, Hassler argued that the Treasury Department should have responsibility for the Survey, as stipulated in the 1807 law, and that having the Navy and the Treasury Department share the responsibility would be unwieldly and unproductive.

Second, Hassler argued that the techniques and instruments he would bring to bear on the Survey would have scientific impacts well beyond the development of charts and maps. Therefore, he proposed that an oversight board should consist solely of scientists with appropriate expertise and training.

In effect, Hassler was making the case for a scientific project funded by the federal government under the control of scientists, rather than bureaucrats, politicians, or the military, even if the military had a dog in the fight. It was an argument scientists would continue to make over the course of many decades.

Hassler got most of what he demanded of Congress in 1832, but the link between science and the military continued to be the subject of debate. Matters came to a head 9 years later when a "select" congressional committee began to examine whether the Survey's spending was justified, and whether its productivity would be improved under Navy oversight. After considerable deliberation, the committee decided simply to shorten Hassler's leash. Starting in 1843, the Survey would be required to clear all new plans with the White House, once its board—consisting of the superintendent (Hassler), his two senior assistants, two naval officers, and four engineers—had signed off. The decision was not what Hassler had pressed for, but the decision for him, turned out to be moot. He fell ill in the summer of 1843 and died the following November.

Alexander Bache, who later became one of the leaders of the Lazzaroni, and the first president of the National Academy of Sciences, assumed the reins of the Coast Survey, which he ran until his death in 1867. Bache was a much smoother operator than Hassler, and was able to keep congressional critics largely at bay by convincing them that the Survey was carrying out its work efficiently, competently, and with an eye toward saving money. As a graduate of the United States Military Academy at West Point, he also had a good feel for the politics of the military, and succeeded in fending off the Navy's continuing attempts to gain the upper hand.

Congress spent 7 years dithering over the weights and measures issue; two decades debating whether to authorize a Naval Observatory; and more than 35 years flip-flopping over military management of the Coast Survey. With hindsight, that fraught history should have been a warning to the Lazzaroni—especially Bache—and their aspirations for science in the post-Civil War era.

The bitter internecine conflict ended on May 9, 1865, leaving a death toll of 750,000 soldiers[9]—far more than the total number of American casualties in all other wars combined—and saddling the nation with a debt exceeding $2 billion. Faced by daunting financial needs for Reconstruction, Congress was fully prepared to put the Army and Navy on a diet. That was not good news for anyone trying to make the case for military stewardship of science.

The failure of the National Academy of Sciences to make any significant contribution to wartime technology certainly did not help the argument that science writ large was indispensable to the armed services. Nor did it help that Lincoln, who had been a strong science advocate, was assassinated less than a month before the war ended. And Vice President Andrew Johnson, who replaced him, was such a divisive figure[10] that even if he had grabbed Lincoln's science baton, it is doubtful he would have had much impact.

For American science and technology policy, the Gilded Age was one of profound transformation and disruption. The military not only shed its responsibilities in managing research and exploration programs, it also altered the curriculum and administration of its flagship training institution at West Point, New York. The United States Military Academy, as it is formally known, traces its origin to George Washington's desire to establish a national military school focused on the art and science of war. Concerned about an elitist image such an institution might project, and the lack of any constitutional justification for establishing it in the first place, Congress ultimately settled simply on creating a "Corps of Artillerist and Engineers" at West Point.[11] The year was 1794.

After becoming president, Thomas Jefferson, who had been an earlier critic of Washington's proposal, reversed course and began to press for the creation of a national university focused on science and engineering.[12] Congress adopted his plan, in part, and in 1802 Jefferson signed a bill establishing a "Corps of Engineers" at West Point that would "constitute a military academy."[13] As its first superintendent, Jefferson chose Jonathan Williams, a grand nephew of Benjamin Franklin, better known for his interest in science than for his minimal military service. But it is to Sylvanus Thayer, the "Father of West Point," who became superintendent in 1817, that credit truly goes for making the Military Academy the center of engineering education in the United States. It remained so until the end of the Civil War.

The eponymous Gilded Age is best known for extreme wealth inequality and excesses of the privileged class, but it was also a period of ascendance for the liberal arts. At West Point, the cultural renaissance meant the end of a science and engineering emphasis. It also meant the transfer of control from the Corps of Engineers to the Secretary of War.

As the military departed from its perch atop the American science and engineering pyramid, science bureaus began to proliferate elsewhere in the federal government, and private institutions and individuals began to vie for a larger piece of the scientific action. At the time, no effective federal structures existed for managing competing interests, avoiding unnecessary duplication of activities, preventing political considerations from trumping scientific judgment, and ensuring that federal funds were being used wisely. The surveys of Western lands, which proliferated after the Civil War ended, illustrate the problems clearly.[14,15]

Unless you've visited the "City of Gnomes," located midway between South Fork and Gunnison deep in the Colorado Rocky Mountains, it is doubtful you've heard of George Wheeler. If you haven't made a pit stop in Glenrock, Wyoming, off Interstate 25, you probably don't know the name Ferdinand Hayden. And even if you've hiked through Kings Canyon National Park in California or boated on Lake Powell in northern Arizona and southern Utah, you probably have no idea who Clarence King and John Wesley Powell were, or why those popular tourist sites are named for them.

Wheeler, Hayden, King, and Powell all played major roles exploring the West during the early years of the Gilded Age. Their expeditions were not for cowards. Powell, for example, took on the Colorado River rapids, rafting six thousand feet below the rim of the Grand Canyon, never knowing what might be around the next bend, and whether he and his team would survive the challenge. His journal entry from August 13, 1869 captures that apprehensive mood:[16]

We are three quarters of a mile in the depths of the earth, and the great river shrinks into insignificance as it dashes its angry waves against the walls and cliffs that rise to the world above; the waves are but puny ripples, and we but pigmies, running up and down the sands or lost among the boulders.

We have an unknown distance yet to run, an unknown river to explore. What falls there are, we know not; what rocks beset the channel, we know not; what walls rise over the river, we know not. Ah, well! we may conjecture many things. The men talk as cheerfully as ever; jests are bandied about freely this morning, but to me the cheer is somber and the jests are ghastly.

With some eagerness and some anxiety and some misgiving we enter the canyon below and are carried along by the swift water through walls which rise from its very edge.

Wheeler, Hayden, and King's adventures were no less fraught. Having met nature's challenges with steely resolve, both they and Powell returned from their respective expeditions primed to challenge each other for preeminence in the public and scientific arenas. They wrangled over federal money, competed for dominance in the territories they explored, and vied for political favors. And they left their collective marks on agronomy, botany, cartography, ethnology, geography, geology, hydrology, minerology, mining, paleontology, and zoology.

But their back stabbing and incessant arguing over which government agency or department should have control over their surveys finally got to be too much for Congress to bear. Matters came to a head in 1878, the same year Joseph Henry died, after having led the National Academy of Sciences for two decades and the Smithsonian for more than three decades. He had been one of the more circumspect leaders of the Lazzaroni, and his cautious approach to science policy and politics had been his trademark during years of public service. He had held the Academy together in very trying times,[17] but under his stewardship, it had become more a forum for scientific discussions among America's most distinguished researchers and less an advisory organ to which government officials could turn.

Henry's death was a great loss for American science, but it opened up the possibility that the National Academy might begin to play the advisory role that Alexander Bache and a number of its other founders had imagined. Enter Othniel Marsh, the renowned Yale paleontologist, who took over as the Academy's interim president. He immediately made it clear that under his stewardship the Academy was open for advisory business.

The timing was fortuitous: Representative Abram Hewitt of New York was already looking for help in sorting out the survey controversies. Hewitt was a force to be reckoned with. He was a lawyer, an industrialist, and had been chairman of the Democratic National Committee in 1876 and 1877. He would become mayor of New York City in 1887.

In short order, Hewitt used his position and power on the House Appropriations Committee to call for an Academy study of the entire mess, covering the activities of Wheeler, Hayden, King, and Powell, as well as the Coast and Geodetic Survey, as the Coast Survey was then known. It was the first time Congress had made a request of the Academy, and Marsh was intent upon getting it done right.[18]

Raymond Canning Cochrane, who has chronicled the first one hundred years of the Academy, describes the seminal report and its reception in Congress as follows:[19]

> The report's principal objective was the attainment of an accuracy and economy impossible in the five surveys. It recommended that the Coast and Geodetic Survey be transferred from the Treasury Department to Interior and that the Survey assume responsibility for all measuration in the public domain. It proposed that Congress establish a new and independent U.S. Geological Survey in the Department of Interior to undertake all study of geological structures and economic resources of the public land areas. The Land Office in Interior would be limited to control of the disposition and sale of public lands. The Academy committee recommended that, when that task had been accomplished, the Hayden, Powell and Wheeler surveys west of the hundredth meridian should be discontinued, except those for military purposes. It also recommended discontinuance of the geographical and geological surveys of the Department of the Interior and the mapping surveys of its General Land Office.

> Finally, the Academy report recommended formation of a commission comprising the Commissioner of the Land Office, the Superintendent of the Coast and Geodetic Survey, the Director of the U.S. Geological Survey, the Chief of the Corps of Engineers, and three others appointed by the President to study and report to Congress a standard of classification and valuation of the public lands and a system of land-parceling survey. Although the public lands in the West totaled 1,101,107,183 acres, for geological and climate reasons the larger portion had no agricultural value; and as the Academy report said, the existing method of parceling out homesteads was therefore impractical and undesirable.

The House committee that requested the study adopted the entire plan of the Academy in a bill reported to the Congress, and a jubilant Marsh wrote his fellow committeeman William B. Rogers: "You will be pleased to know that our Report was as well received in Washington as it was by the Academy…"

The last phrase proved to be more a matter of hope than reality.[20] The House of Representatives did seem poised to accept the Academy's recommendations without any changes, but Hayden, who loathed the findings that stripped him of his survey authority, had other ideas. He found allies among Western members who took strong issue with the public lands provision, and he urged them to kill the bill. They did succeed in having the land language removed, but they fell short in derailing the balance of the legislation. Along the legislative trail, they left their mark on the Academy, accusing its members of passing judgment on an issue beyond the confines of science.

Hayden was not done. As the bill moved to the Senate, he continued his lobbying efforts, and seemingly succeeded, when the upper chamber voted not only to accept the House language that struck the land provisions, but also to discontinue all the surveys except Hayden's. For the moment, it appeared that Hayden had won a significant victory, and that the Academy had come up short in its first attempt to be a major force in science policy.

Hayden was content, but in the House, Abram Hewitt, who had commissioned the Academy report, was not. As anyone familiar with the ways of Washington knows, the power of the purse reigns supreme. And as an appropriator, Hewitt held all the high cards. Without a doubt, he was less than thrilled with Hayden's machinations, and he used his clout to commit the Senate amendments to the dust bin of history. Hewitt inserted language in the appropriations bill that discontinued Hayden's survey, as well as Powell's and Wheeler's, and gave the Interior Department control of the Geological Survey, while keeping the Coast and Geodetic Survey in the Treasury Department. As for the public land provision, Hewitt, who was from New York, knew enough not to pick a fight he could not win, and left the issue out of the bill entirely.

Hayden had lost his very public battle, tarnishing his image in the process, but Powell and Wheeler, who now needed to find other work, were not so tainted. Powell returned to his home at the Smithsonian, and Wheeler contented himself with writing scientific papers, after receiving a promotion to the rank of captain in the Army Corps of Engineers. As for King, fortuitously, he had just about wrapped up his survey activities, and would have been primed for a new assignment, regardless of the congressional outcome. He got his chance shortly after President Rutherford B. Hayes signed the appropriations bill into law on March 3, 1879, becoming—with Powell's backing—director of the newly created U.S. Geological Survey. It was not a match made in heaven, and King resigned 2 years later when James Garfield—who had been an ally of Hewitt's in the House—became president. Garfield immediately named Powell director, and it is Powell who deserves credit for guiding the Survey from an extemporized origin into a permanent civilian science agency.

Congress had dealt with the difficulties the surveys had created, but concerns about science spending, proliferation of scientific bureaus, duplication of federally supported work, political favoritism, and allegations of corruption persisted. In 1884, Congress decided more needed to be done, and established a joint House-Senate commission under the chairmanship of Senator William B. Allison of Iowa.[21] Intended initially to examine the organization of four federal activities—the Signal Service, the Geological Survey, the Coast and Geodetic Survey, and the Navy's Hydrographic Office—the commission's work broadened substantially. It delved into whether the federal government should engage in or support research that does not have any demonstrable practical objective; whether Congress should have a direct voice in evaluating science projects; whether the federal government should establish a Department of Science to coordinate research activities; whether Congress or the president should have a science advisory committee; and whether the federal government more generally was spending too much money on scientific research.

The Allison commission held hearings and deliberated for a year and a half before issuing its recommendations. As the six-member committee began to organize itself, Theodore Lyman, a Massachusetts representative and the only member of the commission with scientific credentials, approached Othniel Marsh, then the president of the National Academy of Sciences. In asking him for the Academy's assistance, Lyman especially wanted advice on how the government should coordinate its scientific activities to achieve the best possible outcomes in the most efficient manner. The Academy acted quickly, reporting its recommendations to the Commission only three months after receiving Lyman's request.

In brief, the Academy report[22] first asserted that "the administration of a scientific bureau or department involves greater difficulty than that of a pure business department," requiring "a combination of scientific knowledge with administrative ability, which is more difficult to command than either of these qualities separately." The report then argued that the difficulties are exacerbated when there is no central authority to coordinate disparate activities. Therefore, the Academy concluded, the government should establish a Department of Science under the direction of a science administrator.

Recognizing that establishing a new department might be a political non-starter, the report stated that in the absence of a Department of Science, all federal science activities should be consolidated into a single existing department, and divided among four bureaus[23] within that department. To oversee the bureaus, the Academy proposed a commission consisting of five government and four non-government members.[24] Finally, the report urged the government not to undertake work that was more appropriate to individual investigators, universities, or the states; but instead should focus its activities on increasing knowledge that would promote the general welfare, thereby implying the importance of practical outcomes in federal science activities.

From a 21st century perspective, the last proposition seems very strange, given the extensive role the federal government plays today in both sponsoring academic research and maintaining major research facilities open to scientists of every stripe.[25] But in 1884, none of today's federal agencies that dole out tens of billions of dollars annually to America's research community were in existence, or could even have been imagined.

To paint the contrasting policy pictures with an even broader brush, the differences between 1884 and modern America are so immense, that if the National Academy of Sciences and its sister organizations, the National Academy of Engineering and the National Academy of Medicine (known collectively as the National Academies), were asked for a set of recommendations today, its list would likely not include a single one of the items Marsh sent Allison at Lyman's request. In fact, a 21st-century list would almost be the antithesis of the Academy's 1884 set of prescriptions.

Lyman had not asked Marsh to assess whether the federal government was spending too much money on research. As a result, the Academy remained silent on that issue.

The National Academy membership included many of America's leading scientific lights, but its role in the policy and political arenas was not well established. In truth, it did not speak for the nation's scientific community, and witnesses at the Allison Commission hearings made that quite clear, often to keep their own ox from being gored.

John Wesley Powell is a prime example. He had replaced Clarence King in 1881 as director of the U.S. Geological Survey, but he still maintained his directorship of the Smithsonian's Bureau of Ethnology. In his testimony, he labeled the Academy's proposal for a nine-member commission representing military, civilian, governmental, and private interests completely unworkable. Instead—as you might guess—he proposed handing over all federal science programs to the Smithsonian, whose board, he asserted, had sufficient breadth and expertise to oversee their conduct.

William E. Chandler, Secretary of the Navy, is another example. He was intent upon keeping naval science under control of the military, and spoke out strongly against any centralized control of government science, either through a new Department of Science, or through consolidation of all activities into an existing one. Instead, he testified in favor of having each government department maintain authority over science programs that best suited its needs. In other words, for Chandler, the status quo was just fine.

Simon Newcombe, who served Chandler as the Navy's first scientist, sided with his boss, but with a caveat. Newcombe proposed that the president be required to appoint a single administrator to oversee and coordinate science activities across the federal government. Newcombe is often forgotten in the annals of American science and technology policy, but his recommendation deserves credit, as the precursor of today's White House structures: The Office of Science and Technology Policy (OSTP), the President's Council of Advisors on Science and Technology (PCAST), and the National Science and Technology Council (NSTC), all of which are chaired or co-chaired by the President's Science Advisor, or more technically, the Assistant to the President for Science and Technology.

Days turned into weeks, weeks into months, and months into a year and a half, as the Allison Commission methodically worked its way through the technical and administrative issues. As more time passed and more witnesses testified, it became apparent that the commission had opened up a Pandora's box. Although political philosophy was not in its original charter, a debate over the proper role of government in the lives of the public became enmeshed in the discussions of financial and structural issues.

Louis Agassiz, the renowned biologist and geologist from Harvard, one of the leaders of the Lazzaroni and a founder of the National Academy of Sciences, was a staunch believer in limited government, and he made his opinion clear. Government should stay out of the lives of its citizens as much as possible, he said, and insofar as science is concerned, it should undertake or sponsor only those activities that lie beyond the capabilities of universities, private citizens, or associations, such as the National Academy of Sciences, the American Academy of Arts and Sciences, and the American Association for the Advancement of Science. To some extent, his position echoed the message in the Academy Report, but unlike the Academy's, it was predicated on *laisssez-faire*, or libertarian beliefs, rather than on practical public policy rationales that underscored the Academy's assessment.

Powell was not willing to let Agassiz's challenge go unanswered. He publicly attacked not only the foundation of Agassiz's philosophy, but also his personal integrity—quite viciously, in fact—accusing his adversary of planning to use the great wealth he had accumulated over many years to establish supreme control over American science.[26] In terms of the nexus between political philosophy, economics, and science policy, Powell made cogent arguments that still have immense resonance more than 130 years later. The following words[27] capture the kernel of his beliefs:

(6) ALL GOVERNMENTAL RESEARCH STIMULATES, PROMOTES AND GUIDES PRIVATE RESEARCH

Possession of property is exclusive; possession of knowledge is not exclusive; for the knowledge which one man has may also be the possession of another. The learning of one man does not subtract from the learning of another, as if there were a limited quantity to be divided into exclusive holdings; so discovery by one man does not inhibit discovery by another, as if there were a limited quantity of unknown truth. Intellectual activity does not compete with other intellectual activity for exclusive possession of the truth; scholarship breeds scholarship; wisdom breeds wisdom, discovery breeds discovery. Property may be divided into exclusive ownership for utilization and preservation, but knowledge is utilized and preserved by multiple ownership. That which one man gains by discovery

*is a gain by other men. And these multiple gains become invested capital, the inter-
est on which is all paid to every owner, and the revenue of new discovery is bound-
less. It may be wrong to take another man's purse, but it is always right to take
another man's knowledge, and it is the highest virtue to promote another man's
investigation. The laws of political economy that relate to property do not belong
to the economics of science and intellectual progress. While ownership pf property
precludes other ownership of the same, ownership of knowledge promotes other
ownership of the same, and when research is properly organized ever man's work
is and aid to every other man's.*

In essence, Powell made intellectual and economic arguments for federal support of scientific research, as well as for open communication of research
results. They are guiding principles for federal science and technology policy
in modern America, although it is likely that few of today's policymakers recognize their connection to Powell.

The Allison Commission, which had been authorized on July 7, 1884,
concluded its work on January 30, 1886. During its 18 months of existence,
it had contended with charges of personal aggrandizement leveled by one witness against another; turf battles between government bureaus; charges of corruption that heated up dramatically after Grover Cleveland—the first Democrat
to be elected since the end of the Civil War—became president on March 4,
1885; and battles over political philosophy it had never envisioned. When it
finally called it a day, the Commission elected not to alter any existing laws
or recommend any changes to the conduct of science. Nor did it propose any
reorganization of research structures within the federal government. In so
doing, it firmly rejected the idea of a Department of Science, but tacitly affirmed
the importance of science to the national interest that Powell had provided.
If there was any loser in the entire affair, it was probably the National Academy
of Sciences, which came under criticism for blundering badly in its failure to
convince any member of the commission of its principal recommendation:
the creation of either a Department of Science, or an alternate centralized
administrative science structure.

The Allison Commission affair left a blemish on the National Academy of
Sciences' advisory reputation. But the Academy's failing would earn it a minor
demerit compared with what was about to unfold for American science, more
generally, in the last decade of the 19th century. The Commission's 1886
decision to accept the status quo and not change the way the nation managed
its science portfolio would prove to be a false indicator of how members of
Congress truly felt.

The unease began to play out 6 years later, and involved America's most
respected science administrator, John Wesley Powell,[28] who had been directing
the U.S. Geological Survey since 1881. He had weathered the survey storm of
1878, and had seen his political philosophy tacitly embraced by the Allison
commission. But for years, he had been unsuccessful in getting Congress to

recognize that water management and irrigation were critical issues in the rapidly developing arid West.

The National Academy had come to his assistance when it submitted its 1878 report to Abram Hewitt. It had endorsed Powell's recommendation that the U.S. Geological Survey develop policies and regulations for land classification and water rights. In spite of the Academy's position, Congress had refused to go along. Instead, it had responded by restricting the Survey's work to data collection and barring it from engaging in all policy and regulatory matters.

Now, 10 years later, members of Congress representing Western states were beginning to grasp the gravity of the water issues. With their prodding, the House and Senate passed a joint resolution directing the Secretary of the Interior to commission an irrigation study.[29] It specifically called for policy recommendations on dams, reservoirs, and waterways, and it surprised no one that Powell's Geological Survey would be assigned the task.

Perhaps Congress had not foreseen the action the Executive Branch would take, but with or without congressional approval, the General Land Office elected to suspend all filings until the survey had been completed. Powell immediately saw the opening he had long been waiting for: to put an end to haphazard development and replace, it with sustainable land use policies and regulations based on scientific data, especially those resulting from comprehensive irrigation and drainage surveys.[30] He emphasized that he would need substantial time to complete his comprehensive studies.

Western developers with major commercial interests would have none of that, and they reacted swiftly. In their view, it was acceptable for scientists to generate data, and even make recommendations on specific projects, but it was completely unacceptable for scientists to hold commercial interests hostage to their findings, more broadly—which, of course, was Powell's intention.

In light of the moneyed opposition, it didn't take Congress long to give Powell his irrigation walking papers. Appropriators quickly eliminated every penny of funding his survey work required.[31] But, having been stung by Powell's overreach, legislators decided to send a stern message to America's scientists: keep your noses out of things political and stick to things scientific. To make sure the recipients got the message, Congress slashed funding in 1892, not only for the U.S. Geological Survey, but also for the Coast and Geodetic Survey, the Smithsonian Institution, the Naval Observatory, and the Ethnology Bureau. Lesson learned!

In spite of the painful rebuke, the National Academy, now under the leadership of Wolcott Gibbs,[32] decided to weigh in on another policy matter, one that was only slightly less fraught. While water was a serious Western development issue, in reality, it only directly affected people who were living in the arid West or who had commercial interests there, such as the powerful railroads. From the distance, there wasn't much for anyone else to see. That was not true about forests.

Images of deforestation were not easy to ignore, and protection of forests had been on the minds of conservationists for many decades. For a number of years, forest preservation had gained little traction, but in 1891, Congress finally authorized the president to dedicate forested lands. During the next 2 years, 18 million acres received the reserve designation, although without any regulations regarding management. Gibbs thought the time was ripe for the Academy to weigh in, and with the backing of a number of prominent conservationists, he approached Hoke Smith, Secretary of the Interior and a noted reformer, with the suggestion that the Academy would be happy to carry out a forestry study if the secretary were to make the request. Hoke welcomed the proposal, and in June 1895 made a formal request—along with $25,000—for an Academy study.[33]

On February 22, 1897, shortly before William McKinley entered the White House, President Cleveland issued a forest proclamation. Using the Academy's preliminary report as the basis for his action, he designated 13 new forest preserves covering more than 21 million additional acres. But the Academy's celebration was to be short lived. Congress challenged the president's authority to make such a designation, and shelved the bulk of the Academy's final report, which contained detailed recommendations on the management of the reserves. Congress did provide funding for fire protection in the final 3 years of the 1890's, but it would wait until the dawn of a new century for the full weight of the Academy's recommendations to finally take hold.

Before we close the curtain on the Gilded Age, we need to take a brief look at the status of medicine and public health policy in the last few decades of 19th century America. The narrative is relatively short.

Except for the armed services, the states and private practitioners had full responsibility for the nation's public health issues.[34] Among them was little, if any, coordination, and until Massachusetts broke the ice in 1869, there was not a single board of health. The American Medical Association, founded in 1847, and the American Public Health Association, established in 1872, led the way on matters of public health. But by 1875, in competing proposals, both organizations had begun to agitate for a centralized board of health, or alternatively, a national department of health.

Nothing focuses the political mind better than a crisis, and 1878 delivered one in the form of a yellow fever epidemic in the Mississippi River Valley. The scourge, which eventually claimed as many as 20,000 lives, put quarantine issues at the top of the public health agenda. It was the only practical option policymakers had, since the cause of yellow fever was not known. (It would be discovered 22 years later by Army Medical Corps physicians.[35])

In 1879, after considerable debate and political jousting, Congress finally agreed[36] to establish a National Board of Health (NBH) charging it with three tasks: "(1) obtaining information on all matters affecting public health; (2) advising governmental departments, the Commissioners of the District of Columbia, and the executives of several states on all questions submitted by them—or whenever in the opinion of the NBH such advice may tend to the pres-ervation and improvement of public health; and (3) with the assistance of the National Academy of Sciences, reporting to Congress on a plan for a national public health organization, with special attention given to quarantine, and espe-cially regulations to be established among the states, as well as a national quar-antine system."[37]

Three months later, Congress enacted additional legislation—"An Act to Prevent the Introduction of Contagious Diseases into the United States"—conferring on the National Board of Health new quarantine powers for a 4-year period. The act removed the quarantine authority from the Marine Health Sys-tem, to which Congress had granted it only 13 months earlier. That action vio-lated a cardinal rule of politics: once you have granted someone something, don't expect to be able to take it away without a fight. The seeds of National Board of Health's failure had been sewn.

The Marine Health Service fought the National Board of Health every step of the way, taking full advantage of congressional states' rights proponents, who objected to the broad federal powers the Board had been given. Four years after it had been established, the National Health Board and the health research activ-ities it fostered were history. By refusing to reauthorize the Board, Congress effectively ended its mandate, although as an entity it remained on the books until 1893, when it met its demise with finality.

Although the Marine Health Service recovered its quarantine responsibili-ties, it did not have a mandate to fulfill any of the medical research objectives of the National Board of Health. It remained for the Army Medical Corps to pick up the pieces. Bailey K. Ashford, John Shaw Billings, Walter Reed, George Sternberg, and J.J. Woodward are names that stand out as leaders of those research efforts. They achieved success, in spite of limited budgets, largely through their brilliance and dedication.[38]

As the 19th century came to a close, American science stood on the cusp of ascendancy, but many policy questions and political considerations remained unresolved. It would take two major wars before the nation's science and sci-ence and technology policy would achieve the prominence we see today. The next chapter will take us through the impacts of those two world conflicts and set the stage for the modern era of American dominance.

Chapter 4

A new century: A new America 1900–1925

William B. Allison left his imprimatur on science policy through the congressional commission he led from 1884 to 1886. Independently wealthy and a political striver, he had reached the top rung of Senate Republicans by 1901. At the time, the current positions of Senate majority and minority leader did not exist, but Allison and his Rhode Island colleague, Nelson W. Aldrich, effectively ran their party from positions as leaders of the Republican Steering Committee.[1]

The Senate had always been a club, and at the turn of the century, its 90 members, representing 45 states, adhered to the well-worn tradition. The more raucous House had 357 members, on average one for every 214,000 U.S. residents.[2] To conduct their daily affairs, all 447 lawmakers had to be content to use their chamber desks or temporarily ensconce themselves in committee rooms that were not in use. For many of them, renting space elsewhere near the Capitol was an unaffordable option.

The cramped mode of doing the people's business changed in 1908 when the Cannon House Office Building opened on the south side of the Capitol, and a year later when the Russell Senate Office Building opened on the north side. Allison and his colleagues would barely recognize the Capitol campus today. Members of the House occupy quarters in three buildings[3] along Independence Avenue; their Senate counterparts have offices in three buildings[4] on Constitution Avenue. On the east side of the Capitol, the Supreme Court, the Library of Congress, and the Capitol Visitor Center all contribute to the sizable federal footprint.

The "Hill" looked very different in 1901. Most of the land around the Capitol was either vacant or not associated with federal functions. One exception was a small area along C Street, now partially occupied by the Longworth House Office Building. A small set of buildings on both sides of the street housed the Office of Weights and Measures.

Article I, Section 8, Clause 5 of the Constitution specifically granted Congress the authority to "fix the Standard of Weights and Measures," but until 1828, lawmakers had shown little appetite for exercising their authority.[5]

Navigating the Maze. https://doi.org/10.1016/B978-0-12-814710-8.00004-8

Different states had different standards, and adherents of states' rights were loath to cede such power to the federal government, even though it was constitutionally mandated. But by 1828, matters had come to a head. The Philadelphia Mint urgently needed direction on the quantity of gold a gold coin should have. On May 19, Congress acted, declaring "…the brass troy pound weight procured by the minister of the United States at London, in the year one thousand eight hundred and twenty-seven, for the use of the mint, and now in the custody of the director thereof, shall be the standard troy pound of the mint of the United States, conformably to which the coinage thereof shall be regulated."

A weight is one thing; a measuring device is quite another matter. The Philadelphia Mint had a precision standard, but it lacked a precision balance, which it needed to measure a quantity of gold precisely. The task of designing and constructing such a device fell to Joseph Saxton, a gifted instrument maker, who became an elected member of both the American Philosophical Society and the National Academy of Sciences, despite having no formal education beyond the age of 12. Saxton finished his work in 1838, and the mint had its standards of weights and measures.

The Philadelphia Mint was not the only federal enterprise in need of standards. Customs houses had similar requirements, and 8 years before Saxton completed his work for the mint, Levi Woodbury, a New Hampshire senator, had called on the Treasury Department to conduct a study of the standards customs houses were using around the country. Ferdinand Hassler, who had been relieved of his duties as head of the Survey of the Coast in 1818, was tapped for the position of Superintendent of the Office Weights and Measures, and from 1830 to 1832, he set about carefully selecting a set of standards. All that remained was supplying them to the customs houses. If that had been his only task, he might have carried it out with alacrity.

But in 1832, after a 14-year hiatus, Hassler was reappointed Superintendent of the Survey—at the same time retaining his leadership post at Weights and Measures—and his standards activities slowed to a crawl. Once again, Congress found itself losing patience with him—as it had when it had relieved him of his Survey duties in 1818—and made its annoyance clear in 1835 with a sharply worded communication to the Treasury Secretary, who by that time, was Levi Woodbury. Hassler got the message and, with his son assisting him, stepped up the pace sufficiently to satisfy his congressional overseers, at least for the time being.

Until that point, he had been satisfied with shared working quarters at the U.S. Arsenal. But now, he apparently recognized the importance of having a site dedicated to his weights and measures assignment. He chose a set of row houses on C Street south of the Capitol for the standards activities, and that's where the Office of Weights and Measures remained for almost seven decades. Neither Saxton, who took over from Hassler in 1844, nor any of the other 19th century superintendents saw any need to move the office to larger quarters.

The dawn of the new century brought with it not only a dramatic total solar eclipse that was visible on May 28, 1900 across the southern United States,[6] but also an array of technological innovations that would propel the American economy for decades to come. Nothing was more transformative than electricity. The 1893 Columbian Exposition, also known as the Chicago World's Fair, gave visitors a preview of how the new technology was about to alter the American landscape.

Thomas Edison, almost every grade school student is taught, was the father of electricity. But on May 1, 1893 it was Nikolai Tesla and George Westinghouse who commanded center stage in Chicago when the exposition opened. For half a dozen years, Edison and Tesla had been fighting over the relative advantages of AC (alternating current) power and DC (direct current) power. The "War of the Currents," as the rivalry is known, was bitter, personal, filled with misinformation—of which Edison was particularly guilty—and replete with publicity stunts, one of which, the first electrocution of a murder convict, went horrifically awry. The public relations and media battle came to a climax in Chicago when President Grover Cleveland threw a switch that lit up the fairgrounds with 100,000 incandescent lightbulbs powered by Westinghouse's generators that used Tesla's AC design.[7]

The die was cast that evening, and despite Edison's continuing vitriolic attack, Tesla's technology eventually came to dominate America's electricity markets. General Electric, which had been a prime promoter of Edison's DC design, capitulated shortly after the Columbian Exposition and joined Westinghouse in the AC corner. The Exposition might have marked the beginning of the end of the War of the Currents, but the battle itself had highlighted the pressing need for electricity standards.

When Henry S. Pritchett took over the helm of the Coast and Geodetic Survey in 1897, he realized that the head of the Office of Weights and Measures had to have technical qualifications that were commensurate with the demands of the new technologies.[8] Samuel Wesley Stratton, a University of Chicago physicist, was his choice, and Pritchett authorized him to develop plans immediately for expanding the office along the lines of equivalent German and British institutions. Stratton proceeded cautiously, and when he presented them with Treasury Secretary Lyman J. Gage's blessing, Congress readily accepted his call for a new National Bureau of Standards.[9] The enabling legislation, which passed on March 3, 1901,[10] established the bureau with the mandate of setting standards for weights and measures. That much was to be expected. But the legislation went even further by declaring the National Bureau of Standards a national physical laboratory after the German and British models Stratton had found so appealing. It is a role the institution continues to play today but under a more appropriate name, The National Institute of Standards and Technology.

President William McKinley, who won reelection as president in 1900, signed the legislation, and the Bureau, or NBS, as it became known, drew its first breath on July 1, 1901 with Stratton at the helm. Administratively NBS was lodged in the Treasury Department, which had been home to the Coast and Geodetic Survey for decades. But in 1903, with Theodore Roosevelt in the White House,[11] Congress (almost as an afterthought) transferred NBS to the newly created Department of Commerce. That same year NBS moved to its new campus in then bucolic northwest Washington, where it would remain for more than half a century.

In 1901, the birth year of NBS, the American science and technology landscape was very fragmented. Practical research, especially in agriculture and mining, was the essential domain of the Land Grant colleges. Basic research—inspired simply by the quest for knowledge—was the province of private universities, learned societies, and philanthropies. And applied research and development thrived on the burgeoning industrial terrain.

The federal government was a relatively minor player, and in many quarters, it was viewed with suspicion. Academic scientists, who treasured the freedom to pursue their own goals, eschewed it. Land Grant colleges were under the control of the states. And industry did not want interference from Washington. In retrospect, it is remarkable that the National Bureau of Standards generated sufficient political support to surmount those obstacles.

While NBS's establishment was groundbreaking in the research arena, it was of little consequence in the policy arena. Congress had charged it with setting standards, testing materials, evaluating processes, and conducting physical research. But neither it nor any other federal body had the authority to develop governmental science and technology strategies based on sound research. The story of the War of the Currents illustrates the extent of the policy vacuum.

By the last decade of the 19th century, it was abundantly clear that electricity was going to revolutionize the nation in ways that few previous technological advances had. To the cognoscenti—meaning anyone with a modicum of physics knowledge—it was clear that AC power held the all the high cards. High voltage transmission lines would be needed to minimize energy losses; lower voltages would be needed in homes and workplaces to minimize danger; and transformers,[12] which could only operate with AC, would be required to move the voltages up and down. Had a federal science and technology policy structure existed, the battle between AC and DC might have been over before it started.

But no such body existed, and the National Academy, which might have played a significant role, carried the stigma of blemished past policy performances. As a result, Westinghouse and General Electric and their intellectual progenitors, Tesla and Edison, were left to spar with each other in the arena of public opinion. Although the flashy success of AC power at the Chicago event succeeded in settling the debate, many decades passed before legislators saw the wisdom of establishing effective science and technology policymaking as a central feature of the federal government.

The rapidly changing industrial terrain, which triggered the creation of the National Bureau of Standards in 1901, would soon motivate lawmakers to expand the federal science bureaucracy further. In 1910, responding to the proliferation of mining and the increasing number of catastrophic mining accidents, Congress established the Bureau of Mines under the purview of the Department of the Interior. Joseph A. Holmes, a geologist from North Carolina, a conservationist who had been active in the Geological Survey, assumed the leadership reins.

The phrasing of the 1910 legislation[13] is highly significant for what it both prescribed and proscribed:

> *Section 2. That it shall be the province and duty of…[the Bureau of Mines]…to make diligent investigation of the methods of mining, especially in relation to the safety of miners, and the appliances best adapted to prevent accidents, the possible improvement of conditions under which mining operations are carried on, the treatment of ores and other mineral substances, the use of explosives and electricity, the prevention of accidents, and other technologic investigations pertinent to said industries, and from time to time make such public reports of the work, investigations, and information obtained…*

> *Section 4. That the Secretary of the Interior is hereby authorized to transfer to the Bureau of Mines from the United States Geological Survey the supervision of the investigation of structural materials and the analyzing of coals, lignites, and other mineral fuel substances and the investigation as to the causes of mine explosions…*

> *Section 5. That nothing in this Act shall be construed as in any way granting to any officer or employee of the Bureau of Mines any right or authority in the inspection or supervision of mines or metallurgical plants in any State.*

A cursory reading shows that the Bureau of Mines had far-reaching authority to conduct research—although not yet in the health arena—but it had no power at all to examine or regulate the conduct of mining companies. Those responsibilities, to the extent they were exercised, remained in the province of the individual states. The mining act reflected growing congressional interest in scientific and technological research, but lingering reluctance to allow scientists or scientific administrators to promote policies or regulations based on the research.

But just 3 years later, Congress expanded the bureau's authority substantially,[14] describing its mission in very broad terms: "mining, metallurgy, and mineral technology." More significantly, Congress charged it with "improving health conditions, and increasing safety, efficiency, economic development, and conserving resources through the prevention of waste in the mining, quarrying, metallurgical, and other mineral industries…" The trajectory of a more expansive federal role in science and technology policy was starting to take shape. It would come into much sharper focus in the coming decades, as the United States found itself—albeit reluctantly at first—playing a crucial role in two world conflicts, both ultimately settled on the basis of technological superiority.

Before we leave the Bureau of Mines, it's worth scrolling forward a dozen years to March 3, 1925. On that date, Congress passed the Helium Act of 1925,[15] authorizing the bureau "to maintain and operate helium production and repurification plants, together with facilities and accessories thereto; to store and care for helium; to conduct exploration for and production of helium on and from lands acquired or set aside under this Act; to conduct experimentation and research for the purpose of discovering helium supplies and improving processes and methods of helium production, repurification, storage, and utilization." As we'll see in Chapter 12, the 1925 act, although never anticipated by the drafters, sowed the seeds of hotly debated helium policies of the 21st century.

The story behind the helium legislation began 10 years earlier during the First World War in the skies over Great Yarmouth and King's Lynn on England's east coast. As the BBC describes the scene, residents reported "an eerie throbbing sound above them, followed shortly afterward by the sound of explosions in the street."[16] What they heard and experienced was a German bombing attack carried out by a lighter than air zeppelin capable of traveling 85 miles per hour and carrying several thousand pounds of explosives. Although dirigibles never proved to be terribly effective in carrying out bombing missions, military planners considered them useful for aerial surveillance, and by 1925, the U.S. Army and Navy had developed an abiding interest in developing them. Their connection to helium is easy to understand.

Dirigibles are essentially big balloons to which a heavier-than-air payload is attached. Helium is a very light, inert, and nonflammable gas—the stuff in party balloons—and is ideal for inflating dirigibles. But helium is so light that once freed from any confine, it can escape the pull of Earth's gravity and leave the atmosphere. Obtaining it requires capturing it as it emerges from deep underground rock formations, where, as the product of radioactive decay of uranium, it is sometimes trapped along with natural gas or oil.

In 1914, the Bureau of Mines had established a Petroleum Division, so it was well positioned in 1925 to take on an associated helium mission. The driving force in Congress was military in nature, but the helium legislation also gave the bureau the authority to "lease" any surplus gas to "American citizens or American corporations." The language is significant because it reflected a willingness of the federal government to engage with the general public and private business in ways that might have been anathema just a few decades earlier.

The guns of war had been blazing in Europe, in the Middle East, and on the high seas of the Atlantic for almost 3 years before the United States entered the world conflict on April 6, 1917. America's noninterventionist World War I stance belied its muscular ventures into the global arena during the preceding decades. Two foreign exploits stand out: the Spanish American War in 1898, which gave the United States control of the Philippines, Puerto Rico, and Guam, and the Hay-Bunau-Varilla Treaty of 1903, which gave the United States perpetual control of the planned Panama Canal for a purchase price of $10 million and an annual payment of $250 thousand.

The successful prosecution of the two ventures presaged America's eventual transformation into the global power that both McKinley and Roosevelt had envisioned. They also cast a spotlight on the importance of science and technology. Military engagements required development of the instruments of war, and the Panama Canal[17] project—one of the largest and most daunting endeavors ever undertaken—required development of civil engineering capabilities almost unimaginable. Both propelled American science and technology policies of the early 20th century and laid the foundation of America's rise to global science and engineering preeminence in the latter part of the century.

Theodore Roosevelt was passionate about establishing the United States as a world power. He was equally passionate about preserving the environment, and during his charismatic presidency, he set aside more than 200 million acres for public use, more than 150 million of them as national forests. Management of forested lands had been a thorny policy problem for decades, and finally in 1905, Congress decided to act, creating the U.S. Forest Service[18] to administer them.

It was a heady era not only for environmental preservationists, but also for progressive conservationists of another stripe. John Wesley Powell had made irrigation and land management a *cause célèbre* in the 1880s, one that cost him dearly when he lost congressional funding for his survey work in 1892. Now, two decades later, it wasn't commercial interests that were lining up against the environmental dicta, but rather socioeconomic progressives devoted to public ownership of utilities, including water and electricity.

The bitter clash occurred again in the West, this time in California. The battlefield was the watershed of Yosemite National Park, a public treasure Congress had created in 1890 at the behest of the famed naturalist John Muir.[19] The issue was the construction of a dam in the Hetch Hetchy Valley to address San Francisco's growing need for water, which the catastrophic 1906 earthquake fires had brought into sharp relief.

The debate attracted national attention, with Muir and his preservationist allies from the Sierra Club—the environmental organization he had founded in 1892—leading the charge against the project. Their mantra was "safeguard nature at all costs." On the other side were the San Francisco dam promoters, who predicated their advocacy on the proposition that nature should be conserved, but in ways most beneficial to society.[20]

After the dust had settled, Congress came down on the side of San Francisco, and passed the "Raker Act" in 1913, authorizing construction of the dam.[21] Muir died the following year, never witnessing the flooding of the valley he held so dear.

The construction of the O'Shaughnessy dam finally began on August 1, 1919, and by the time it was completed in 1923, 1200 acres of Yosemite National Park had fallen victim to the vast Hetch Hetchy Reservoir. It would take another decade to fill the entire valley.

Setting aside the efficacy of the outcome, the contentious debate highlighted a glaring deficiency: No federal policymaking apparatus existed to regulate the use of national park acreage. Three years after passing the "Raker Act," Congress remedied the shortcoming by establishing the National Park Service as an Interior Department bureau in 1916,[22] granting it broad authority to manage the system, which had grown to 14 parks and 21 national monuments by that time.

If we were to pin a policy label on the early 20th century, it would probably be bureau building. That is particularly true in the science and technology arena. Electricity, mining, oil, forests, land management, and water resources all called attention to the policy vacuum. The National Bureau of Standards, the Bureau of Mines, the Forestry Service, and the National Park Service were federal bodies established to fill the voids. But that was only the beginning.

In 1902, Congress passed the Newlands Reclamation Act,[23] creating a "reclamation fund" to support the development of irrigation projects in arid lands. The Reclamation Service administered the program as part of the U.S. Geological Survey until 1906, when the Department of the Interior granted it status as the Bureau of Reclamation. Originally hailed as a major breakthrough in land and water management, the 1902 Act was the target of environmentalists almost from the outset. They criticized it as a commercial boondoggle, which, they claimed with some validity, lined the pockets of developers and despoiled millions of acres of land.

Criticism aside, the Act was, without question, a prime enabler of western development, much as John Wesley Powell had envisioned in 1878.[24] He might well have regarded it as the capstone of his career had he been alive to see its implementation. But whether the enduring consequences of dam construction on most of the major rivers in the West would have troubled him is hard to know. He died in 1902, three months after Congress passed the Newlands Act, long before the evidence was in.

America's population migrated westward, but it also migrated from farms to cities. And it grew dramatically in size, from about 50 million in 1880 to more than 90 million just 30 years later. Driven by waves of immigrants, the demographic complexion was changing dramatically, as well. For lawmakers and policymakers, population data was invaluable.

The Constitution recognized the importance of the size of the population and how it was distributed among the states for determining the composition of the House of Representatives and for levying taxes, but nothing more. In the words of Article I, Section 2,

Representatives and direct Taxes shall be apportioned among the several States which may be included within the Union, according to their respective Numbers, which shall be determined by adding to the whole Number of free persons, including those bound to Service for a Term of Years, and excluding Indians not taxed, three fifths of all other Persons. The actual Enumeration shall be made within the three Years of the first Meeting of the Congress of the United States, and within every subsequent Term of ten Years, in such a manner as they shall by Law direct.

As the population grew in size and diversity, conducting a census and analyzing the data every 10th year was becoming a daunting, if not impossible, task. To remedy the logistical problem and provide a more reliable assessment of the data, Congress established a permanent Census Office in 1902.[25] It acquired the name Bureau of the Census a year later when it moved from the Department of the Interior to the newly created Department of Commerce and Labor.[26]

One of the most popular sites along the Mall in Washington is the National Air and Space Museum.[27] Established in 1946 as the National Air Museum, it houses the largest collection of historic airplanes and space craft in the world. The museum is part of the Smithsonian Institution, and it is the rare guide who would be able to recount the unsung role the Smithsonian played in the history of aeronautics and aviation.

The story begins in 1887, the year Samuel Pierpont Langley,[28] an astronomer and a physicist, became secretary of the Smithsonian Institution. By that time, the institution had begun to extend its reach from curation to research, and in 1890, at Langley's urging, it established the Smithsonian Astrophysical Observatory in Washington.[29] In addition to his scientific passions, Langley was consumed with aviation and the possibility of developing a piloted aircraft that was heavier than air. He received support from the War Department and achieved an initial success with an unpiloted model in 1896. But he struggled to realize his original goal, and finally conceded defeat after two piloted airplanes crashed. The second and final failure occurred on December 8, 1903, just nine days before the Wright Brothers accomplished the feat in Kitty Hawk, North Carolina, without any federal support.

Langley's failure to deliver the goods he had promised the War Department put the government's interest in aviation on ice. But the hold would prove temporary. After Langley died in 1906, Charles Doolittle Wolcott,[30] who had drawn Theodore Roosevelt's attention during his tenure as director of the Geological Survey, became secretary of the Smithsonian. In that capacity, and with his government *bona fides*, Wolcott donned the mantle of federal scientific guru, keeping the Smithsonian's interest in aviation alive in the process.

By 1912, federal interest in aviation had grown, and President William Howard Taft, who had been elected 4 years earlier, was persuaded that the time was ripe for addressing the opportunities and challenges of the new technological arena. He appointed a commission,[31] chaired by the president of the Carnegie Institution of Washington, Robert S. Woodward, specifically charging it with reporting on the need for an aviation research laboratory. Wolcott, as a member of the commission, was poised to make the case for Smithsonian leadership.

In shades of John Wesley Powell's 1884 testimony before the Allison Commission when he was the director of the Smithsonian's Bureau of Ethnology and had argued for consolidation of all federal science programs under the Smithsonian's aegis, Wolcott overreached, as well. He pressed for a Smithsonian aviation laboratory funded by the federal government, and not surprisingly,

Wolcott met with the same stony silence as Powell had before. Congress was not disposed to hand over the research purse strings to a private entity. Undeterred, the Smithsonian proposed what in hindsight seems like an even more ambitious plan: coordinating aeronautical research across all federal agencies.[32] The plan to establish a new advisory committee for that purpose was stillborn when the comptroller of the treasury determined that it was illegal for any federal employee to serve on such a body that was not sanctioned by Congress.[33]

The year was now 1914, and hostilities had broken out in Europe. Woodrow Wilson was president, and the American posture was one of nonintervention. Nonetheless, military planners in the War Department became concerned about America's weak air capability. On July 28, 2014, the day Austria-Hungary declared war on Serbia, the United States military had just 23 airplanes. Elsewhere, the numbers stacked up this way:[34] France, 1400; Germany, 1000; Russia, 800; and the United Kingdom, 400. But the War Department's concerns did not resonate well with isolationists in Congress and a president who had pledged to keep the nation out of war.

Wolcott took another page out of Powell's playbook, and with the support of Acting Secretary of the Navy Franklin Roosevelt, he persuaded congressional aeronautic proponents to add a rider to the fiscal year 1916 Navy appropriations bill. The legislation, which passed on March 3, 1915, established the National Advisory Committee for Aeronautics.[35] The groundbreaking legislation set the course for American aeronautics and aerospace policy for the century that followed. The two short paragraphs of 1915 are as vital to modern American science and technology policy as any that have ever appeared in the Congressional Record and are worth quoting in full:

An Advisory Committee for Aeronautics is hereby established, and the President is authorized to appoint not to exceed twelve members, to consist of two members from the War Department, from the office in charge of military aeronautics; two members from the Navy Department, from the office in charge of naval aeronautics; a representative each of the Smithsonian Institution, of the United States Weather Bureau, and of the United States Bureau of Standards; together with not more than five additional persons who shall be acquainted with the needs of aeronautical science, either civil or military, or skilled in aeronautical engineering or its allied sciences: Provided, That the members of the Advisory Committee for Aeronautics, as such, shall serve without compensation: Provided further, That it shall be the duty of the Advisory Committee for Aeronautics to supervise and direct the scientific study of the problems of flight, with a view to their practical solution, and to determine the problems which should be experimentally attacked, and to discuss their solution and their application to practical questions. In the event of a laboratory or laboratories, either in whole or in part, being placed under the direction of the committee, the committee may direct and conduct research and experiment in aeronautics in such laboratory or laboratories: And provided further, That rules and regulations for the conduct of the work of the committee shall be formulated by the committee and approved by the President.

That the sum of $5,000 a year, or so much thereof as may be necessary, for five
years is hereby appropriated, out of any money in the Treasury not otherwise
appropriated, to be immediately available, for experimental work and investiga-
tions undertaken by the committee, clerical expenses and supplies, and necessary
expenses of members of the committee in going to, returning from, and while
attending meetings of the committee: Provided, That an annual report to the Con-
gress shall be submitted through the President, including an itemized statement of
expenditures.

The key words appear near the end of the first paragraph: "That it shall be the duty of the Advisory Committee for Aeronautics to supervise and direct the scientific study of the problems of flight, with a view to their practical solution, and to determine the problems which should be experimentally attacked and to discuss their solution and their application to practical questions. In the event of a laboratory or laboratories, either in whole or in part, being placed under the direction of the committee, the committee may direct and conduct research and experiment in aeronautics in such laboratory or laboratories…" Wolcott failed in his ill-conceived attempt to consolidate aeronautics under the Smithsonian, but after licking his wounds, he paved the way for legislation that set the stage for NASA's establishment in 1958 half a century later[36] and some of the most awe-inspiring exhibits in the Smithsonian National Air and Space Museum.

The Air and Space Museum occupies a striking building six blocks from the Capitol, and although NASA's headquarters is just a few blocks away, the nuts and bolts of the space agency are flung far across the country. That is not true for the National Institutes of Health (NIH), which occupies a 300-acre campus with more than 75 buildings[37] in Bethesda, Md. just inside the Capital Beltway, across the way from Walter Reed National Military Medical Center[38] and ten miles from the White House. But the path that took NIH from a Washington vision to a Bethesda reality in the 20th century was a long and tortuous one. It required, as is often the case with a grand idea, a combination of policy imperatives, political expediency, personal devotion, and a soupçon of serendipity.

The story begins with several events prior to 1900 that bear repeating. Internecine battles within the embryonic federal bureaucracy during the last two decades of the 19th century had led to the demise the National Board of Health after only 4 years of existence. Its failure reflected, in part, the public policy weakness of the National Academy of Sciences, which had been instrumental in establishing the agency. And when Congress elected not to reauthorize the board's budget in 1883, the Marine Hospital Service, which had led the fight against it, and the Army Medical Corps found themselves the primary beneficiaries. The Marines regained control of quarantine matters, and the Army assumed much of the board's research mandate.

The division of responsibility might have remained that way if hadn't been for Joseph Kinyoun, a name often forgotten in the annals of science policy.

Kinyoun[39] was born to a slave-owning family in Dan'l Boone country just before the Civil War began. East Bend, a remote town of about 11,000 residents in Yadkin County, North Carolina, where the Kinyouns lived, was probably best known in 1860 for its two dozen or so liquor stores. In today's political parlance, it was a swing county, split between Unionists and Confederates. Kinyoun's father, a lawyer and surgeon who had trained at University Medical College of New York (now New York University School of Medicine),[40] was one of the latter. He joined the Confederate Army as a captain when the war began. Four years later, like many of the lucky ones who had fought for the Confederacy and survived the horrors of the battlefield, he found his East Bend home in ashes and his possessions gone.

With little to keep them in the war-ravaged hill country, the Kinyouns left North Carolina, eventually settling in rural Missouri, where they began their lives anew. There, Joseph excelled in his studies and, despite the family's now modest circumstances, decided to follow in his father's footsteps and enroll in a New York medical school. The year was 1881, and medicine had been undergoing a steady transformation since the elder Kinyoun had received his degree 22 years earlier. Research was in the ascendancy: understanding the scientific cause of a disease was slowly being recognized as central to developing treatments for it.

During his 2-year stint at Bellevue Hospital Medical College[41]—which ironically would merge with his father's alma mater, University Medical College of New York in 1898—Kinyoun encountered firsthand the excitement of the new spotlight on medical research, especially through what would soon become known as the fields of microbiology, bacteriology, and infectious diseases. So, it was not surprising that his return to Missouri in 1882 as a physician in sole practice would be short lived.

Kinyoun's fortunes soon became intertwined with Andrew Carnegie's fortune, after the preeminent Scottish-American industrialist turned his attention from accumulating wealth to philanthropic giving.[42] The Bellevue laboratory for pathology and bacteriology research was one of his first beneficiaries, and 1885 found Kinyoun back at his old stomping grounds, becoming the new Carnegie Laboratory's first student of bacteriology. His specialty was to be cholera.

A year later, led by its surgeon general, John Hamilton, the Marine Hospital Service, perhaps smarting from the loss of its research mandate first to the National Board of Health and then to the Army Medical Corps, decided to establish a small research program of its own. And small it was—a one-room "Hygienic Laboratory"[43] in the attic of the Hospital Service's quarantine facility on Staten Island, New York. The laboratory's first priority was to be cholera.

The synchronicity could not have been more apparent. Yet it was—Hamilton's trusted assistant and facilitator was Preston Heath Bailhache, Kinyoun's uncle.[44] Whether or not Bailhache served as Kinyoun's facilitator, as well, is not clear, but in October 1886 the Marine Hospital Service offered Joe Kinyoun a position, and shortly thereafter named him director of the new Hygienic Laboratory.

When the Marine Health Service moved to Washington in 1891, Kinyoun found himself in the lap of luxury: his laboratory occupied the entire top floor of the four-story Butler Mansion at the corner of B Street (now known as Independence Avenue) and 3rd Street, SE, three blocks from the grounds of the Capitol on a site now occupied by the John Adams Building of the Library of Congress. In the succeeding years, he created a research capability that would serve as a model for the National Institutes of Health decades later. Kinyoun's efforts were heroic for his time, but he could not have imagined how they would eventually evolve. There were still many twists and turns to be navigated along the road to NIH's creation and its remarkable Bethesda campus.

Two years after the Health Service's move, the Army, not to be upstaged, established a medical school in Washington,[45] also on B Street, a block away from the Smithsonian Castle. It was there, incidentally, that Walter Reed began his distinguished bacteriology research career. As the 19th century was drawing to a close, the Marine Health Service and the Army Medical Corps had both staked out prime real estate in the nation's capital and were positioning themselves to lead an effort in medical research. But it would take a few more years before any effort crystalized.

Just as the 1878 Mississippi Valley yellow fever scourge had spurred the creation of the National Board of Health—albeit short-lived—the dire consequences of the Spanish American War of 1898 thrust medicine into the national spotlight. It was not battle-related deaths that created the impetus, but rather what American troops encountered in the tropics of Cuba, the Philippines, and Puerto Rico. The war, which lasted six months, took the lives of only 332 fighters,[46] but tropical diseases the Navy and Marines encountered produced an additional 2597 fatalities.[47]

That disparity set the stage for a new national agenda focused on research into diseases. What had been strictly a public health issue in 1878 now became a national security issue. Quarantine was not a viable solution on the battlefield; finding the causes of disease and their cures was a necessity as the United States expanded its global reach in the aftermath of the Spanish American War, and especially during Theodore Roosevelt's presidency. The impetus was just what the Marine Health Service's Hygienic Laboratory and the Army's Medical School had been waiting for.

The Hygienic Laboratory walked off with the prize, in large part due to Kinyoun's visionary efforts. Although he left his director's post in 1899, he had provided the intellectual heft, policy rationale, and organizational plan that led Congress to adopt the following legislative language in 1901:[48]

Be it enacted by the Senate and House of Representatives of the United States of America in Congress assembled, *That the following sums be, and are hereby, appropriated, for the objects hereinafter expressed for the fiscal year ending June thirtieth, nineteen hundred and two, namely:*

UNDER THE TREASURY DEPARTMENT
Pᴜʙʟɪᴄ Bᴜɪʟᴅɪɴɢs

....

Mᴀʀɪɴᴇ Hᴏsᴘɪᴛᴀʟs: *For building for laboratory, Marine-Hospital Service: For the erection of the necessary buildings and quarters for a laboratory for the investigation of infectious and contagious diseases, and matters pertaining to public health, under the direction of the Supervising Surgeon General, thirty-five thousand dollars; and the Secretary of the Navy is authorized to transfer to the Secretary of the Treasury, for the use as a site for said laboratory, five acres of the reservation now occupied by the Naval Museum of Hygiene.*

Congress also changed the name of the laboratory's parent agency a year later to the Public Health and Marine Hospital Service. And under its broadened public health mandate, the Hygienic Laboratory established a Division of Scientific Research comprising bacteriology, chemistry, pathology, pharmacology, and zoology.[49] The course for a new, greatly expanded medical research enterprise was being charted.

In 1906, the American Association for the Advancement of Science (AAAS) responded to a proposal of one of its members, Yale economist J.P. Norton, and created the Committee of One Hundred on National Health.[50] Its primary objective was to make the case for a national department of health, which among its prime mandates, would be promoting research in preventive medicine and public hygiene. The committee, chaired by Irving Fisher, another Yale economist, aimed high, arguing for consolidation of all federal health and medicine programs into a Cabinet level department.

The AAAS campaign lasted half a dozen years, but in the end, its bold plan proved to be too big a lift for the committee's Washington allies. In 1912, Congress simply passed legislation that codified the work of the Public Health and Marine Hospital Service. The bill shortened its name to the Public Health Service, and with a single phrase, broadened the mandate of the Hygienic Laboratory's research programs,[51] another milestone on the road to Bethesda. The relevant language reads, "The Public Health Service may study and investigate the diseases of man and conditions influencing the propagation and spread thereof…" Brevity is not only the soul of wit,[52] it is often the essence of the most effective public policy.

Epidemics and wars shake the political establishment in ways that no other events can. Between 1912 and 1930, there were three significant tremors that spurred policymakers to ramp up America's medical research capabilities. World War I, which had provided an impetus for creating the National Advisory Committee for Aeronautics, the precursor to NASA, was one. Weaponizing chemicals and biological agents—which had yet to be banned[53]—was a military matter. But understanding how to combat them was an issue for public health officials, as well as the War Department. Before 1917, the policy imperative might have been theoretical, but when America entered the world conflict on

April 7, the new technological threats to human health and life created a sense of urgency.

A year later, an influenza pandemic struck continents around the globe.[54] With the cause of the disease poorly understood and no effective treatment available, 50 million people, or about 5% of the world's population, succumbed. The American death toll ran as high as 675,000. In the nation's capital, especially, the impacts were profound: public schools, universities, churches, and libraries were closed, and indoor public gatherings were banned. Yet, despite such precautions, the disease persisted there largely unabated for months. From October 1, 1918 to February 1, 1919, the height of the contagion, Washington, then a city of 418,000 residents, reported almost 38,000 cases and nearly 3,000 fatalities.

For elected officials, who witnessed the scourge first hand, it should have been a wakeup call, but it wasn't. A serious legislative initiative would have to wait until another significant influenza outbreak in the winter of 1928–29.[55] Although the incidence and death rates were much lower than they had been 10 years before, health professionals were alarmed that so little progress had been made in controlling and treating the disease. Two years later, Congress finally decided to make medical research a priority.

The Ransdell Act[56] of 1930 changed the name of the Hygienic Laboratory to the National Institute of Health,[57] providing money to build new scientific facilities and establishing a system of research fellowships. The sweeping legislation was a landmark in the history of NIH, and it established the principles for future federal science agencies. For those reasons, its wording is worth repeating verbatim.

Be it enacted by the Senate and House of Representatives of the United States of America in congress assembled, *That the Hygienic Laboratory of the Public Health Service shall hereafter be known as the National Institute of Health…The secretary of the Treasury is authorized to utilize the site now occupied by the Hygienic Laboratory and the land adjacent thereto owned by the government and available for this purpose, or when funds are available therefor, to acquire sites by purchase, condemnation, or otherwise, in or near the District of Columbia, and to erect thereon and to furnish and equip suitable and adequate buildings for the use of such institute. In the administration and operation of this institute the Surgeon General shall select persons who show unusual aptitude in science…*

Section 2. The Secretary of the Treasury is authorized to accept on behalf of the United States gifts made unconditionally by will or otherwise for study, investigation, and research in the fundamental problems of the diseases of man and matters pertaining thereto, and for the acquisition of grounds or for the erection, equipment, and maintenance of buildings and premises…

Section 3. Individual scientists, other than commissioned officers of the Public Health Service, designated by the Surgeon General to receive fellowships may be appointed for duty in the National Institute of Health... Scientists so selected may likewise be designated for the prosecution of investigations in other localities and institutions in this and other countries during the term of their fellowships...

Secton 5. The facilities of the institute shall from time to time be made available to bona fide health authorities of States, counties, or municipalities for purposes of instruction and investigation.

Approved, May 26, 1930.

The language in Sections 3 and 5 established the operating principles of the modern National Institutes of Health: extramural grants, fellowships, and intramural research facilities in Bethesda and elsewhere around the country.

Without question, influenza provided the political lubricant that finally led lawmakers to establish NIH in 1930. But momentum had been building ever since American soldiers had confronted the horrors of chemical and biological weapons on the European killing fields of World War I. The war had also focused congressional interest airpower, leading to the creation of NASA's precursor, the National Advisory Committee for Aeronautics. Yet, for the United States military, the conflict's most significant technological impact might have been on the open seas.

Even though U.S. entry into the war would not come until 1917, Naval concerns about Germany's use of submarine warfare were already palpable in 1915. That year, Navy Secretary Joseph Daniels created the Naval Consulting Board, appointing Thomas Edison as its chairman and filling out seats at the table with some of the most eminent American engineers, technologists, and inventors.[58] Conspicuously missing were any members of the National Academy of Sciences. That should not have been surprising, given the institution's pitiable advisory track record during the Civil War and the Spanish American War, and Edison's bona fides as an inventor.

As the War of the Currents illustrated, Edison was certainly not shy about making his case for DC electricity, even to the point of distortion of the essential science and defamation of the character of his principal adversary, Nikolai Tesla. In the end, his scorched earth campaign did not help him win his cause. As chairman of the Naval Consulting Board, he had a perfect opportunity to recapture any influence he might have lost in his previous battle. He boldly moved to expand the board's purview to every facet of military technology—not just naval—in the process, embracing virtually all scientific and engineering disciplines. But the board, even if it boasted some of the most creative minds in the country, could not carry out a comprehensive mission of examining, evaluating, and developing military technologies unless it had the requisite facilities. In 1916, led by Edison, the Naval Consulting Board made its case to the House Naval Affairs Committee, chaired by Lemuel Padgett, a Tennessee attorney, well respected for his knowledge of all matters naval.

The committee found the Consulting Board's case compelling and authorized $1 million for a new military research laboratory. But the board members fell victim to infighting over where to site the laboratory, and by 1917, when America entered World War I, the opportunity for building the new research facility had passed. It would not be the last time scientists or engineers forfeited a grand project opportunity because of siting deliberations. What happened to the Superconducting Super Collider (SSC)[59] project in 1993 is an excellent example, and worth a brief digression.

The particle physics—sometimes called high-energy physics—community and its Department of Energy allies had agonized for more than 2 years over where to build the massively expensive SSC facility, dangling one possibility after another in front of salivating members of Congress, who craved the construction money, scientific infrastructure, and jobs that would flow to their district and state. Eventually, legislators whose home bases had lost out in the site selection competition lost their enthusiasm for the $11-billion project—which, to be fair, suffered from escalating costs, management missteps, technical problems, and the consequences of the end of the Cold War—and sank it in 1993.

The sad saga of the SSC and its object policy lessons bear further narrative. But peeling back the curtain on one of America's biggest science stumbles will have to wait until we complete the account of how two world wars enabled the United States to become the international leader in science and technology and the dominant economic and military power globally. So now, back to business.

World War I placed a hold on the Naval Consulting Board's plan for a new military science and technology facility, but as events would prove, the hold was only temporary. At 11:00 a.m. on July 2, 1923, the Naval Research Laboratory (NRL)[60] opened its doors in Washington on a site along the east bank of the Potomac River, one of the locations the board had actually considered. In line with Edison's vision, NRL developed a broad research program, spanning a panoply of disciplines and targeting not only issues of clear military relevance, but also those of fundamental science. Its wide reach echoed the mandate Congress had given the National Bureau of Standards in 1901.

The 7-year interregnum between NRL's 1916 congressional authorization and its 1923 ribbon cutting opened a door for the National Academy of Sciences, which was still seeking ways to be relevant in Washington's corridors of power. The story[61,62] of how it began to rehabilitate its image and reinsert itself into policymaking highlights the nexus between policy and politics, and the role of personal relationships in both arenas.

Five actors played principal roles during the war years of 1914 through 1918: George Hale, an astronomer who was director of the Mount Wilson Observatory; Robert Millikan, a physicist who was a member of the University of Chicago faculty; Elihu Root, a former Secretary of both War and State under McKinley and Theodore Roosevelt, who was president of the Carnegie Corporation; Major General George Squier, a graduate of West Point with a Ph.D. in electrochemistry from Johns Hopkins, who was the Army Signal Corps chief officer; and William Welch, a physician and pathologist, who was president of the National Academy of Sciences. Of the five, only Root lacked a background in science or engineering. But his position at Carnegie and his political experience would prove useful, if not vital, in pumping new life into the Academy's moribund ability to engage in federal science and technology matters.

Hale got the ball rolling shortly after fighting broke out in Europe, a conflict President Woodrow Wilson considered strictly a European matter. Roosevelt, who had opposed Wilson in the 1912 election as nominee of the "Bull Moose" Progressive Party, and Root, who had been a member of Roosevelt's Cabinet from 1901 through 1908, saw things quite differently. Germany, in their eyes, was a threat to American ideals and needed to be countered. Although Roosevelt's unconventional, disruptive, and ultimately unsuccessful presidential run in 1912 had left him with few friends in either major party, Elihu Root's stellar reputation remained intact. He had not been involved in the 1912 election, and from his perch at Carnegie, he was well positioned to assist Hale and the National Academy.

Hale was both an internationalist and admirer of the august science societies in France and Britain, which he viewed as models for the U.S. National Academy of Sciences. He also regarded World War I as a technology war. Whoever had the better science had the better chance of winning.

Driven by his twin desires of revitalizing the Academy and assisting the Allies in Europe, Hale pressed Welch to offer the Academy's services to Wilson, if and when America joined the conflict. Welch, a pragmatist, saw little chance of sparking any White House interest, and set Hale's proposition aside. But in 1916, with German submarine activity ratcheting up in the North Atlantic, Hale requested that his proposal be brought to a vote of the Academy membership. And at the annual meeting that April it passed unanimously. The resolution read in full,[63] "*Resolved*, That the President of the Academy be requested to inform the President of the United States that in the event of a break in diplomatic relations with any other country the academy desires to place itself at the disposal of the Government for any service within its scope."

Perhaps even more significantly, the Academy membership empowered its Council "to organize the Academy for the purpose of carrying out the resolution most effectively." That language sent a clear signal that the Council was free to break new ground. Despite Welch's fears, Wilson readily accepted the offer of assistance, perhaps recognizing the inevitability of American entry into the war in the near future.

It was now up to the Academy to avoid embarrassing itself as it had twice before. To set a new course, Hale immediately formed an organizing committee, which in short order recommended establishing a National Research Council[64] under the Academy's auspices. Providing the breadth and depth of the needed expertise, the committee decided, would require more than the Academy membership could deliver. Accordingly, the committee stipulated that

NRC participants did not have to be members of the Academy, but they should reflect all relevant institutional players: universities, industry, government, and foundations.

To make the sale and generate the revenues the NRC would need, Hale brought on board University of Chicago physicist Robert A. Millikan, a renowned scientist with an exceptional research pedigree who would eventually win a Nobel Prize in physics. But it was Millikan's other attributes that captured Hale's attention. He was, according to National Academy records,[65] "a likable and energetic physicist and ... a consultant and a supplier of trained physicists to Western Electric and AT&T." He was ideally suited to the role in which Hale cast him, straddling "the worlds of academia and high-tech industry" and "effective at working with officers in the military's technical bureaus."

Millikan certainly had the contacts and charisma to attract top-flight researchers to the NRC, which he did successfully. But neither he, nor anyone else in the Academy, had the deep pockets needed to sustain its work. They needed help, and Hale knew where to get it. His relationship with Elihu Root—which stemmed from Carnegie's support of big astronomical instruments—and their shared antipathy toward Wilson's noninterventionist posture, opened the door to the philanthropy's coffers.[66] And by May 1918, the Carnegie Corporation had bankrolled the NRC to the tune of $150,000. Other foundations added to the sum, as did the federal government eventually, but Root's timely support was vital to the NRC's ability to survive in more than name only.

Money is essential, but so, too, is a mandate to do something. And in that respect, the NRC was fortunate. In early 1917, a few months before the United States entered the war, the Council of National Defense, which Wilson had established a year earlier to coordinate national security resources, requested the NRC to serve as its research arm. And in mid-summer, a few months after America's entry into the world conflict, the NRC received an extraordinary request. It came from George Squier, the tenacious head of the Army Signal Corps, who had both a personal passion for science and technology and a grand vision for American research, especially, but not exclusively, in support of the military.

Squier asked the NRC to be the research arm of the Signal Corps and requested Millikan to place himself under direct military command by enlisting as a major in the Officer's Reserve Corps. Today, that would seem like a strange request, but in 1917, no federal instrument existed for transferring funds from a military service to a civilian organization, such as the NRC. Of course, by acceding to Squier's request, Millikan also placed himself under Squier's direct control, which could not have gone unnoticed by either of them.

Their relationship evokes a World War II parallel involving Lieutenant General Leslie Groves, who was responsible for the Manhattan Project—

as the top-secret atomic bomb effort was called—and J. Robert Oppenheimer, the theoretical nuclear physicist whom Groves chose to lead the team building the plutonium "Gadget" and its uranium counterpart at Los Alamos, a desolate area atop a high mesa in New Mexico.[67] But there are significant differences.

Unlike Squier, Groves, who was an officer in the Army Corps of Engineers, had no background in science, and didn't hide his dislike and mistrust of physicists. Unlike Millikan, Oppenheimer had shown little prior evidence of having any executive flair and, although not a Communist, his association with Communist sympathizers had led the FBI to question his allegiance to the United States. Oppenheimer, even if he had been asked to trade in his civvies for a uniform, is unlikely to have done so.

But like Millikan, Oppenheimer grew into his management responsibilities quickly. And just as Squier had developed an easy rapport with Millikan, Groves eventually came to respect Oppenheimer, and even regard him as indispensable to the Manhattan Project. Both wartime episodes illustrated the growing bond between the civilian science community and the military that continued through much of the 20th century. For supporters of scientific research, that bond would prove to be indispensable during the Cold War era, which lasted roughly from the end of World War II in 1945 until the collapse of the Soviet Union in 1991.

George Hale's assessment of World War I as a technology war would prove to be true on many counts: on the battlefield, in the air, and on the sea. It would also be borne out in university and industrial laboratories, corporate and philanthropic board rooms, government agencies, the White House, and congressional offices. For the duration of those years, the National Research Council, although not a government agency, became a de facto national scientific clearinghouse and research coordinator. It carried out those tasks in spite of having a skeleton administrative staff and little financial support from the federal government. Even with those disadvantages, it animated the vision the Lazzaronis had for the National Academy of Sciences when they established it half a century earlier.

The impact of the war experience extended well beyond the Academy and its newly formed operational branch, the National Research Council. It gave the public a foretaste of the growing importance of science and technology in American life. And perhaps even more significantly, it reinforced a growing awareness among policymakers that the federal government had a huge stake in fostering a healthy research enterprise.

There remained one science problem the war's impact failed to resolve. It did little to spur the development of policies that captured the importance of fundamental research—the exploration of nature for knowledge's sake—in nourishing technological progress. It failed to do so, even though the connection was more than implicit in Woodrow Wilson's executive order of May 11, 1918, which requested the Academy "to perpetuate the National Research Council."

In Executive Order 2859, Wilson enumerated the NRC's functions as follows:[68,69]

To stimulate research in the mathematical, physical, and biological sciences, and the application of these sciences to engineering, agriculture, medicine, and other useful arts...

To survey the larger possibilities of science, to formulate comprehensive projects of research, and to develop effective means of utilizing the scientific and technical resources of the country...

To promote cooperation in research at home and abroad...

To serve as a means of bringing American and foreign investigators into active cooperation with the scientific and technical services of the War and Navy Departments and with those of the civil branches of government.

To direct the attention of scientific and technical investigators to the present importance of military and industrial problems in connection with the war...[and]

To gather and collate scientific and technical information at home and abroad, in cooperation with governmental and other agencies...

The nexus between the fundamental and the practical is quite clear, but Wilson's 1918 message had little impact on how lawmakers and policymakers actually saw things during the next three decades. It would take another war and the efforts of Vannevar Bush, an engineer and administrator extraordinaire with political savvy and connections, to make the case compellingly.[70] We'll get to Bush's World War II story and the foundation he laid for modern American science and technology policy in due course. But, we have a few more threads to weave into the 20th century fabric before we get there.

Policies can promote technological change, but frequently it's the other way around. The history of the American highway system[72] during the early part of the 20th century is a good example. The episode also shows how competing interest groups can impede effective policymaking to the detriment of everyone involved. It also demonstrates that the most politically achievable outcome is not necessarily the best one. With that teaser, let's get on with the story.

At the close of the 19th century there were about 8000 automobiles in the entire country, primarily owned by the wealthy. Trains, river boats, horses, stage coaches, and horse-drawn carriages accounted for almost all intercity travel. And existing roads reflected the state of the nonmechanized technology. They were mostly improved dirt or gravel wagon trails and were a hodgepodge across state lines.

In 1908, a technological revolution occurred that would change the American landscape forever. It was the year Henry Ford unveiled the Model T, the first automobile made of interchangeable parts. His goal was manufacturing "a car for the great multitude."[71] Interchangeability got him part way to his goal, but not far enough. Five years later, though, he achieved his objective when the first moving assembly line opened in Highland Park, Michigan. American life would never be the same. By 1927, the last year of its production, 15 million Model Ts had rolled off the line. At the time, just over 119 million people lived in the United States, constituting 26 million households.[73] Henry Ford's Model T had become ubiquitous on the American landscape.

The two technological advances—interchangeability and assembly line production—changed the face of personal travel, as well as manufacturing more generally. They also made interstate journeys more commonplace, and existing American roads antiquated. Highways had to be either upgraded or constructed, and just as important, routes had to be harmonized across state boundaries.

Until the turn of the century, road management had been the sole responsibility of states and localities. If they neglected maintenance, so be it. If local citizens didn't care about road conditions, it was their business. But Henry Ford's ability to give the multitudes a car at a price they could afford and to afford them the ability to travel longer distances in shorter times was poised to alter attitudes dramatically.

The pace of the technological change caught policymakers off guard. Prior to 1900, the only federal highway program—if indeed it merited the label "program"—was the Office of Road Inquiry, which was authorized in 1893 for the purpose of advising states and localities on how they might best improve their roads. A 1902 bill that would have created a Bureau of Public Roads and authorized $20 million in federal matching grants to local jurisdictions for improving their roads failed almost at the outset.

Most members of Congress did not believe the legislation passed constitutional muster, because highway concerns were not among the federal powers enumerated in Article I, Section 8. Accordingly, based on the Tenth Amendment,[74] they argued, such matters fell strictly under the jurisdiction of the states. It took 5 years, but the Supreme Court eventually dispensed with the states' rights argument in 1907, holding in Wilson v. Shaw[75] that the "Commerce Clause"[76] of the Constitution gave the federal government the authority to fund and regulate interstate highways. The ruling cleared the way for new federal legislation.

But, the policy dithering and political bickering continued until well after the Model T had made its appearance on the scene in 1908. Special-interest arguments about priorities poisoned the well repeatedly. Motorist organizations, such as the American Automobile Association (AAA), founded in Chicago in 1902, pushed for investments in long-distance paved highways. Farmers argued for a focus on the local roads they used for transporting their produce.

In 1912, Congress finally did more than simply slog through the murky highway mud. In August of that year, after failing to pass a $25 million highway "rental" plan for postal service use, it appropriated $500,000 for a trial program to improve roads used by the service. States and counties would be on the hook for two thirds of the cost of any project—still a pretty good deal—but they would have to abide by federal labor rules. That proved to be too much of a lift for many states, although the program continued to muddle along for 3 years, until 1915 when the Justice Department, siding with the states and counties, declared the federal regulations an overreach. In fairly short order, the program collapsed.

A year later, Congress took another run at the issue. The House overwhelmingly passed legislation providing $25 million to improve post roads with the federal government assuming 30%–50% of the cost, depending on the project. The money would be split among the states according to a formula that gave equal weight to population size and post road miles, thereby striking a balance between interstate and local rural travel.

Unlike the 1912 legislation, the 1916 House bill would have allowed states to choose the projects (subject to federal review) and manage them without further federal oversight. In addition to the funding provisions, the bill, more significantly, would have required participating states to have a state highway agency up and running by 1920. It seemed as though a federal highway program—limited though it might be—was back on track.

But before the Senate could act, opposition began to build: the apportionment formula did not provide funding where it was most needed, the bill's foes argued; preparing for war was more important than building roads, they said; projects would simply reward political supporters, they contended. At this point, the American Association of State Highway Officials (AASHO)—which had only been in existence for a little over a year—jumped in.

The AASHO proposed that the federal government triple the commitment to $75 million over 5 years, split among the states according a three-part formula with population, mileage, and total area given equal weights. As is often true with legislative language, it is important to read the words carefully to see who benefits most. By including "total area" in the formula, the AASHO tilted its proposal heavily toward the farmer and the local rural roads at the expense of advocates of long-distance highways, a bias that would continue for a number of years.

Its plan upped the federal share to half the cost of each project, but capped spending at $10,000 per mile. Although it required the Secretary of Agriculture to sign off on every project, it followed the lead of the stalled House bill, granting states full responsibility for each one, and requiring them to establish a state highway agency for that purpose.

On June 27, 1916, the House and Senate accepted the AASHO language almost verbatim and passed the Federal Road Act of 1916. Although the legislation established the principle of federal-state highway cooperation, it failed to address the interstate needs of an automotive revolution that was already visible on the horizon. Technological change was happening so fast that federal policymakers didn't even appreciate how inappropriate the existing state-by-state patchwork of road names was for the new era. It would be private automobile clubs, such as the AAA, and the American Association of State Highway Officials that would ultimately provide the impetus for assigning uniform numbers to routes that stretched across the nation. But that wouldn't happen until 9 years later.

The early history of American highways demonstrates that disruptive technological change can easily outpace the formulation of policies needed to manage it, partly because policymakers can be slow to recognize the exigency, and partly because disagreements among interest groups can retard the process. As we have seen, it took Congress more than a decade to enact the skeletal 1916 road bill. It would take almost five decades more for the House and Senate to pass the 1956 National Interstate and Defense Highways Act[77] and bring the nation's infrastructure up to par with automotive technology.

Chapter 5

From depression to global engagement 1925–1945

Herbert Hoover is probably best known as the president who was in the Oval Office on October 29, 1929. That day, known ever since as Black Tuesday, began a deep and protracted stock market decline, ushering in the Great Depression. Within 2 days, the Dow Jones Industrial Average lost about a quarter of its value, and by 1932, the last year of Hoover's presidency, shares on the New York Stock Exchange had fallen cumulatively by an average of a staggering 90%. The financial collapse that began on Wall Street in 1929 spread across the world, leaving few areas untouched. In the 3 years that followed, the global economy contracted by almost 15%. Hoover's reputation would forever be tarnished by the economic misery that touched tens of millions of people. By the time he left office, one in four Americans was unemployed, and most of those who were fortunate to have jobs were barely getting by.

Were it not for the Depression, however, Hoover might be remembered for his work promoting science and technology. He came to it naturally, having earned degrees in mining and civil engineering, and believing in progressive Republican principles. He served as Secretary of Commerce from 1921 until 1928 during the presidencies of Warren Harding and Calvin Coolidge at a time that was challenging financially for America's scientific enterprise. Federal support of wartime research had not translated into peace time dollars, and any hope the National Research Council harbored for extending the central role it had played during the war was dashed within months of the treaty signing at Versailles that ended the war.

With federal underwriting of the science and technology enterprise existing in little more than name only, industry became the principal player on the research stage. Several giants dominated the scene in the 1920s, among them American Telephone & Telegraph (AT&T), Dow Chemical, DuPont, General Electric, General Telephone and Electronics, Kodak, and Westinghouse. They had one thing in common: they were "vertically integrated." They conducted their own research (R), developed (D) their own technologies, tested (T) their own innovations, evaluated (E) the results of their endeavors, and marketed

Navigating the Maze. https://doi.org/10.1016/B978-0-12-814710-8.00005-X

their products. In modern Defense Department parlance, they conducted their own RDT and E.

With few exceptions, the "central laboratories" of 1920s American industry focused on applied research directed at improving their company's product lines. The quest to understand nature was not part of the industrial research portfolio. Basic or fundamental research, as it is commonly called, simply didn't align with industry's principal goal of generating profits—and it never will.

As Secretary of Commerce, Herbert Hoover well understood and appreciated industry's research priorities, and he harbored no illusion of being able to alter them. But as Secretary of Commerce, he also understood the importance of scientific research in a broader context. In 1925, his department housed the National Bureau of Standards, the Bureau of the Census, the Bureau of Fisheries, the Bureau of Mines, the Coast and Geodetic Survey, and the U.S. Patent Office. It was, at the time, and in most respects, the federal hub for scientific research.

Heading the Commerce Department gave Hoover a first-hand look at the integral nature of research. Compartmentalizing it, as an individual company did by focusing on applied programs, was understandable from the company's viewpoint but he believed the future of American industry and the future of the country required more. They depended on a steady flow of scientific discoveries, and if industry was ill equipped to perform that mission, then others must. But who should they be, and where would the needed resources come from?

For his time and position, Hoover showed extraordinary breadth and depth of knowledge of the science and technology enterprise. In a December 1926 speech before a joint meeting in Philadelphia of the scientific honor society, Sigma Xi, and the American Association for the Advancement of Science, he answered these questions with logic and clarity. His words bear close reading, because they remain just as relevant today in the 21st century as they were in the third decade of the 20th century.[1]

> I should like to discuss with you for a few moments certain relationships of pure and applied science research to public policies and above all the national necessity for enlarged activities in support of pure scientific research…
>
> …Men in the scientific world will have no difficulty in making a distinction between the fields of pure and applied science. It is, however, not so clear in industry or in our governmental relations and sometimes even in our educational institutions.
>
> At least for the practical purposes of this discussion I think we may make this definition – that pure science research is the search for new fundamental natural law and substance – while applied science is clearly enough the application of these discoveries to practical use. And the two callings depart widely in their motivating impulses, their personnel, their character, their support and their economic setting. And these differences are the root of our problem.

As a nation we have not been remiss in our support of applied science. We have contributed our full measure of invention and improvement in the application of physics, in mechanics, in biology and chemistry and we have made contributions to the world in applied economics and sociology.

Business and industry have realized the vivid values of the application of scientific discoveries. To further it in twelve years our individual industries have increased their research laboratories from less than 100 to over 500... But all these laboratories and experiment stations are devoted to the application of science, not fundamental research. Yet the raw material for these laboratories comes alone from the ranks of our men of pure science whose efforts are supported almost wholly in our universities, colleges and a few scientific institutions.

We are spending in industry, in government, national and local, probably $2000,000,000 a year in search for applications of scientific knowledge – with perhaps 30,000 engaged in the work...

...Yet the whole sum which we have available to support pure science is less than $10,000,000 a year, with probably less [sic] than 4,000 men engaged in it, most of them dividing their time between it and teaching...

...Teaching is a noble occupation, but other men can teach and few men have that quality of mind which can successfully explore the unknown in nature. Not only are our universities compelled to curtail the resources they should contribute in men and equipment for this patient groping for fundamental truth because of our educational pressures, but the sudden growth of industrial laboratories themselves and the larger salaries they offer have in themselves endangered pure science by drafting men from universities...

Some scientific discoveries and inventions have in the past been the result of the genius struggling in poverty. But poverty does not clarify thought, nor furnish laboratory equipment... Discovery nowadays must be builded [sic] upon a vast background of scientific knowledge, of liberal equipment...

The day of the genius in the garret has passed, if it ever existed... We do have the genius in science; he is the most precious of all our citizens. We cannot invent him; we can, however, give him a chance to serve.

...How are we to secure the much wider and more liberal support to pure science research? It appears to me that we must seek it in three directions – first, from the government both national and state; second, from industry; and third, from an enlargement of private benevolence... And the point of application is more liberal appropriations to our National Bureaus for pure science research instead of the confinement as to-day of these undertakings of applied science work. And we must have more liberal support of pure science research in our state universities and other publicly [sic] –supported institutions.

Our second source of support must come from business and industry. You are aware of the appeal in this particular from the National Academy of Sciences a year ago – that they might be entrusted with a fund largely for the better support of proved [sic] men now engaged in such research in our universities and elsewhere... That appeal has been met generously by some of our largest industries; it is under consideration by others; it has been refused by one or two largely because they have not grasped the essential differences between the applied science investigations upon which they are themselves engaged and the pure science which must be the foundation of their future inventions. A nation with an output of fifty billion annually in commodities which could not be produced but for the discoveries of pure science could well afford, it would seem, to put back a hundredth of one per cent [sic] as an assurance of progress...

From benevolence we have had the generous support of some individuals to our universities and scientific institutions, but this benevolence has come from dishearteningly meager numbers... In a nation of such high appreciation of the value of knowledge, and of such superabundance of private wealth, we can surely hope for that wider understanding which is the basis of constructive action...

...Our nation must recognize that its future is not merely a question of applying present-day science to the development of our industries, or to reducing the cost of living, or eradication disease and multiplying our harvests, or even to increasing the diffusion of knowledge. We must add to knowledge, both for the intellectual and spiritual satisfaction that comes from widening the range of human understanding and for the direct practical utilization of these fundamental discoveries.

There are few occupants of the White House who have had Herbert Hoover's appreciation for science—Barack Obama being a notable contender—as well as the complex nature of the science and technology enterprise and the policies required for the enterprise to thrive. Sadly, Hoover's words have largely been relegated to the footnotes of history. There are probably few modern-day policymakers who are aware of them, and even fewer scientists. Yet the arguments Hoover made remain applicable today. If anything, the gap between the nature of research performed in industrial laboratories and the nature of research performed elsewhere has grown even larger. And that gap has the capacity to stifle American innovation in the 21st century. More of that later.

Hoover wasn't just venting or proselytizing at the December 28, 1926 joint meeting of the scientific research honor society Sigma Xi and the AAAS. And he certainly wasn't bragging. He had been hard at work for months, pressing industry to pony up money for pure science research. But when he spoke about the National Academy of Sciences' appeal to industry, he didn't reveal to his audience that he had led it. If any of them had seen a front-page New York Times article on April 21, 1926, they certainly would have known. The headline read in full:

HOOVER LEADS GROUP RAISING $20,000,000 TO AID PURE SCIENCE

Heads of Great Corporations Enter Campaign to Endow Research in Universities

$3,000,000 PLEDGED SO FAR

Educators Hail Fulfillment of Their Needs – Industrialists See Only Hope of Progress

10-YEAR PROGRAM PLANNED

Trustees of National Academy Fund Include Root, Hughes, Mellon John W. Davis, Owen D. Young

Educators might have seen only hope of progress, but the hue of their crystal ball was far too rosy, as 3-year's worth of collections would show. By 1930, of the $3,000,000 pledged, only $379,660 had materialized.[2] The optimism of Hoover and the trustees of the fund had proven woefully misguided.

It is understandable why corporate contributions would have dried up following the 1929 stock market crash, but lack of financial support prior to that can only be attributed to the self-interest of industrial leaders who were focused on near-term results at the expense of long-term growth. If anything, those attitudes are far worse today, for reasons we will come back to.

In the end, Hoover's dream of a National Research Fund dedicated to pure science turned out to be a fantasy. American support for fundamental research and scientific discovery would remain tepid until the end of World War II. It would take the science policy acumen and political savvy of Vannevar Bush,[3] as well as the existential threats of the Soviet Union during the Cold War, to weave the quest for scientific knowledge for knowledge's sake into the American fabric.

Halfway between P and Q Streets in northwest Washington, D.C. on the east side of 35th Street stands a nondescript neoclassic yellow brick and sandstone building that rarely sees any tourist traffic. Opposite Georgetown University, it's a site that played a formative role in the story of America's best-known industrial research facility, Bell Laboratories. During its 71 years of existence, Bell Labs, as scientists and techies called it, was AT&T's innovation powerhouse, and an exemplar of industrial research facilities, bar none.

It all began on March 10, 1876. That was the day a Scottish immigrant, Alexander Graham Bell, used a telegraph modified with a liquid transmitter and spoke these famous words to his assistant: "Mr. Watson, come here. I want to see you."[4] Just 3 days prior to the legendary demonstration, Bell had received patent 174,465 for his invention,[5] which his filing detailed as "The method of, and apparatus for, transmitting vocal or other sounds telegraphically, as herein described, by causing electrical undulations, similar in form to the vibrations of the air accompanying the said vocal or other sounds, substantially as set forth."[6]

Sometimes misfortune can be fortune in disguise, and what happened next is a perfect example. Bell and his partners, Thomas Sanders, a wealthy Boston businessman, and Gardner Greene Hubbard, a friend, financial backer, and soon to be father-in-law, offered to sell their patent to Western Union, for $100,000, but the telegraph company turned them down.[7] That rejection would ultimately make Bell, Sanders, and Hubbard millionaires, although it would take two decades for their fledgling company, Bell Telephone,[8] formed in 1877, to produce big returns.

In the meantime, Bell received a small financial reward, when L'Académie Française chose him as the winner of the 1880 Volta prize[9] for his invention of the telephone. He used the 50,000 francs to found Volta Associates[10] with the goal of developing devices that could record and transmit sound. His success continued, and 7 years later he sold his new patent rights to American Gramophone.[11]

With his new-found wealth, he turned some of his attention to something and someone close to his heart. Mabel Hubbard, his wife and daughter of his partner, had lost her hearing to scarlet fever when she was five, and for Bell, now 40 years old, helping the deaf had special meaning. In 1887, he opened the Volta Bureau, setting up shop in his parents' Washington carriage house, and for the next 6 years, in addition to working on inventions, he conducted research on the cause of deafness.

As his work expanded, his need for expanded space grew, and in 1893, he moved into his newly constructed building near his parent's home on what is now the corner of 35th St. and Volta Place. Although primarily remembered for its work on deafness, the Volta Bureau is often considered the precursor to Bell Labs, because the structure served as Alexander Graham Bell's first dedicated research laboratory.

Bell and his partners were quick to recognize the potential of long-distance telephone service, and in 1885 they formed AT&T Long Lines as a Bell Telephone subsidiary. AT&T's rapid success led it to be the dominant corporate entity, and in 1899 it traded places with Bell Telephone, becoming Bell's parent company. With that change, AT&T's monopolistic die was cast. It would have major ramifications for science and technology—and the policies that guide them—extending almost 100 years into the future. It would also have significance for the operation, support, and extraordinary success of Bell Labs, which saw first life in lower Manhattan after several of AT&T's engineering groups merged with Western Electric Research Laboratory in 1925. Bell Labs would ultimately generate eight Nobel prizes and a slew of blockbuster innovations, among them the transistor, the maser (the precursor to the laser), the solar cell, data networking, and cellular phone technology.

Alexander Graham Bell could never have dreamed of the technologies Bell Labs produced and the impact they would have on American life. Bell and his eponymous research laboratory owe their success not only to human creativity and the quest for knowledge, but also to the pillars of American innovation: patent protection, the availability of capital investment, and a free market of ideas. In the case of Bell Labs, itself, the benefits of being part of a regulated monopoly effectively guaranteed it robust, enduring financial support.

The Depression was devastating for America, and, indeed, the entire world. It left few sectors of the economy unscathed, and it damaged the lives of millions of people. Hoover's plans for a National Research Fund lay in ruin, and by 1934, the National Bureau of Standards had lost 55% of the funding it had only 4 years earlier. Tens of thousands of scientists and technical workers were without jobs, and scores of research projects foundered.[12]

On March 4, 1933, Franklin Delano Roosevelt took office as the 32nd president of the United States. That same day, Henry A. Wallace was sworn in as Secretary of Agriculture, following in the tradition of his father Henry C. Wallace, whom Warren G. Harding had tapped for the position in 1921. But Wallace, the elder, known as Harry, and Wallace, the younger, had dramatically different political philosophies. Harry was a Republican and a member of the protectionist school of trade policy, while Henry A. was a progressive Democrat in the mold of his boss, FDR, and an ardent internationalist.

If science had any advocate in the early years of the Roosevelt administration, it was Henry A., although his views differed sharply from those of the traditionalists who populated such august bodies as the National Academy of Sciences and the American Association for the Advancement of Science. Wallace came to those views quite naturally. He had an undergraduate degree in animal husbandry, was an expert in statistics and econometrics, and had developed a modest reputation for his work in agricultural research.

From Wallace's perspective, the full measure of the natural sciences extended well beyond the intellectual worth of discovery and the growing contributions to military technologies, industrial innovation, and agricultural productivity. Wallace believed that natural science's human impacts—mostly overlooked, as he saw it—constituted an essential part of the measure. His view that the social sciences were as important as the natural sciences was not widely held by the scientific elite, but they meshed well with Roosevelt's vision of the New Deal. Wallace's aim was to make certain that economics, sociology, psychology, and political science received as much attention as agronomy, astronomy, biology, chemistry, math, and physics. He hoped that natural scientists would accept his proposition, but he was highly skeptical they would.

The verdict was not long in coming. Once again, not surprisingly, it would involve the National Academy of Sciences.

At the time, the chairman of the National Research Council was Isaiah Bowman,[13] a prominent geographer and director of the American Geographical Society. Later, his reputation would be sullied by charges of anti-Semitism when he was president of Johns Hopkins University. But in 1933, Bowman was well respected scientifically and well connected politically, leading policy-makers to take his views seriously.

From its inception, the NRC had organized its activities around individual scientific disciplines, much the way its parent, the National Academy of Sciences, did, in the tradition of universities from which both drew most of their members. Bowman recognized that the government's science and technology portfolio did not fit neatly into such a mold, and that any work the NRC carried out for the government needed a much broader, more encompassing perspective. He proposed that the federal government create a board under the auspices of the Academy and the NRC, mandating it to examine and advise the government across the breadth of its science and technology activities. Wallace, who viewed such a construct as consistent with his own vision of the integration of the natural and social sciences, put Bowman's proposal on Roosevelt's desk. The result was the Science Advisory Board, established by executive order in the summer of 1933 with Karl Taylor Compton serving as its chair.

Compton seemed like an ideal choice. His pedigree extended well beyond his physics research in atomic spectroscopy and electronic and atomic collisions. He had been elected to both the National Academy of Sciences and the American Philosophical Society. He was a member of the American Chemical Society, the American Physical Society, the Optical Society of America, and the Franklin Institute. He had served in the State Department and had carried out military research for the Army Signal Corps during World War I. At the time Roosevelt named him to head the Science Advisory Board, Compton was chairman of the board of the American Institute of Physics and president of one of the leading engineering schools in the world, the Massachusetts Institute of Technology.

To be sure, there were a few gaps in Compton's résumé: agriculture, biology, and the social sciences were missing, for example. But Roosevelt would have been hard pressed to find anyone else with Compton's science and technology breadth who could match his enthusiasm for the entire enterprise.

Both Compton and Bowman saw the Board's strength in its arms-length relationship with the government, which gave it an ability to provide unbiased assessments of the strengths and weaknesses of federal programs, and the bureaus that oversaw them. Under Compton's stewardship, the Board fully bought into Wallace's vision of science for the social good, proposing that a portion of public works funds be allocated to research in support of the industrial recovery and relief programs that were key elements of Roosevelt's "New Deal."

It seemed like a good idea on paper, but it ran into a major legal obstacle. Funds allocated for public works projects could only be used for construction. As we will see in the narrative about the 2009 America Recovery and Reinvestment Act in Chapter 11, that object lesson was not lost on proponents of the successful science stimulus in 2009. Before Compton and his colleagues could find a legal work-around, the Advisory Board encountered new headwinds. Even though it had adopted Wallace's worldview, it had not reached out to the social science community. Nor had it brought into its fold scientists who worked for the government. Both mistakes would prove to be a significant impediment to the Board's success.

But the infighting didn't end there. Some members of the National Academy, including its president, William Wallace Campbell, believed they had been ignored, as well, and began attacking Compton and Bowman. For anyone who might think of science as a monolithic endeavor, the behavior of the community in 1934 should give great pause. Such backbiting is more the rule rather than the exception to it, and it carries a strong message for policymakers: do not count on the unconditional support of scientists for strategies they do not clearly recognize as benefitting their own self-interests.

Roosevelt's decision to expand the Board with members whose science pedigrees were either weak or absent entirely only made matters worse. But Compton was undeterred, and later in the year, perhaps recognizing that money might unify the science community, he made a pitch to Roosevelt. Boost federal spending on natural science research by $15 million annually for 5 years, and commit one third of the money to non-governmental research. The amount of money, the non-governmental nature of a large chunk of it, and the multiyear spending plan were all unheard of at the time. Compton did get one thing right, the natural science community swung behind his plan. It was clearly in their own self-interest.

Roosevelt was more circumspect in his response and tasked his Secretary of the Interior, Harold Ickes, with reviewing the proposition. Ickes, Roosevelt's main go-to man, also chaired the recently constituted National Resources Planning Board, which served as the New Deal's central coordinating committee. Compton's proposal reached the highest echelons of Roosevelt's policy apparatus. All it needed was the thumbs up from Ickes. But on the advice of Frederic Delano, the president's uncle who served on the Resources Planning Board, Ickes gave it the thumbs down.

There were a number of problems with Compton's proposal, as Delano and Ickes saw it. First, the magnitude of spending on science was far too large. Second, the multiyear commitment had no legal precedent and no implementation mechanism through the annual appropriations process. Third, there was no provision for continuing activities beyond the 5-year term. Fourth, the social sciences—which were essential parts of Wallace's vision for the New Deal—had been left out. And fifth, setting up a fund to support unspecified research projects was more than the Roosevelt Administration could countenance.

In a last gasp, the Science Advisory Board scaled its proposal back from 5 to 2 years and slashed the spending total to $1.75 million. But it met with the same fate as the original one.

Compton had given it his best shot, but in the end, Wallace's skepticism proved correct. Getting all the natural sciences and all the social sciences on the same page and getting them to set goals that were politically realizable proved to be an insurmountable task. With Compton's proposals consigned to the dustbin, Roosevelt notified the Science Advisory Board in February 1935 that its services were no longer needed. The Science Board would close its doors at year's end.

Compton's efforts were not entirely for naught. He had raised the profile of research on the national stage, and the with the demise of the Science Board in the offing, the National Resources Board saw fit to establish its own science committee, but making sure that the social sciences and education were accorded the same stature and representation as the natural sciences. Two years later, at the committee's behest, Roosevelt called for a reexamination of the "role of the federal government in supporting and stimulating research," noting that research was "one of the Nation's greatest resources."

The study group, chaired by University of Chicago psychologist, Charles Judd, cast its net widely. It's report, "Research – A National Resource,"[14] which appeared in November 1938, provided the first true look at the full array of America's science and technologies activities. It's 21 findings and 8 recommendations provided a framework for the federal government's role in scientific research. They are certainly worth reading for their historical perspective. But there is a more compelling reason. The extent to which they continue to have relevance is truly remarkable, as the following selection illustrates:

Findings

1. From the earliest days of national history the Government of the United States has conducted scientific investigations in order to establish a sound basis for its legislative and administrative activities, Government agencies were pioneers in this country in carrying on research...

3. Research is at the present time universally recognized as highly important. Universities, the foundations, special research institutions, and industrial and commercial concerns are all engaged in the encouragement and prosecution of research in many lines.

4. Competition for research workers and the demand for large funds to support research have created a situation which calls for better coordination of the research facilities of the Nation than now exists.

5. Research is of many different types. There is ample opportunity for all the agencies, private and public, engaged in research to make valuable contributions, especially if further cooperation can be developed.

6. The Congress engages directly in many lines of inquiry through its special committees and special commissions. It often carries on through these

committees and commissions researches which contribute significant findings to both the natural sciences and social sciences.

7. When research projects become elaborate and the necessity for continued investigations of particular types arises, the Congress creates permanent research agencies to make these investigations...

10. Most branches of the government are supplied with a research division...

12. Research agencies in the Government have taken advantage of the aid which can be contributed by able scientists not in the employ of the Government...

14. The solution of the problem of the utilization of the research facilities of the country as aids to research in the Government is rendered possible by the existence of national councils made up of specialists in the major lines of research.

15. The Government has developed a pattern of cooperation with research agencies outside the Government by making contracts for the prosecution of special research projects with responsible institutions and national organizations...

17. Research supported by States and municipal governments is now common, the possibilities of cooperation between the Federal Government and the other units of government in the country should be more fully explored than they have been up to the present time.

18. Similarly, the universities in some areas have organized with a view to cooperating with one another and with the Government.

19. International cooperation in scientific research now exists on a large scale. It could be encouraged to the great advantage of this Nation if the Federal Government would adopt the practice which is common among Governments of other nations of according official recognition and, wherever necessary, financial support to international gatherings of scientists.

20. The methods of securing the funds necessary for research in the Government can be improved. Clear and explicit statements as to the purposes of research projects should be prepared by research agencies. The equipment of the Bureau of the Budget for the consideration of research proposals should be substantially increased...

Recommendations

On the basis of the survey which it has sponsored, the Science Committee makes the following recommendations:

4. That research agencies of the Government be authorized and encouraged to enter into contracts for the prosecution of research projects with the National Academy of Sciences, the National Research Council, the Social Science Research Council, the American Council on Education, the American Council of Learned Societies, and other recognized research agencies.

5. That official recognition and, where necessary, financial support be given by the Government to international meetings of scientists, and that American participation in international organizations and projects be encouraged.

6. That research within the Government and by nongovernmental agencies which cooperate with the Government be so organized and conducted as to avoid the possibilities of bias through subordination in any way to policy-making and policy-enforcing.

7. That research agencies of the Government extend the practice of encouraging decentralized research in institutions not directly related to the Government and by individuals not in its employ...

"Research – A National Resource" might have had a greater impact had it not been for the gathering storm clouds that presaged the outbreak of the Second World War. The timing for a U.S. science renaissance could not have been worse. Charles Judd's ad hoc committee released its report on November 21, 1938. Less than ten months later, World War II began when Germany invaded Poland on September 1, 1939.

About the only area of research that managed to gain a broader mandate before the nation once again turned its attention to military matters, was public health. From its inception, the National Institute of Health, reflecting its Hygienic Laboratory genesis, had been focused solely on infectious diseases. On August 5, 1937, Congress broke new ground when it authorized the creation of a new entity, the National Cancer Institute, as part of the Public Health Service.

The legislation[15] not only expanded the boundaries of federal health research, it explicitly gave the government the ability to sponsor research outside the confines of federal laboratories. The 1937 Act created a National Cancer Advisory Council, authorizing it: "To review applications from any university, hospital, laboratory, or other institution, whether public or private, or from individuals, for grants-in-aid for research projects relating to cancer, and certify to the Surgeon General its approval of grants-in-aid in the cases of such projects which show promise of making valuable contributions to human knowledge with respect to the cause, prevention, or methods of diagnosis or treatment of cancer..."

Had the war not intervened, it is possible that other areas of scientific research might have received equally expansive recognition. But a greater federal role in the science and technology arena would have to wait for the world conflict to run its course. The dawn of a new American research era would not occur until the end of the 1940s. Much would transpire before that time.

Although America's entry into World War II did not occur until Japan attacked Pearl Harbor on December 7, 1941, the U.S. had already been supplying the Allies with military equipment for many months. If World War I had been a war of technology, it was becoming abundantly clear to military planners and policymakers that World War II was going to be a technology conflict on steroids. That posed a big problem for the United States, because the work of the Science Advisory Board, the National Resources Planning Board, and Judd's ad hoc committee had all pointed to significant shortfalls in America's research capabilities. The technological demands of a new world war would reveal just how far those deficits extended.

Vannevar Bush was among those who well understood the precarious condition of the nation's research establishment. An engineer by training, he had left his post as vice president of MIT to become president of the Carnegie Institution of Washington in 1939. From his perch in the nation's capital and with his ties to MIT, he was in a perfect position to assess both how the war might affect the nation's science enterprise, and whether the nation's science enterprise was capable of assisting the impending war effort. Both assessments alarmed him greatly.[16]

Striking the correct balance between military research and "civilian" research, both pure and applied, also weighed heavily on his mind. He elaborated on the thorny matter in his 1939 "Report of the President" of the Carnegie Institution, writing:[17]

THE EMERGENCY

Much of the world is at war. We are fortunately able to stand aside, but no evaluation of the condition and program of an institution can be completely divorced from the stress of the times in which it operates. Even in these fortunate United States all plans are thus conditioned, and every individual is thus affected.

The scientist in particular is faced with a quandary. The same science which saves life and renders it rich and full, also destroys it and renders it horrible. Is it then possible to remain in a detached atmosphere, to cultivate the slowly growing body of pure knowledge, and to labor apart from the intense struggle in which the direct application of science now implies so much for good or ill?...

The quandary may be immediate and direct. Science and its applications have produced the aircraft and the bomb. Entirely apart from all questions of national sympathies, from all opinion concerning political ideologies, we fear to witness the destruction of the treasures of civilization and the agony of peoples, by reason of this new weapon. As science has produced a weapon, so also can it produce in time a defense against it. Science is dedicated to the advance of knowledge for the benefit of man. Here is a sphere where the benefit might perhaps indeed be immediate, real and satisfying. Can a scientist, skilled in a field such that his efforts might readily be directed to the attainment of applications which would afford protection to his fellow men against such an overwhelming peril, now justify expending his effort for any other more remote cause?

Every individual scientist must of course render his own answer. Only a very small percentage are in fields that their efforts could in any case be suddenly altered so as to become immediately effective, and these only are directly faced with the problem...

Throughout the Institution most of our work is, and should remain, far afield from the techniques of the present struggle. Yet if some of our laboratories can contribute, as they did twenty-two years ago, to the solution of immediate problems in the national interest, they should do so…

Yet we should not become stampeded. There is still a duty to keep the torch of pure science lit, and this duty is only the greater under stress… If it is really good that man should look at the stars and should contemplate his great destiny, then it is imperative that in those regions which enjoy the blessings of peace the search for the eternal verities should continue.

The dual character of science influences much of our outlook. We look at the stars, and we build yet greater machines to aid our vision for two reasons. The stars are a laboratory, wherein are pressures and temperatures far beyond those we can artificially produce… We also look at the stars for the same reason that inspired the shepherd on the ancient hill, because we are bound to think of greater things than the comforts or dangers of the morrow…

The same thread runs through all our research…

Our detachment makes it possible for us to keep the even tenor of our way and largely to devote our efforts to inquiries which are the most fascinating that engage the scientific mind, and which will require long and continuous effort by many men for their solution. Some of us certainly should depart at least temporarily from this sustained effort if we see a way in which our science may definitely aid in mitigating some great immediate ill that threatens humanity. Most of us can continue along the familiar path, with clear conscience, toward a distant goal…

Vannevar Bush would reprise the theme at the end of the war in *Science the Endless Frontier – A Report to the President on a Program for Postwar Scientific Research, July 1945*,[18] where he would lay out his vision for American leadership in science and technology that would endure for decades.

Bush's defense of pure science notwithstanding, America's science community, including Bush, came to the defense of the nation in major ways during World War II. The war needs largely shaped America's science and technology policies of that era, and resources for research largely reflected military exigencies. In 1940, Roosevelt was facing both re-election and the inevitability of America's entry into the war. He had learned two lessons from the Depression: the value of preparedness and the importance of acting boldly. Drawing on both, he established the National Defense Research Committee to help mobilize America's scientific talent in support of weapons research. The NDRC[19] grew out of a series of meetings the president had convened with four leaders of the science community. James B. Conant, a chemist who was president of Harvard University at the time, and Frank B. Jewett, a physicist who had been the first head of Bell Labs and was then serving as the National Academy's president, had joined Bush and Compton in helping Roosevelt develop plans to confront the Nazi technological war machine. Strikingly, no social scientists or representatives of the military had seats at the planning table. Nor did anyone from any of the federal agencies.

The composition of the eight-member NDRC was somewhat broader. Rear Admiral Harold C. Bowen, director of the Naval Research Laboratory, and Brigadier General George V. Strong, who represented the Army, complemented the original gang of four, along with Conway P. Coe, Commissioner of Patents, and Richard C. Tolman, dean of the California Institute of Technology graduate school and professor of physical chemistry and mathematical physics. But the NDRC, even with Bush at its helm, still lacked sufficient engineering and industrial expertise on which the military technologies depended. It also lacked both the money and the authority to translate research outcomes into the development and production of instruments of war.

At Bush's suggestion, Roosevelt established a new organization in June 1941 with far broader representation, more extensive authority, and significantly expanded funding, naming Bush as its chairman. The Office of Scientific Research and Development or OSRD,[20] as it became known, would manage all wartime science and technology activities for the military. And with the war effort marshalling most of the nation's scientific talent, OSRD would become the de facto coordinator of virtually all American research, from the physical sciences and engineering to the agricultural sciences and health. As its chairman, Bush became the most powerful voice of American science and technology. His other Washington responsibilities only enhanced his clout.

Bush was a member of the National Advisory Committee for Aeronautics— which he had chaired for 2 years prior to 1940—and in his advisory role to the military's Joint Chiefs of Staff, he led the Joint Committee on New Weapons and Equipment. He was the point person for OSRD's Section S-1, its top-secret Committee on Uranium, and, as Roosevelt's confidant on all matters technological, he was in effect the first presidential science advisor, even if he didn't have the title.

In 1945, as the war was winding down, Bush would use his influence and stature to shape American science and technology policy according to his own vision. His paradigm would endure for half a century or more. If he had gotten it wrong, the outcome could have been ruinous. But he didn't, and the result, as we will see, turned out to be nothing short of remarkable.

From 1941 to 1945, military needs largely dictated America's science and technology policy. Of the many wartime efforts, the Manhattan Project[21] undoubtedly is the best known and best documented. It began with discussions among three Hungarian-born Jewish émigrés, Leo Szilárd, Eugene Wigner, and Edward Teller. Among the most talented nuclear physicists in the world, they had sought refuge in the United States as Hitler began to crank up the Nazi war machine. Believing firmly that an "atomic bomb" was not only feasible, but already on Hitler's radar screen, they sought Albert Einstein's help in warning both the Belgian government—which was innocently supplying Germany with uranium from the Congo—and President Roosevelt of the German threat. It was early July 1939. Two months later Hitler would attack Poland.

Szilárd, Wigner, and Teller knew that Einstein's name would attract more attention than their own, and ultimately, they would be proved right. Belgium was the easy part: Einstein wrote a letter directly to the country's ambassador explaining the danger.[22] But getting a similar message to Roosevelt proved more complicated. To be certain the president would actually see it and grasp its significance, someone had to deliver it and explain it. Szilárd settled on the economist Alexander Sachs, who was a quick study and very close to Roosevelt.

Szilárd drafted the letter, Einstein signed it, and on August 15, they sent it to Sachs. Had Germany not invaded Poland two weeks later, Sachs might have been able to get a meeting with Roosevelt promptly, but with the world in turmoil, he had to wait until October 11. Once in the Oval Office, he made the case in a lengthy discourse, so lengthy, in fact, that the president finally interrupted him, saying, "Alex, what you are after is to see they don't blow us up."[23]

Roosevelt acted quickly, establishing a small Advisory Committee on Uranium. To chair it, he chose Lyman Briggs, then director of the National Bureau of Standards, adding two military men as liaisons to the Army and Navy. Even though Briggs had no background in nuclear physics, Roosevelt's choice seemed reasonable, because NBS had been the government's principal research agency from its inception in 1903. To provide the needed nuclear physics expertise, Briggs invited Szilárd, Wigner, and Teller to the committee's first meeting on October 21—just 10 days after Sachs had spoken with Roosevelt. Later he added Enrico Fermi, an Italian émigré, and without question, the preeminent American expert in nuclear fission, the essence of the atomic bomb.

The committee suffered no lack of scientific brilliance, but the backgrounds of its immigrant advisors were not in sync with American norms. Fearful of German scientific proficiency, they veered toward extreme secrecy. Whatever progress American researchers might make on a nuclear weapon, they feared, would find its way into Hitler's bomb development plans. Moreover, their European experience had not prepared them for the more egalitarian nature of American science. Embracing a larger community to build support in order to get things done was simply not part of their DNA. As a result, outside their small circle, only a few physicists understood the urgent nature of the fission program.

Ernest Lawrence, who had won the physics Nobel Prize in 1939 for inventing the cyclotron, was one. Arthur Compton, who had won the prize previously in 1927 for demonstrating the particle nature of light, and was Karl Compton's younger brother, was another. By the spring of 1940, both of them had become extremely worried that the Briggs committee had made far too little progress during the six months of its existence. Despite their stellar reputations, however, they were unable to do much about it.

That June, the newly formed NDRC, chaired by Vannevar Bush, took ownership of the advisory group, renaming it the Committee on Uranium. But even that transfer and the new oversight that came with it had little impact on the sluggish pace. The committee simply had the wrong leadership and the wrong mix of members and advisors. It is a lesson that policymakers need to take to heart.

Inside the committee, secrecy remained a serious issue, but outside, it was business as usual. The contradiction might have remained, but for a plea from the British government that specifically targeted work at Lawrence's Berkeley cyclotron laboratory. Britain's intervention into American nuclear affairs reflected three realities. First, British bomb research was far ahead of the U.S. effort. Second, Britain was already in a death struggle with the Nazis, while America was still debating whether to join the battle. Third, with its resources stretched thin by the war, the British program would benefit from a collaboration with U.S. scientists. The secrecy request and the collaboration proposal both found ready audiences.

Unexpectedly, it was the science community rather than the federal government that reacted first to the British call for greater secrecy. The National Academy of Sciences seized the day. It established a committee to review scientific manuscripts and censor those that might give Germany insight into critical nuclear physics research results. Executing a collaboration agreement was more complicated, but from a scientific perspective, the timing was right.

When the Office of Scientific Research and Development subsumed the NDRC in June 1941, physicists already understood that two kinds of atomic bombs were potentially realizable. One would require uranium-235, an "isotope" that occurs in natural uranium ore only at a level 0.7%, and the other would need plutonium-239, an isotope of a synthetic element that does not occur in nature at all. Producing a sufficient amount of either fissile material was a monumental task and far beyond the scope or funding of the Uranium Committee.

British physicists had been looking into the feasibility of an atomic bomb since April 10, 1940. In early 1941, they tried to get Briggs to share their preliminary findings with his committee. But Briggs, consistent with his peculiarly plodding personality, simply sat on the material, not letting anyone else see it. Finally, out of frustration, the British team—code named "MAUD"—decided to bypass Briggs entirely, and that August they sent one of their members, Mark Oliphant, to meet with Lawrence at his Berkeley, California cyclotron laboratory.

The meeting was successful: the findings were solid. And on October 3, 1941, the British delivered a copy of their final report directly to Vannevar Bush, once again bypassing Briggs. Before the week was out, Bush had the report in Roosevelt's hands and requested the president's approval for an American program to validate the British findings. Roosevelt gave it a green light, and on December 18, 11 days after Japan had attacked Pearl Harbor, Bush convened the first meeting of OSRD's Section S-1, dedicated to the development of an atomic bomb. A month later Roosevelt formally authorized the project.

Briggs remained chairman of the committee, but the addition of new members injected a new dynamism into its operations in spite of him. Plutonium production work began under Fermi at the University of Chicago in January. And with an executive committee comprising Bush, Compton, Conant, Lawrence, and Harold Urey (a professor of physical chemistry at Columbia University) the committee authorized construction of a uranium isotope separation facility in Tennessee that June. Two months later, the Army Corps of Engineers became operational arm of the project, taking on its massive construction needs.

The "Manhattan Engineering District," as it was called in recognition of the central role New York physicists had been playing, took on a distinctly military flavor. Brigadier General Leslie Groves assumed command of the project in September, and a month later appointed J. Robert Oppenheimer to coordinate the scientific research. Now known as the "Manhattan Project," or simply "Manhattan," the main research and development effort moved from New York to Los Alamos, a remote mesa in New Mexico, which the scientists and engineers who worked on the top-secret project called, "The Hill."

But the Los Alamos facility did not have responsibility for producing the highly-enriched (U-235) uranium or the plutonium needed for the weapon. That task fell to other sites: Oak Ridge, Tennessee; Hanford, Washington; Chicago, and later Argonne, Illinois; and Ames, Iowa. In all, the Manhattan Project was an immense undertaking, employing 130,000 workers and costing almost $2 billion in 1945 dollars, or nearly $20 billion in 2016 dollars.[24]

At Los Alamos, Oppenheimer assembled a brilliant team, inspiring them with a fanaticism rarely seen in any research arena. Groves kept tabs on their activities, ever watchful for breaches of security. On July 16, 1945, their work came to fruition at the "Trinity Test Site" in Alamogordo, New Mexico. The plutonium bomb, they called "The Gadget," worked better than they had imagined. It exploded with an equivalent energy of 20 kilotons of TNT, producing a mushroom cloud that reached an altitude of 7.5 miles in the sky, melted the desert sand, and sent a shock wave felt more than a hundred miles away.[25]

Three weeks later, Harry S Truman, who had assumed the presidency after Roosevelt's death on April 12, authorized nuclear strikes on two Japanese industrial cities. On August 6, a B-29 bomber, carrying the inscription, "Enola Gay," dropped a uranium-235 bomb called "Little Boy" on Hiroshima. Three days later, "Bockscar," another B-29, flew over Nagasaki and released "Fat Man," a replica of the Trinity Gadget. The devastation was terrifying. Although the exact death toll will never be known, estimates run as high as 200,000–300,000.

The two bombings, which ushered in the nuclear age, demonstrated the destructive power of modern physics. They elevated killing to a new and frightening level. But they also revealed the extraordinary capabilities of American scientists and showcased the effectiveness of wartime science and technology policy—at least once the kinks were ironed out. Science had won the war, not only with the atomic bomb, but with radar, sonar, aviation, and medicine. Whether the policies that produced such scientific and technological breakthroughs could be adapted to advance an America at peace was yet to be determined.

Chapter 6

Donning the mantle of world leadership 1945–1952

Hiroshima was obliterated on August 6, 1945; Nagasaki, on August 9. Unaware the United States did not have any more atomic bombs in its arsenal, Japan surrendered unconditionally on August 14 (V-J Day), signing the formal documents on September 2 aboard the U.S.S. Missouri in Tokyo Bay. World War II was over. The nation celebrated.

Nothing captured the country's euphoric mood more than Alfred Eisenstaedt's iconic *Life* magazine photograph above a V-J Day caption that read, "In the middle of New York's Time Square, a white-clad girl clutches her purse and skirt as an uninhibited sailor plants his lips squarely on hers."[1] Two thousand miles away at Los Alamos, the initial joy of victory quickly turned sour. On V-J Day, the scientists and engineers danced and hooted. But as the horror of what they had unleashed began to sink in, their mood took a decidedly gloomy turn.

On October 16, the Army presented a commendation to the laboratory for its remarkable work. Accepting the award, J. Robert Oppenheimer, who had led his scientific troops for 16 months at the isolated high-desert outpost, expressed the thoughts many of his team members were undoubtedly harboring: "If atomic bombs are to be added to the arsenals of a warring world or to the arsenals of nations preparing for war, then the time will come when mankind will curse the names of Los Alamos and Hiroshima."[2]

Vannevar Bush's outlook was far rosier. In fact, it was positively upbeat. It's possible he was just an optimist, or had a much broader perspective than Oppenheimer. It's possible his distance from the Manhattan Project allowed him to be more positive. It's also possible his monograph, *Science: The Endless Frontier*,[3] did not reflect the dangers Oppenheimer saw on the horizon simply because his essay appeared after the war had ended in Europe, but still a month before Hiroshima and Nagasaki were decimated. We will never know, but Bush's vision of a future filled with scientific largess struck a resonant chord with President Truman and a soon-to-be peacetime Congress.

For quite some time, Bush had been contemplating what science could do for America and what government needed to do to make it happen. The official record shows that Franklin Roosevelt, anticipating the end of the war, sent Bush a letter dated November 17, 1944, requesting his views on science policies for a nation at peace. Some historians have suggested that Bush, whose relationship

Navigating the Maze. https://doi.org/10.1016/B978-0-12-814710-8.00006-1

89

with Roosevelt had grown quite close, planted the request in Roosevelt's mind, or even suggested some of the language. You can judge for yourself after you have read Roosevelt's letter.

THE WHITE HOUSE
WASHINGTON

November 17, 1944

Dear Dr. Bush:

The Office of Scientific Research and Development, of which you are the Director, represents a unique experiment of team-work and cooperation in coordinating scientific research and in applying existing scientific knowledge to the solution of the technical problems paramount in war. Its work has been conducted in the utmost secrecy and carried on without public recognition of any kind; but its tangible results can be found in the communiques coming in from the battlefronts all over the world. Someday the full story of its achievements can be told.

There is, however, no reason why the lessons to be found in this experiment cannot be profitably employed in times of peace. The information, the techniques, and the research experience developed by the Office of Scientific Research and Development and by the thousands of scientists in the universities and in private industry, should be used in the days of peace ahead for the improvement of the national health, the creation of new enterprises bringing new jobs, and the betterment of the national standard of living.

It is with that objective in mind that I would like to have your recommendations on the following four major points:

First: What can be done, consistent with military security, and with the prior approval of the military authorities, to make known to the world as soon as possible the contributions which have been made during our war effort to scientific knowledge?

The diffusion of such knowledge should help us stimulate new enterprises, provide jobs for our returning servicemen and other workers, and make possible great strides for the improvement of the national well-being.

Second: With particular reference to the war of science against disease, what can be done now to organize a program for continuing in the future the work which has been done in medicine and related sciences?

The fact that the annual deaths in this country from one or two diseases alone are far in excess of the total number of lives lost by us in battle during this war should make us conscious of the duty we owe future generations.

Third: What can the Government do now and in the future to aid research activities by public and private organizations? The proper roles of public and of private research, and their interrelation, should be carefully considered.
Fourth: Can an effective program be proposed for discovering and developing scientific talent in American youth so that the continuing future of

scientific research in this country may be assured on a level comparable to what has been done during the war?

New frontiers of the mind are before us, and if they are pioneered with the same vision, boldness, and drive with which we have waged this war we can create a fuller and more fruitful employment and a fuller and more fruitful life.

I hope that, after such consultation as you may deem advisable with your associates and others, you can let me have your considered judgment on these matters as soon as convenient — reporting on each when you are ready, rather than waiting for completion of your studies in all.

Very sincerely yours,

/s/

Franklin D. Roosevelt.

Dr. Vannevar Bush
Office of Scientific Research and Development
Washington, D.C.

The letter, although brief, is extremely specific. Had Herbert Hoover been the writer, the detailed nature of the request would not raise any eyebrows. He had a background in geology, mining engineering, and civil engineering, and he was a student of science. But Roosevelt was not in Herbert Hoover's league in that respect. The specificity of his questions lends credence to the possibility that Bush was the true author—either that or Roosevelt had someone else with science credentials upon whom he was relying. The speculation is moot, because by the time Bush completed his assignment, Roosevelt had died, and Truman was in the Oval Office.

After summarizing Roosevelt's request, of which Truman probably was unaware, Bush, as OSRD director, continued with the following words in a letter dated July 5, 1945:

It is clear from President Roosevelt's letter that in speaking of science he had in mind the natural sciences, including biology and medicine, and I have so interpreted his questions. Progress in other fields, such as the social sciences and the humanities, is likewise important; but the program for science presented in my report warrants immediate attention.

In seeking answers to President Roosevelt's questions I have had the assistance of distinguished [OSRD] committees specially qualified to advise in respect to these subjects. The committees have given these matters the serious attention they deserve; indeed, they have regarded this as an opportunity to participate in shaping the policy of the country with reference to scientific research. They have had many meetings and have submitted formal reports. I have been in close touch with the work of the committees and with their members throughout. I have examined all of the data they assembled and the suggestions they submitted on the points raised in President Roosevelt's letter.

Although the report which I submit herewith is my own, the facts, conclusions, and recommendations are based on the findings of the committee which have studied these questions. Since my report is necessarily brief, I am including as appendices the full reports of the committees.

A single mechanism for implementing the recommendations of the several committees is essential. In proposing such a mechanism I have departed somewhat from the specific recommendations of the committees, but I have since been assured that the plan I am proposing is fully acceptable to the committee members.

The pioneer spirit is still vigorous within this Nation. Science offers largely unexplored hinterland for the pioneer who has the tools for his task. The rewards of such exploration both for the Nation and the individual are great. Scientific progress is one essential key to our security as a nation, to our better health, to more jobs, to a higher standard of living, and to our cultural progress.

Respectfully yours,

/s/

V. Bush, Director

The President of the United States
The White House
Washington, D.C.

Bush opened his report with the attention-grabbing headline, "Scientific Progress is Essential," followed by an exposition that reflected his keen understanding of how to spark a politician's interest. Start with human health and the military, and then move on to jobs. It's a prescription for success every science policy advocate should strive to emulate.

The introduction to the report and the chapters that follow reveal Bush's keen understanding of the scope and organization of federal science programs and the rationale for the government's role in science and technology. What he saw and what he proposed in 1945 remain the cornerstone of science and technology policy today. And for that reason, they bear close scrutiny. The following excerpts from his remarkable work highlight his discernment and enduring vision. Even if some of his ideas were flawed, is impossible to overstate the importance of the document in the annals of American science and technology policy.

SCIENCE

THE ENDLESS FRONTIER

A Report to the President
by
Vannevar Bush
Director of the
Office of Scientific Research and Development

Part One
INTRODUCTION
Scientific Progress Is Essential

We all know how much the new drug, penicillin, has meant to our grievously wounded men on the grim battlefronts of this war—the countless lives it has saved—the incalculable suffering its use has prevented. Science and the great practical genius of this Nation made this achievement possible.

Some of us know the vital role which radar has played in bringing the Allied Nations to victory over Nazi Germany and in driving the Japanese steadily back from their island bastions. Again it was painstaking scientific research over many years that made radar possible.

What we often forget are the millions of pay envelopes on a peacetime Saturday night which are filled because new products and new industries have provided jobs for countless Americans. Science made that possible, too...

Advances in science when put to practical use mean more jobs, higher wages, shorter hours, more abundant crops, more leisure for recreation, for study, for learning how to live without the deadening drudgery which has been the burden for the common man for ages past. Advances in science will also bring higher standards of living, will lead to the prevention or cure of diseases, will promote conservation of our limited natural resources, and will assure means of defense against aggression. But to achieve these objectives—to secure a high level of employment, to maintain a position of world leadership—the flow of new scientific knowledge must be both continuous and substantial...

Science Is a Proper Concern of Government

It has been basic United States policy that Government should foster the opening of new frontiers. It opened the seas to clipper ships and furnished land for pioneers. Although these frontiers have more or less disappeared, the frontier of science remains. It is in keeping with the American tradition—one that has made the United States great—that new frontiers shall be accessible for development by all American citizens...

Government Relations to Science—Past and Present

From the early days the Government has taken an active interest in scientific matters... Since 1900 a large number of scientific agencies have been established within the Federal Government, until in 1939 they numbered more than 40.

Much of the scientific research done by Government agencies is intermediate in character between two types of work commonly referred to as basic and applied research. Almost all Government scientific work has ultimate practical objectives but, in many fields of broad national concern, it commonly involves long-term investigations of a fundamental nature. Generally speaking, the scientific agencies of Government are not so concerned with immediate practical objectives as are the laboratories of industry nor, on the other hand, are they free to explore any natural phenomena without regard to possible applications as are the educational and private research institutions...

We have no national policy for science. The Government has only begun to utilize science in the Nation's welfare. There is no body within the Government charged with formulating or executing a national science policy. There are no standing committees of the Congress devoted to this important subject. Science has been in the wings. It should be brought to the center of the stage—for in it lies much of our hope for the future.

There are areas of science in which the public interest is acute but which are likely to be cultivated inadequately if left without more support than will come from private sources... To date, with the exception of the Office of Scientific Research and Development, such support has been meager and intermittent.

For reasons presented in this report we are entering a period when science needs and deserves increased support from public funds.

Freedom of Inquiry Must Be Preserved

The publicly and privately supported colleges, universities, and research institutes are the centers of basic research. They are the wellsprings of knowledge and understanding. As long as they are vigorous and healthy and their scientists are free to pursue the truth wherever it may lead, there will be a flow of new scientific knowledge to those who can apply it to practical problems in Government, in industry, or elsewhere.

Many of the lessons learned in the war-time application of science under Government can be profitably applied in peace... But we must proceed with caution in carrying over the methods which work in wartime to the very different conditions of peace. We must remove the rigid controls we have had to impose, and recover freedom of inquiry and that healthy competitive scientific spirit so necessary for expansion of the frontiers of scientific knowledge...

Part Two

THE WAR AGAINST DISEASE

In War

The death rate for all diseases in the Army, including overseas forces, has been reduced from 14.1 per thousand in the last war to 0.6 per thousand in this war...

The striking advances in medicine during the war have been possible only because we had a large backlog of scientific data accumulated through basic research in many scientific fields in the years before the war.

In Peace

In the last 40 years life expectancy in the United States has increased from 49 to 65 years largely as a consequence of the reduction in the death rates of infants and children; in the last 20 years the death rate from the diseases of childhood has been reduced 87 percent...

These results have been achieved through a great amount of basic research in medicine and the preclinical sciences...

Progress in combating disease depends upon an expanding body of new scientific knowledge.

Unsolved Problems

As President Roosevelt observed, the annual deaths from one or two diseases are far in excess of the total number of American lives lost in battle during this war...

Notwithstanding great progress in prolonging the span of life and in relief of suffering, much illness remains for which adequate means of prevention and cure are not yet known. While additional physicians, hospitals, and health programs are needed, their full usefulness cannot be attained unless we enlarge our knowledge of the human organism and the nature of disease. Any extension of medical facilities must be accompanied by an expanded program of medical training and research.

Broad and Basic Studies Needed

...Progress in the war against disease results from discoveries in remote and unexpected fields of medicine and the underlying sciences.

Coordinated Attack on Special Problems

...Government initiative and support for the development of newly discovered therapeutic materials and methods can reduce the time required to bring the benefits to the public.

Action is Necessary

...It is clear that if we are to maintain the progress in medicine which has marked the last 25 years, the Government should extend financial support to basic medical research in the medical schools and in the universities, through grants both for research and for fellowships...

Part Three

SCIENCE AND THE PUBLIC WELFARE

Relation to National Security

In this war it has become clear beyond all doubt that scientific research is absolutely essential to national security. The bitter and dangerous battle against the U-boat was a battle of scientific techniques—and our margin of success was dangerously small. The new eyes which radar supplied to our fighting forces quickly evoked the development of scientific countermeasures which could often blind them...

The Secretaries of War and Navy recently stated in a joint letter to the National Academy of Sciences:

This war emphasizes three facts of supreme importance to national security: (1) Powerful new tactics of defense and offense are developed around new weapons created by science and engineering research; (2) the competitive time element in developing those weapons and tactics must be decisive; (3) war is increasingly total war, in which the armed services must be supplemented by active participation of every element of civilian population.

To insure continued preparedness along farsighted technical lines, the research scientists of the country must be called on to continue in peacetime some substantial portion of those types of contribution to national security which they have made so effectively during the stress of the present war...

Military preparedness requires a permanent independent civilian-controlled organization, having close liaison with the Army and Navy, but with funds directly from Congress and with clear power to initiate military research which will supplement and strengthen that carried on directly under the control of the Army and Navy.

Science and Jobs

One of the hopes is that after the war there will be full employment, and that the production of goods and services will serve to raise our standard of living. We do not know yet how we will reach that goal, but it is certain that if can be achieved only by releasing the full creative and productive energies of the American people...

More and better scientific research is essential to the achievement of our goal of full employment.

The Importance of Basic Research

Basic research is performed without thought of practical ends. It results in general knowledge and an understanding of nature and its laws. This general knowledge provides the means of answering a large number of important practical problems, though it may not give a complete specific answer to any one of them. The function of applied research is to provide such complete answers. The scientist doing basic research may not be at all interested in the practical applications of his work, yet the further progress of industrial development would eventually stagnate if basic scientific research were long neglected...

Today, it is truer than ever that basic research is the pacemaker of technological progress. In the nineteenth century, Yankee mechanical ingenuity, building largely upon the basic discoveries of European scientists, could greatly enhance the technical arts. Now the situation is different.

A nation which depends upon others for its new basic scientific knowledge will be slow in its industrial progress and weak in its competitive position in world trade, regardless of its mechanical skill.

Centers of Basic Research

Publicly and privately supported colleges and universities and the endorsed research institutes must furnish both the new scientific knowledge and the trained research workers. These institutions are uniquely qualified by tradition and by their special characteristics to carry on basic research... It is chiefly in these institutions that scientists may work in an atmosphere which is relatively free from the adverse pressure of convention, prejudice, or commercial necessity...

Industry is generally inhibited by preconceived goals, by its own clearly defined standards, and by the constant pressure of commercial necessity. Satisfactory progress in basic science seldom occurs under conditions prevailing in the normal industrial laboratory. There are some notable exceptions, it is true, but even in such cases it is rarely possible to match the universities in respect to the freedom which is so important to scientific discovery...

If the colleges, universities, and research institutes are to meet the rapidly increasing demands of industry and Government for new scientific knowledge, their basic research should be strengthened by use of public funds.

Research Within the Government

Although there are some notable exceptions, most research conducted within governmental laboratories is of an applied nature...

Research within the Government represents an important part of our total research activity and needs to be strengthened and expanded after the war. Such expansion should be directed to fields of inquiry and service which are of public importance and are not adequately carried on by private organizations...

In the Government the arrangement whereby the numerous scientific agencies form parts of large departments has both advantages and disadvantages. But the present pattern is firmly established and there is much to be said for it. There is, however, a very real need for some measure of coordination of the common scientific activities of these agencies, both as to policies and budgets, and at present no such means exist.

A permanent Science Advisory Board should be created to consult with these scientific bureaus and to advise the executive and legislative branches of Government as to the policies and budgets of Government agencies engaged in scientific research.

The board should be composed of disinterested scientists who have no connection with the affairs of any Government agency.

Industrial Research

The simplest and most effective way in which the Government can strengthen industrial research is to support basic research and to develop scientific talent...

One of the most important factors affecting the amount of industrial research is the income tax law. Government action in respect to this subject will affect the rate of technological progress in industry...

The Internal Revenue Code should be amended to remove present uncertainties in regard to the deductibility of research and development expenditures as current charges against net income.

Research is also affected by the patent laws. They stimulate new invention and they make it possible for new industries to be built around new devices and new processes...

Yet uncertainties in the operation of the patent laws have impaired the ability of small industries to translate new ideas into processes and products of value to the Nation. These uncertainties are, in part, attributable to the difficulties and expense incident to the operation of the patent system as it presently exists. The uncertainties are also attributable to the existence of certain abuses which have appeared in the use of patents. The abuses should be corrected...

International Exchange of Scientific Information

International exchange of scientific information is of growing importance. Increasing specialization of science will make it more important than ever that scientists in this country keep continually abreast of developments abroad. In addition, a flow of scientific information constitutes one facet of general international accord which should be cultivated...

The Government should take an active role in promoting the international flow of scientific information.

The Special Need for Federal Support

We can no longer count on ravaged Europe as a source of fundamental knowledge...

New impetus must be given to research in our country. Such new impetus can come promptly only from the Government...

In providing government support, however, we must endeavor to preserve as far as possible the private support of research both in industry and in the colleges, universities and research institutes.

Part Four

RENEWAL OF OUR SCIENTIFIC TALENT

Nature of the Problem

The responsibility for the creation of new scientific knowledge rests on that small body of men and women who understand the fundamental laws of nature and are skilled in the techniques of scientific research... I cannot improve on [Harvard] President Conant's statement that:

...in every section of the entire area where the word science may properly be applied, the limiting factor is a human one. We shall have rapid or slow advance in this direction or in that depending on the number of really first-class men who are engaged in the work in question... So in the last analysis, the future of science in this country will be determined by our basic educational policy.

A Note of Warning

It would be folly to set up a program under which research in the natural sciences and medicine was expanded at the cost of the social sciences, humanities, and other studies so essential to national well-being...

The Wartime Deficit

…With mounting demands for scientists both for teaching and for research, we will enter the postwar period with a serious deficit in our trained scientific personnel.

Improve the Quality

Confronted with these deficits, we are compelled to look to the use of our basic human resources and formulate a program which will assure their conservation and effective development…

Remove the Barriers

Higher education in this country is largely for those who have the means…

If ability, and not the circumstances of family fortune, is made to determine who shall receive higher education in science, then we shall be assured of constantly improving quality at every level of scientific activity.

The Generation in Uniform Must Not Be Lost

…The Armed Services should comb their records for men who, prior to or during the war, have given evidence of talent for science, and make prompt arrangements, consistent with current discharge plans, for ordering those who remain in uniform as soon as militarily possible to duty at institutions here and overseas where they can continue their scientific education…

A Program

…To encourage and enable a larger number of young and women of ability to take up science as a career, and in order to reduce the deficit of trained scientific personnel, it is recommended that provision be made for a reasonable number of (a) undergraduate scholarships and graduate fellowships and (b) fellowships for advanced training and fundamental research. The details should be worked out with reference to the interests of the several States and of the universities and colleges; and care should be taken not to impair the freedom of the institutions and individuals concerned.

Part Five
A PROBLEM OF SCIENTIFIC RECONVERSION

Effect of Mobilization of Science for War

We have been living on our fat. For more than 5 years many of our scientists have been fighting the war in the laboratories, in the factories and shops, and at the front. We have been directing the energies of our scientists to the development of weapons and materials and methods on a large number of relatively narrow projects initiated and controlled by the Office of Scientific Research and Development and other Government agencies. Like troops, scientists have been mobilized and thrown into action to serve their country in time of emergency. But they have been diverted to a greater extent than is generally appreciated from the search for answers to the fundamental problems—from the search on which human welfare and progress depends...

Security Restrictions Should be Lifted Promptly

Our ability to overcome possible future enemies depends upon scientific advances which will proceed more rapidly with diffusion of knowledge than under a policy of continued restriction of knowledge now in our possession...

Part Six

THE MEANS TO THE END

New Responsibilities for Government

...The Federal Government should accept new responsibilities for promoting the creation of new scientific knowledge and the development of scientific talent in our youth...

In discharging these responsibilities Federal funds should be made available. We have given much thought to the question of how plans for the use of Federal funds may be arranged so that such funds will not drive out of the picture funds from local governments, foundations, and private donors. We believe that our proposals will minimize that effect...

It is also clear that the effective discharge if these responsibilities will require the full attention of some over-all agency devoted to that purpose... Such an agency should furnish the funds needed to support basic research in the colleges and universities, should coordinate where possible research programs on matters of utmost importance to the national welfare, should formulate a national policy for the Government toward science, should sponsor the interchange of scientific information among scientists and laboratories both in the country and abroad, and should ensure that all the incentives to research in industry and the universities are maintained.

The Mechanism

There are within Government departments many groups whose interests are primarily those of scientific research... But nowhere in the governmental structure receiving funds from Congress is there an agency adapted to supplementing the support of basic research in the universities, both in medicine and the natural sciences; adapted to supporting research on new weapons for both Services; or adapted to administering a program of science scholarships and fellowships.

A new agency should be established, therefore by the Congress for the purpose. Such an agency, moreover, should be independent agency devoted to the support of scientific research and advanced scientific education alone... Separation of the sciences in tight compartments, as would occur of more than one agency were involved, would retard and not advance scientific knowledge as a whole.

Five Fundamentals

There are certain basic principles which must underlie the program of Government support for scientific research and education... The principles are as follows:

(1) ...there must be stability of funds over a period of years so that long-range programs may be undertaken.

(2) The agency to administer such funds should be composed of citizens selected only on the basis of their interest in and capacity to promote the work of the agency. They should be persons of broad interest in and understanding of the peculiarities of scientific research and education.

(3) The agency should promote research through contracts or grants to organizations outside the Federal Government. It should not operate any laboratories of its own.

(4) Support of basic research in the public and private colleges, universities, and research institutes must leave the internal control of policy, personnel, and the scope of the research to the institutions themselves.

(5) While assuring complete independence and freedom for the nature, scope, and methodology of research carried on in the institutions receiving public funds, and while retaining discretion in the allocation of funds among such institutions, the Foundation proposed herein must be responsible to the President and the Congress...

Basic research is a long-term process... Methods should therefore be found which will permit the agency to make commitments of funds from current appropriations for programs of 5 years duration or longer...

National Research Foundation

   I. *Purposes*

      The National Research Foundation should develop and promote a national policy for scientific research and scientific education, should support basic research in nonprofit organizations, should develop scientific talent in American youth by means of scholarships and fellowships, and should by contract support long-range research on military matters.

   II. *Members*

      1. Responsibility to the people, through the President and Congress, should be placed in the hands of...persons not otherwise connected to the Government and not representative of any special interest...

      3. The members shall serve without compensation...

      5. The chief executive officer of the Foundation should be a director appointed by the Members...

   III. *Organization*

      1. In order to accomplish the purposes of the Foundation, the Members should establish several professional Divisions... At the outset these Divisions should be:

        a. Division of Medical Research...

        b. Division of Natural Sciences...

c. Division of National Defense...
d. Division of Scientific Personal and Education...
e. Division of Publications and Scientific Collaboration...

IV. *Functions*

1. The Members of the Foundation should have the following functions, powers, and duties:

a. To formulate overall policies for the Foundation...

f. To review the financial requirements of the several Divisions and to propose to the President the annual estimates for the funds required by each Division. Appropriations should be earmarked for the purposes of specific Divisions, but the Foundation should be left discretion with respect to the expenditure of each Division's funds.

g. To make contracts or grants for the conduct of research by negotiation without advertising for bids...

i. To enter into contracts with or make grants to educational and non-profit institutions for support of scientific research.

j. To initiate and finance... research on problems related to national defense.

k. To initiate and finance... research projects for which existing facilities are unavailable or inadequate.

l. To establish scholarships and fellowships in the natural sciences and medicine.

m. To promote the dissemination of scientific and technical information and to further its international exchange.

n. To support international cooperation in science...

V. *Patent Policy*

...In making contracts with or grants to...organizations [outside the Government] the Foundation should protect the public interest adequately and at the same time leave the cooperating organizations with adequate freedom and incentive to conduct scientific research. The public interest will normally be adequately protected if the Government receives royalty-free license for governmental purposes under any patents resulting from the work financed by the Foundation... There should certainly *not* be any absolute requirement that all rights in such discoveries be assigned to the Government, but it should be left to the discretion of the Director and the interested Division whether in special cases the public interest requires such an assignment...

Action by Congress

The National Research Foundation herein proposed meets the urgent needs of the days ahead...

Legislation is necessary. It should be drafted with great care. Early action is imperative, however, if this Nation is to meet the challenge of science and fully utilize the potentialities of science. On the wisdom with which we bring science to bear against the problems of the coming years depends on huge measure our future as a Nation.

Vannevar Bush saw America's future intertwined with science, and he was forceful in expressing his conviction. He was also forceful in proposing a blueprint for organizing America's post-war research activities. Despite its defects, *Science: The Endless Frontier* is a remarkable treatise. It provides compelling rationales for basic research, the benefits of science and technology in American life, and the importance of federal support of the science and technology enterprise. It is remarkable for its clarity and vision.

But it is also remarkable for its lack of any significant historical context, except for the backdrop World War II provided. It is difficult to know whether Bush was not a student of history or whether he simply saw the often unchoreographed and ineffective policy gambits of the past as irrelevant distractions to his visionary proposals for the future. Perhaps, if Roosevelt hadn't died, the omission would have been immaterial, because the 32nd president had a close and trusted relationship with Bush, and it's reasonable to assume that the two had spoken about the importance of science in an America at peace.

But *Science: The Endless Frontier* landed on Harry S. Truman's desk after he had been in office for a scant 84 days. And as David McCullough recounts in his noted biography, *Truman*,[5] the 33rd president assumed his office "unprepared, bewildered. And frightened." And for good reason as Truman later revealed. McCullough quotes him telling in his wife, Margaret, privately that Roosevelt "never did talk to me confidentially about the war, or about foreign affairs, or what he had in mind for peace after the war." The only science policy perspective Truman probably had would have come from his brief but highly visible tenure as chairman of a special Senate committee on U.S. war production and from the cover letter Bush provided with his monograph. And in a world of complexity, both were scant.

Science policy could not have been terribly high on Truman's "to do" list. He had monumental decisions to make with little time to make them, and little preparation for the task—how to prosecute the final days of the war in the Pacific theater, whether to drop nuclear bombs on Japan, how to work with Churchill and the Allies on rebuilding Europe, how to contain Stalin and the Soviet Union's expansionist objectives, whether to treat Communism as a domestic threat, how to place a wartime economy on a peacetime footing, and how to transition returning warfighters to a civilian workforce. Taking Vannevar Bush's vision for American science and technology in the post-war period and turning it into an immediate reality simply could not compete with the other challenges Truman faced, especially because he was largely ignorant of many science policy essentials when he entered the Oval Office. He would prove to be a quick study, but his late entry into the science policy arena couldn't prepare him for the battles that would soon erupt over competing ideas for the new era of American science.

Science and technology policy is never clear-cut, rational, or scientific, as many modern-day authorities would have you believe. It would be nice if it were, but it isn't. It has always been thick with politics, intrigue, and petty grievances, often driven by special interests and the well-connected and not infrequently purely by serendipity in its outcomes. Vannevar Bush might have thought his professional bona fides, intellect, administrative stature, and grasp of science and technology issues would carry the day, but he didn't fully account for one uninvited guest at his policy table. And the protracted and grueling birth of the National Science Foundation (National Research Foundation, in Bush's language) would reflect his misfortune, as well as his gross miscalculation. It's one of the most illuminating stories in the chronicles of science policy and certainly worth recounting.

With Roosevelt's death, Bush's path to implementing his postwar science policy agenda was no longer smooth. But his troubles actually began 5 years earlier, when Harley Kilgore won election to the United States Senate.

Kilgore was a liberal Democrat from West Virginia, a staunch Roosevelt adherent who drew strong support from organized labor. In 1940, he challenged the incumbent, Rush Holt, Sr., who had run as a New Deal Democrat in 1934, but had turned conservative and isolationist, on one occasion even associating himself with the America First Committee,[5] known for its anti-Semitic, pro-fascist propaganda. Holt found himself out of step with his West Virginian base, especially with John L. Lewis, president of the United Mineworkers Union. Kilgore swept through the open door, defeating Holt in a primary and winning the general election handily.

Kilgore took his seat in the Senate chamber in 1941 and accepted an appointment to the Special Committee to Investigate the National Defense Program. Charged with investigating waste and corruption in war production, the eponymous committee was chaired by Harry Truman. The committee's success in rooting out war profiteering proved to be decisive in Roosevelt's decision to name Truman his vice-presidential running mate in 1944. It also provided Kilgore with an entree to Truman's inner circle and, just as significantly, led to his appointment in 1942 as chairman of the Military Affairs Subcommittee on War Mobilization.

Thrust into a leadership position, Kilgore promptly began to use his power to probe whether the nation was using its scientific and technical capabilities most effectively in prosecuting the war.[6] And at the prompting of Herbert Schimmel,[7] a physicist and congressional staffer, Kilgore took dead aim at Bush's stewardship of OSRD. As Kilgore saw it, Bush was focusing too much on the scientific and industrial elites and ignoring the potential contributions of many others outside the select circle.

Several scientists and engineers who had not made Bush's cut testified before the Kilgore Committee in its early days, feeding the chairman's narrative and tacitly endorsing his draft legislation[8] that would establish a centralized Office of Technological Mobilization and broaden the participation of the scientific community in the war effort. Kilgore also drew support from the American Association of Scientific Workers, which argued that chemists, biologists, clinical doctors and earth scientists had been shunted aside by OSRD.[9] Regardless of its merits, in the politics of science and technology policy, the bill, S. 2721, has to be seen as a direct assault on OSRD and its director.

Like Bush, Kilgore recognized the importance of science, but as a non-scientist, he saw the enterprise though a dramatically different lens. In *Science: The Endless Frontier*, Bush gave strong voice to the importance of scientific research for knowledge's sake—basic research as policy wonks call it—trusting that scientific discovery emanating from it would eventually lead to technological advances, a strong military, and improvement in the human condition. And he believed that scientists would be faithful stewards of their mission if given sufficient latitude. After all, he considered himself one of them.

Kilgore had a more transactional view, as the purpose of "The Technology Mobilization Act" of 1942 revealed: "To regain, maintain, and surpass our previous technical preeminence and attainments; and to make forever secure America's world leadership in the practical application of scientific discoveries…"[10] Basic research that did not have a specific connection to utility was not a priority for Kilgore, at least not in 1942. Foreshadowing legislation he would submit 3 years later, the Technology Mobilization Act also hinted at Kilgore's lack of trust in scientists to manage their own affairs—at least where federal funds were involved—and his conviction that greater central planning at the federal level was needed. In the bill's language, "The Office of Technological Mobilization is authorized and directed to review all projects for research and development, including practical development of inventions which may be brought to its attention; and it shall promote such projects as it deems appropriate…"[11]

As Kilgore saw it, scientists were actors rather than directors on the research stage. And the federal government, as the producer of the science theater, needed to have full control of the show.

Not surprisingly, Bush, with the scientific and industrial elites' backing, parried Kilgore's thrust, and after extensive hearings, Kilgore agreed to modify the legislation. His new bill, "The Science Mobilization Act," introduced in the Senate as S. 702 in February 1943 and accompanied in the House of Representatives by Wright Patman's bill, H.R. 2100,[12] provided a grander vision for a postwar scientific enterprise and, as a concession to his critics, modestly reduced the degree of centralized planning and control. But it retained the emphasis on utility and societal benefits of science, creating an Office of Scientific and Technical Mobilization and specifically authorizing it to "develop comprehensive programs for the maximum use of science and technology in the national interest in periods of peace and war;…to promote the full and speedy introduction of the most advanced and effective techniques—for the benefit of agriculture, manufacturing, distribution, transportation, communication, and other phases of productive activity;…to promote full employment and higher standards of living after the war…"[13]

The Kilgore-Patman legislation did little to mollify critics of the earlier version. The American Association for the Advancement of Science (AAAS), for example, asserted the new bill was not mobilizing science as much as regimenting it.[14,15] In 1943, Bush still had the upper hand, and, of course, Roosevelt had his back. But Kilgore was not done. In late 1944, he made another run at the issue, this time focusing on science in an America once again at peace. In an early 1945 report, he elaborated on his proposal for establishing a National Science Foundation as an independent federal agency. It was to have a director chosen by the president subject to advice and consent by the Senate, but to keep the reins of government tight and scientists in check, it was to adhere to the following directives:[16]

> In exercising his authority and duties, the Director should consult with a National Science Board on all matters of major policy or program. The Board should consist of the Director, acting as Chairman, the Secretaries of War, Navy, Interior, Agriculture, Commerce, and Labor, the Attorney General, and the head of the Federal Security Agency, or their representatives, and eight members at large appointed by the President.

> In general the administrative powers should be vested in the Director, but the allocation of funds to specific fields of research and development, the appointment of members to special advisory research committees, and similar duties or authority of primary importance should depend upon the approval of the Board. Thus, by providing guidance and acting as a check, the Board would share responsibility with the Director for the efficient operation of the Foundation.

> The Foundation should not itself, as a general rule, perform any research or development work. Instead, it should make funds for this purpose available to other organizations, public or private, who are already staffed or equipped to do so. Wherever possible, these other organizations, including private individuals, should be encouraged to participate jointly in formulating, promoting, and carrying through the programs and projects which are deemed desirable in the public interest.

> The National Science Board should be responsible for determining the allocation of research and development funds within the limits appropriated annually by Congress. As a guide, the proposed bill requires particular attention to be given to these categories of research and development: National defense; health and medical care; basic sciences; natural resources; methods, products, and processes which may be valuable for small business enterprises; and peacetime uses for wartime research and wartime facilities...

> To protect the taxpayer's interest, all research and development projects financed in whole or in part by the Federal Government should be undertaken only upon the condition that any invention or discovery resulting therefrom would become the property of the United States.

Although much of the report was administratively prescriptive and clearly intended to constrain the autonomy of the nation's elite scientists and the academic institutions they inhabited—mostly in California and the Northeast—it did include protections for the research community. Individual scientists and technologists, the report stated, "should be encouraged to exercise their creative talents and to develop promising new ideas, and, moreover, ... they should not be prevented in any way from expressing their personal beliefs on scientific and technical matters (except when in violation of national security)." And it had something to say about the future workforce, directing the Foundation "to discover and develop scientific talent, particularly in American youth. To this end it should be empowered to grant fellowships and scholarships in various fields of science."

It's worth a moment's pause to summarize the science policy debate that Truman encountered when he took office. Both Bush and Kilgore agreed that American science needed to thrive in the post-war era, and both saw the need for a new science agency to achieve that outcome. But their visions for a new structure diverged in highly significant ways.

Bush wanted scientists to control the new agency. They would populate the agency's board and select the agency's director. Kilgore wanted a director chosen by the president, along with a collaborative board, populated half by Cabinet officials (or their designees) and half by presidential appointees.

Bush wanted the eminent research institutions to retain their leadership in research. Kilgore labeled Bush's approach undemocratic, inconsistent with the New Deal agenda on which he, Kilgore, had run for office. He wanted to open up the research enterprise to the have-nots, not only as a matter of democratic practice, but as a means of achieving greater economic parity throughout the country.

Bush wanted the new agency to be restricted to the natural sciences. As a true New Dealer in the tradition of Henry A. Wallace, Kilgore insisted that the social sciences be included.

Bush wanted researchers and their institutions to retain the rights to any patents emanating from government-sponsored research, allowing the federal government to have a royalty-free license for government use. Kilgore, again fearful of the concentration of power among the elites, wanted the patents to become the property of the government to maximize their utility throughout the country.

Bush drew his support from the bastions of research and development: the premier academic institutions, such as MIT, Berkeley, and Columbia, and the major industrial laboratories, such as Westinghouse, General Electric, and Bell. He also had an ally in Warren Magnuson, who had been elected to the Senate from Washington in 1944. A first-termer seeking recognition, Magnuson submitted legislation on July 19, 1945, two weeks after *Science: The Endless Frontier* appeared. The Magnuson bill specifically called for establishing a National

Research Foundation, closely following the model Bush had proposed. It garnered little enthusiasm from other senators, and Magnuson let it die.

Kilgore's support came from a more diverse group of scientists and businesses, with whom he had a greater affinity, but few of them had the credentials of Bush's adherents, at least at the outset. Nonetheless, Kilgore was a savvier politician than Bush, and he knew that an agency not subject to customary federal budgetary and administrative protocols was likely to run into severe headwinds. He probably also knew Truman well enough to surmise that the president would share his views.

Truman, who had little time to get up to speed following Roosevelt's death, showed how much he had synthesized in just five months, when he presented a "21-Point Program for the Reconversion Period" in a Sept. 6, 1945 Special Message to Congress[17] four days after Japan had surrendered. Science was prominently on the list, and the words are worth reading.

12. RESEARCH

Progress in scientific research and development is an indispensable condition to the future welfare and security of the Nation. The events of the past few years are both proof and prophecy of what science can do.

Science in this war has worked through thousands of men and women who labored selflessly and, for the most part, anonymously in the laboratories, pilot plants, and proving grounds of the Nation.

Through them, science, always pushing forward the frontiers of knowledge, forged the new weapons that shortened the war.

Progress in science cannot depend alone upon brilliant inspiration or sudden flights of genius. We have recently had a dramatic demonstration of this truth. In peace and in war, progress comes slowly in small new bits, from the unremitting day-by-day labors of thousands of men and women.

No nation can maintain a position of leadership in the world of today unless it develops to the full its scientific and technological resources. No government adequately meets its responsibilities unless it generously and intelligently supports and encourages the work of science in university, industry, and in its own laboratories.

During the war we have learned much about the methods of organizing science, and about the ways of encouraging and supporting its activities.

The development of atomic energy is a clear-cut indication of what can be accomplished by our universities, industry, and Government working together. Vast scientific fields remain to be conquered in the same way.

In order to derive the full profit in the future from what we have learned, I urge upon the Congress the early adoption of legislation for the establishment of a single Federal research agency which would discharge the following functions:

1. Promote and support fundamental research and development projects in all matters pertaining to the defense and security of the Nation.

2. Promote and support research in the basic sciences and in the social sciences.
3. Promote and support research in medicine, public health, and allied fields.
4. Provide financial assistance in the form of scholarships and grants for young men and women of proved scientific ability.
5. Coordinate and control diverse scientific activities now conducted by the several departments and agencies of the Federal Government.
6. Make fully, freely, and publicly available to commerce, industry, agriculture, and academic institutions, the fruits of research financed by Federal funds.

Scientific knowledge and scientific research are a complex and interrelated structure. Technological advances in one field may have great significance for another apparently unrelated. Accordingly, I urge upon the Congress the desirability of centralizing these functions in a single agency.

Although science can be coordinated and encouraged, it cannot be dictated to or regimented. Science cannot progress unless founded on the free intelligence of the scientist. I stress the fact that the Federal research agency here proposed should in no way impair that freedom.

Even if the Congress promptly adopts the legislation I have recommended, some months must elapse before the newly established agency could commence its operations. To fill what I hope will be only a temporary gap, I have asked the Office of Scientific Research and Development and the Research Board for National Security to continue their work.

Our economic and industrial strength, the physical well-being of our people, the achievement of full employment and full production, the future of our security, and the preservation of our principles will be determined by the extent to which we give full and sincere support to the works of science.

It is with these works that we can build the highroads to the future.

It's clear from the message that Truman was enthusiastically on board with the importance of science in the nation's future and with the need for a federal research agency. But other than his reference to the social sciences (cited in Function 2) and to the importance of sharing the fruits of research widely (emphasized in Function 6), he didn't signal whether he was leaning toward Bush or Kilgore, especially on the matter of the proposed agency's operational independence. Given all that was on his plate, it's quite likely he hadn't had time to focus on it.

Truman wanted Congress to act, that much is certain, but absent clear White House guidance, it was almost inevitable that the Bush-Kilgore dispute was likely to fester. And that it did. The bickering extended well beyond Capitol Hill and well beyond Washington: the nation's science community, itself, was sparring.[18]

Magnuson, who had been carrying Bush's water in the Senate, began to collaborate with Kilgore, and the two finally came to an agreement in the summer of 1946, co-sponsoring a bill, S. 1850, that would establish a National Science Foundation mostly along the lines Kilgore had proposed. On July 3, the Senate, by a vote of 48 to 18, passed the compromise legislation, which specified Kilgore's administrative structure—a director appointed by the president and a collaborative board—but it rejected, by a vote of 46 to 26, an amendment that would have included the social sciences.[19]

During the debate, Magnuson refuted an accusation that the legislation was not what the science community wanted, declaring,[20] "Certainly there are portions of the bill with which some one scientist does not agree; there are other portions of the bill to which another group does not agree; but by and large, they all agree to this bill." He then introduced into the record a joint statement of support from 11 prominent scientists, whom he asserted were representative of "the cream and the bulk – at least 98 percent – of the educators and scientists of this country." Leading the list were James B. Conant, president of the AAAS and president of Harvard University; George F. Zook, president of the Council on Education; Morris Fishbein, representing the American Medical Association; Thomas P. Cooper, president of the Association of Land-Grant Colleges and Universities; and, most significantly, Isaiah Bowman, chairman of the committee supporting the Bush report and president of Johns Hopkins University.

Within short order, it became clear that Magnuson had overstated his case significantly. The 11 signatories were among the "cream" of the American science community, but they fell far short of living up to the senator's claim that they represented at least 98 percent of the educators and scientists. No sooner had the Senate passed the bill than scientists of every ilk began to attack the legislation. Many, standing up for Bush's plan, wanted less government control over the Foundation. Others, such as Frank Jewett, who was president of the National Academy of Sciences and president of Bell Labs, and Robert Millikan, who had just stepped down as president of Cal Tech, opposed the concept of the Foundation, period. They saw major federal funding of science as a threat to the autonomy of the enterprise and an excuse for universities, philanthropies, and industries to step back from their support of research.

The splintering of the science community torpedoed the bill when it reached the House of Representatives a few weeks later. John Heselton, a representative from Massachusetts, put it this way: "...the feeling in the House was that until the scientists themselves got together on the kind of organization they wanted, the members of Congress should do nothing."[21] And nothing is what they did. The bill's demise exasperated Truman, who, by that time, had become a staunch supporter of Kilgore's model for the Foundation. No doubt his relationship with Kilgore from his Senate days predisposed him to that view, but now having had a taste of presidential power, he unquestionably also believed in greater White House control over the Foundation's operations.

Truman truly wanted to see science elevated on the national stage, but he needed to find an acceptable path forward.

At the suggestion of several members of his staff, who worried that Congress would now simply adopt Bush's model, Truman turned to John Steelman to help resolve the issue. An economist and sociologist by training, Steelman enjoyed Truman's total trust, so much so that by the time his new committee, officially known as the President's Scientific Research Board, had completed its work in August of 1947, Steelman would have the title, Assistant to the President, in effect White House Chief of Staff.

The Steelman Committee was a who's who of the upper echelons of the federal government. Its members ranged from officials in Truman's Cabinet to heads of agencies that had any whiff of science and technology in their portfolios: The Departments of War, Navy, Commerce, Interior, and Agriculture; the Federal Communications Commission, the Federal Works Agency, and the National Advisory Committee of Aeronautics; and more. The Steelman Committee was about as inclusive as any federal committee could get. That it would be able to generate a comprehensive report in less than a year, as Truman desired, was a reach, especially by today's standards.

It was now October 1946. More than a year had passed since World War II officially ended, and Congress had moved swiftly to transition America's wartime research and development programs to a peacetime footing. On August 1, 1946, it had established both the Office of Naval Research[22] and the Atomic Energy Commission.[23] It had added new institutes at the National Institutes of Health[24] and was expanding activities at the National Bureau of Standards (later known as the National Institute of Standards and Technology). But a new national science agency, the jewel of Vannevar Bush's policy case, was still a diamond in the rough.[25]

The waters had become muddier since Bush submitted his report to Truman in July of 1945. Up to that point—as our historical journey through American science and technology policy has amply demonstrated—the federal government's interest in research was far from passionate. Except during health crises or times of war, it was politically more a dalliance than an imperative. But America's postwar domestic prosperity and global leadership, Bush argued, would require stronger ties between the federal government and the nation's burgeoning research enterprise. It would require spending more federal money on science and creating new administrative structures. His proposed new science agency would be a guiding light.

But now that Congress had expanded research activities throughout the federal government, it was far from clear that Bush's new agency was really needed. It still had support on Capitol Hill, but Truman believed Steelman's committee needed to reexamine its rationale before tackling the question of how it should be organized and run. Truman also saw a glaring omission in

Bush's report: It had failed to frame science and technology policies in political terms. Truman's astute observation carries a message for anyone wanting to wrestle in the policy arena. Ignore politics at your peril.

The results of the 1946 election would further complicate Steelman's work. Truman's popularity had been sinking almost from the day he took office, and in the months following the war's end, his handling of a rash of labor strikes tanked his approval ratings. Steel workers, coal miners, auto workers, meat-packers, railroad engineers, and electrical workers all walked off their jobs. At one point in 1946, more than a million workers were out on strike at the same time.[26] The public rendered its judgment on Truman's performance by turning over control of both houses of Congress to the Republicans.

The Republican takeover did not mean that Congress would sideline science on its policy agenda. Quite the contrary, the imperative for action was strong on both sides of the aisle—science would not become a partisan punching bag until Donald Trump became president in 2017—but a different cast of players would command the spotlight. A flurry of bills emerged. In early 1947, H. Alexander Smith, a New Jersey Republican with Ivy League credentials, introduced a carbon copy of Magnuson's bill in the Senate (S. 526) that adhered to Bush's prescription. Elbert D. Thomas, a Democrat from Utah, sponsored (S. 525) a reprise of the Kilgore-Magnuson compromise.[27,28]

Four separate bills quickly emerged in the House, all echoing Smith's Senate version. Thus began what Milton Lomask has called "the follies of 1947."[29] Mindful of the 1946 debacle, the American Association for the Advancement of Science convened a meeting of an Inter-Society Committee on Science Foundation Legislation on February 23, hoping to arrive at a consensus.[30] Within a month, the group reported that of 140 or so committee members, 63 percent opted for the Kilgore-Magnuson compromise, now Thomas's bill S. 525 and its House counterpart, H.R. 1850, sponsored by Arkansas Representative Wilbur Mills. The remaining 37 percent were evenly split between Bush's plan and one modeled on the newly established Atomic Energy Commission, which had a presidentially appointed director and a small, nine-member presidentially appointed board.[31]

Had the nation's scientists all lined up behind the Inter-Society majority, the National Science Foundation might have been two votes and a presidential signature away from becoming a reality. But they didn't. A vocal minority seemed bent on supporting Bush's plan, regardless of the efficacy. As the science community had proved many times before and would continue to prove many times over, it falls far short on political acumen. Science and technology policy might be part science, but it is also art, artfulness, and politics. Science, scientists understand viscerally; but art, artfulness, and politics are largely alien instincts.

The legislation that emerged from the House-Senate conference was essentially the Smith bill, which Truman had already indicated he would not sign. Why Congress went ahead and sent him a bill he opposed is not certain, but it might have reflected a desire to be done with the issue and send scientists packing. Let Truman deal with the irascible bunch, might have been the thinking on Capitol Hill. If he ultimately agreed to sign it, that would be fine, and if he didn't, the disgruntled brainiacs would be his problem.

What ensued lends credence to that supposition. Congress sent the legislation to the president on July 27, and in accordance with the Legislative Reorganization Act of 1946, adjourned the same day. If Truman failed to sign the bill within 10 days, it would fail to become law by virtue of a "pocket veto."[32] And that's exactly what happened. Two years of thrashing about had ended without any fanfare and without any indication whether there would ever be a National Science Foundation, although Truman encouraged Congress to return to the matter in the future.

Despite his encouragement, Truman was unequivocal in his rejection of the 1947 bill. His criticism was scathing. He concluded his veto justification with these words:

> …If the principles of this bill were extended throughout the Government, the result would be utter chaos. There is no justification in this case for not using sound principles for normal governmental operations. I cannot agree that our traditional democratic form of government is incapable of properly administering a program for encouraging scientific research and education.

But he ended on a positive and very encouraging note:

> I am convinced that the long-range interests of scientific research and education will be best served by continuing our effort to obtain a Science Foundation free from the vital defects of this bill. These defects in the structure of the proposed Foundation are so fundamental that it would not be practicable to permit its establishment in this form with the hope that the defects might be corrected at a later date. We must start with a law which is basically sound.

> I hope that the Congress will reconsider this question and enact such a law early in its next session.

Congress would reconsider the question, but a positive outcome would require another two-and-a-half years of deliberation.

Three weeks after Truman used his pocket veto on August 6, 1947, the Steelman Committee issued its report.[33] Although often overlooked, it is an important document in the annals of American science and technology policy. It is a more focused report than *Science: The Endless Frontier*, and more in tune with the way politicians look at policy matters, especially with regard to the administrative mechanisms. Its major recommendations are worth reading.

In the light of the world situation and of the position of science in this country, this report will urge:

1. That, as a Nation, we increase our annual expenditures for research and development as rapidly as we can expand facilities and increase trained

manpower. By 1957 we should be devoting at least one percent of our national income to research and development in the universities, industry, and the Government.

2. That heavier emphasis be placed upon basic research and upon medical research in our national research and development budget. Expenditures for basic research should be quadrupled and those for health and medical research tripled over the next decade, while total research and development expenditures should be doubled.

3. That the Federal Government support basic research in the universities and nonprofit institutions at a progressively increasing rate...

4. That a National Science Foundation be established to make grants in support of basic research, with a Director appointed by and responsible to the President. The Director should be advised by a part-time board of eminent scientists and educators, half to be drawn from outside the Federal Government and half from within it.

5. That a Federal program of assistance to undergraduate and graduate students in the sciences be developed...

6. That a program of Federal assistance to universities and colleges be developed in the matters of laboratory facilities and scientific equipment as an integral part of a general program of aid to education.

7. That a Federal Committee be established, composed of the directors of the principle Federal research establishments, to assist in the coordination and development of the Government's own research and development programs.

8. That every effort be made to assist in the reconstruction of European laboratories as a part of our program of aid to peace-loving countries. Such aid should be given on terms which require the maximum contributions toward the restoration of conditions of free international exchange of scientific knowledge...

The task of policy formulation for the Federal research and development program requires establishment of a number of coordinating centers within the executive branch of the Government. These would be called upon to make determinations upon a number of interrelated problems, of which the most important are:

1. An over-all picture of the allocations of research and development functions among the Federal agencies, and the relative emphasis placed upon the fields of research and development within the Federal Government must be available.

2. A central point of liaison among the major research agencies to secure the maximum interchange of information with respect to the content of research and development programs and with respect to administrative techniques must be provided.

3. There must be a single point close to the President at which the most significant problems created in the research and development program of the Nation as a whole can be brought into top policy discussions.

Setting up an organization to handle these diverse functions is not a simple task that can be solved, for example, by establishment of a Department of Science. Such an approach was considered in the course of these studies, and, after consultation with scientists and administrators, was rejected...

The three existing mechanisms in the Executive Branch for policy formulation with respect to research and development are inadequate when measured against the policy problems that must be more effectively dealt with.

The following steps should be taken:

1. An Interdepartmental Committee for Scientific Research should be created.
2. The Bureau of the Budget [now called the Office of Management and Budget] should set up a unit for reviewing Federal scientific research and development programs.
3. The President should designate a member of the White House staff for scientific liaison.

Aside from Steelman's recommendations for the National Science Foundation's structure, five items deserve special attention. First, reprising *Science: The Endless Frontier*, Steelman's report emphasized the importance of basic research and recognized the global nature of science. It also highlighted the linkage between research and education. Finally, it called for coordination of research and development across federal agencies, as Bush recommended, but it went one step further. It stressed the need for a single point of contact within the White House for science and technology policymaking.

Vannevar Bush had been that White House point person for six and a half years, serving Roosevelt and Truman as chairman of OSRD from June 21, 1941, when the wartime office was established, until December 31, 1947, when it was disbanded. In 1951, Truman created the Science Advisory Committee (SAC), naming Oliver Buckley, an electrical engineer, as its chairman. But neither Buckley, nor his successors, Lee DuBridge and Isadore Rabi, both physicists, were full-time White House employees. Only after the Soviet Union launched Sputnik in 1957, and at Rabi's suggestion, did Steelman's implicit recommendation for a staff-level presidential science advisor become a reality.

In November that year, President Dwight Eisenhower, a strong but generally underappreciated promoter of scientific research, upgraded the SAC to the President's Science Advisory Committee (PSAC), naming MIT's president James R. Killian as its full-time director and assigning him an office in the White House. Killian also wore the hat of Special Assistant to the President, becoming the first official Presidential Science Advisor. It had taken a decade for Steelman's recommendation to be realized. Once again, it was a military threat—this time from space—that proved essential to overcoming the bureaucratic inertia.

By the time Eisenhower won election to his first term in 1952, the National Science Foundation had become a reality. It was hardly a sprint to the finish: it had taken three more years following Truman's 1947 veto for the new agency to obtain a congressional authorization and a White House sign-off. A split in the science community had no longer been the issue. Truman had made it clear in his veto message that he wanted Congress to try again, but 1948 was a presidential election year, not an easy time to get significant legislation passed. It was especially true that year because Truman was widely expected to lose in November.

Bills were written, and committees considered them. But most, even if they passed committee muster, never made it to the floor, in at least one of the chambers. National Science Foundation legislation fell victim to such inaction, in spite of Truman's support for the compromise that had been worked out.[34] There wasn't much he could do. His political capital, at that point, was very limited, even among some Democrats.

Roosevelt New Dealers had never felt comfortable with Truman, and early in the nominating process, they tried to recruit Dwight Eisenhower to lead the ticket, believing his popularity as the victorious commander of the Allies could unite the party. Eisenhower declined, and following a contentious convention in Philadelphia, the party split three ways: Strom Thurmond, a conservative Dixiecrat, ran from the right; Henry A. Wallace, the New Deal progressive, ran from the left; and Truman competed from the center. Republicans had come out of their convention united, nominating New York Governor Thomas Dewey for president and California Governor Earl Warren as his vice-presidential running mate. They seemed unbeatable, right up to the close of the last polling booths on Election Day.

Reading the tea leaves well in advance of the nominating conventions, Republicans, who controlled both houses of Congress, had no incentive to move any legislation, especially anything Truman might have wanted. Even had they desired to do so, they would have found the road to passage of bills virtually impassable. True, the Democrats were split. But so, too, were the Republicans. A sizable fraction—those who lined up behind Dewey and Warren—was comfortable with many of Roosevelt's programs. But a large contingent found most, if not all, of the New Deal programs abhorrent. The result was gridlock. And Truman made the most of it, first calling Congress back into a special session following the conventions, and then on the campaign trail labeling Capitol Hill Republicans as "The Do-Nothing Congress."[35]

Truman won with 303 electoral votes to Dewey's 189, capturing 24.1 million popular votes to Dewey's 22.0 million. He had defied the odds, the pundits, the pollsters and, most notably, the *Chicago Daily Tribune*, which just as the polls were closing, ran the infamous banner headline, "Dewey Defeats Truman." Democrats regained control of both the House and the Senate, setting the stage for a productive legislative session. Civil rights, labor laws, price controls, housing, education, medical care, farming, and recognition of the new state of Israel were campaign issues.[36] If science was ever mentioned, there is no record of it. It was, and, with rare exceptions, almost always is a non-starter during any political campaign.

Science generally generates few votes among the lay public, and scientists are rarely single-issue voters. Those realities pose significant challenges for anyone involved in science advocacy, or more generally, science and technology policy. Even the 2017 "March for Science," which attracted tens of thousands of participants in Washington and other major cities and garnered significant media coverage, seems to have had little enduring impact on the general public.

———

The National Science Foundation (NSF) legislative log jam broke early in the 81st Congress. The Senate easily passed S. 247, a compromise bill, once again introduced by Democratic Sen. Elbert Thomas from Utah. The companion House legislation (H.R. 4846), sponsored by J. Percy Priest, a Tennessee Democrat, was not a mirror image: it granted dual authority to a presidentially appointed NSF director and a 24-member volunteer National Science Board with representatives from the private sector, also presidentially appointed. The White House signed off on it. The Senate concurred. And Vannevar Bush gave it his blessing, remarking, according to Edmund Day, chairman of the Inter-Society Committee, "…we will get a distinguished and well known scientist, such as Dr. Conant [James B. Conant, a chemist and president of Harvard University], to serve as chairman of the board and someone like Alan Waterman [a Yale University physicist, named chief scientist of the Office of Naval Research in 1946] to serve as director…"[37]

All the lights were green, or so it seemed. The science community had ceased its troublesome bickering; the 1948 election had removed partisan obstacles; Truman and Bush were on the same page. And then a new wrinkle appeared. A number of Republicans had gotten religion on budgetary matters. Led by James Wadsworth of New York, they bottled up the legislation in the House Rules Committee, and there it sat. Bush proposed a way out: language should be added, stating that the legislation would not increase the total of federal dollars spent on scientific research. (In today's budgetary parlance, such language is known as an "offset.") But the White House, through the budget director's office, nixed the idea. Funding basic research at the expense of applied research and development was not in Truman's DNA. The first session of the 81st Congress ended as the three previous ones had, with no NSF resolution in sight.

Shortly after the second session began on January 3, 1950, Priest and his allies exploited a loophole in the House rules, which allowed the chairman of the committee of jurisdiction to bring the bill out of the Rules Committee and onto the House floor 21 days after having requested action. Robert Crosser, the Ohio Democrat who chaired the House Committee on Interstate and Foreign Commerce, to which Priest's bill had been referred, agreed to use the parliamentary procedure, and on April 27, 1950, the House passed the legislation. The Senate followed suit the day after, and Truman signed it into law on May 10.[38]

The National Science Foundation torture was finally over. Almost most 5 years had passed since Bush had sent *Science: The Endless Frontier* to Truman. And following Bush's prescient rumination, Truman named James B. Conant in 1950 to chair the National Science Board, and in 1951, Alan T. Waterman to be the NSF director.

There is no better illustration of the essentials of science and technology policy than the saga of the National Science Foundation's birth. Major policy changes are difficult to accomplish, especially if there is no national security concern or health crisis. They require commitment, patience, fortitude, and a keen appreciation for the political landscape. They require getting different voices to harmonize. They require understanding that overcoming one barrier doesn't mean there won't be another one around the next bend in the road.

The saga also reveals several striking truths about the policy landscape: Science constituencies are fickle; egos are large. The same is true for politicians. For both, compromise often does not come easily.

And in the end, serendipity may be more important than the best laid plans and their most painstaking execution.

The speed with which Congress and the president came to agreement on establishing the Atomic Energy Commission (AEC) offers a striking contrast to the NSF's exceptional labor pains. To be sure, it was not without controversy, but from start to finish, the AEC's gestation period was nine months, very much on a human scale.

The Enola Gay had dropped the first atomic bomb, a uranium device known as Little Boy, on Hiroshima on Aug. 6, 1945, and Bockscar had followed up on Aug. 9, releasing a plutonium bomb, known as Fat Man, above Nagasaki. World War II was effectively over, but the atomic age was just beginning. Less than two months later, on Oct. 3, President Truman sent a message to Congress on atomic energy, urging the creation of an Atomic Energy Commission. He began his message with the following words:

To the Congress of the United States:

Almost two months have passed since the atomic bomb was used against Japan. That bomb did not win the war, but it certainly shortened the war. We know that it saved the lives of untold thousands of American and Allied soldiers who would otherwise have been killed in battle.

The discovery of the means of releasing atomic energy began a new era in the history of civilization. The scientific and industrial knowledge on which this discovery rests does not relate merely to another weapon. It may some day prove to be more revolutionary in the development of human society than the invention of the wheel, the use of metals, or the steam or internal combustion engine.

Never in history has society been confronted with a power so full of potential danger and at the same time so full of promise for the future of man and for the peace of the world. I think I can express the faith of the American people when I say that we can use the knowledge we have won, not for the devastation of war, but for the future welfare of humanity.

To accomplish that objective we must proceed along two fronts—the domestic and the international.

The first and most urgent step is the determination of our domestic policy for the control, use and development of atomic energy within the United States...

He continued with a call for the new agency:

The powers which the Congress wisely gave to the Government to wage war were adequate to permit the creation and development of this enterprise as a war project. Now that our enemies have surrendered, we should take immediate action to provide for the future use of this huge investment in brains and plant...

I therefore urge, as a first measure in a program of utilizing our knowledge for the benefit of society, that the Congress enact legislation to fix a policy with respect to our existing plants, and to control all sources of atomic energy and all activities connected with its development and use in the United States.

The legislation should give jurisdiction for these purposes to an Atomic Energy Commission with members appointed by the President with the advice and consent of the Senate...

The Commission should...be authorized to conduct all necessary research, experimentation, and operations for the further development and use of atomic energy for military, industrial, scientific, or medical purposes. In these activities it should, of course, use existing private and public institutions and agencies to the fullest practicable extent.

Under appropriate safeguards, the Commission should also be permitted to license any property available to the Commission for research, development and exploitation in the field of atomic energy. Among other things such licensing should be conditioned of course upon a policy of widespread distribution of peacetime products on equitable terms which will prevent monopoly.

In order to establish effective control and security, it should be declared unlawful to produce or use the substances comprising the sources of atomic energy or to import or export them except under conditions prescribed by the Commission.

Finally, the Commission should be authorized to establish security regulations governing the handling of all information, material and equipment under its jurisdiction...

And he concluded his message with recommendations to control proliferation
of atomic weapons and advance the peaceful use of atomic energy:

> In international relations as in domestic affairs, the release of atomic energy con-
> stitutes a new force too revolutionary to consider in the framework of old ideas.
> We can no longer rely on the slow progress of time to develop a program of control
> among nations. Civilization demands that we shall reach at the earliest possible
> date a satisfactory arrangement for the control of this discovery in order that it
> may become a powerful and forceful influence towards the maintenance of world
> peace instead of an instrument of destruction...

> The hope of civilization lies in international arrangements looking, if possible, to
> the renunciation of the use and development of the atomic bomb, and directing and
> encouraging the use of atomic energy and all future scientific information toward
> peaceful and humanitarian ends. The difficulties in working out such arrange-
> ments are great. The alternative to overcoming these difficulties, however, may
> be a desperate armament race which might well end in disaster. Discussion of
> the international problem cannot be safely delayed until the United Nations Orga-
> nization is functioning and in a position adequately to deal with it.

> I therefore propose to initiate discussions, first with our associates in this discov-
> ery, Great Britain and Canada, and then with other nations, in an effort to effect
> agreement on the conditions under which cooperation might replace rivalry in the
> field of atomic power.

Congress had already been working with the White House on draft legislation. And the same day Truman delivered his message, a pair of Democrats,
Andrew May from Kentucky, who chaired the House Military Affairs Committee, and Edwin Johnson from Colorado, who occupied the number two slot on
the Senate Military Affairs Committee, introduced legislation establishing an
Atomic Energy Commission (AEC), which largely followed the wording
proposed by the War Department.[40]

May, by that time, was embroiled in several high-profile controversies,
including a bribery charge that ultimately led to his re-election defeat in
1946 and an ignominious hitch in federal prison. But it wasn't May's notoriety
that caused the May-Johnson bill to run into trouble. Even though Manhattan
Project heavyweights J. Robert Oppenheimer, Ernest Lawrence, and Enrico
Fermi came out in support of the legislation, other scientists were far more critical. Like the National Science Foundation, the AEC was to have a presidentially appointed administrator and a presidentially appointed board—in this
case a commission with a revolving part-time membership of nine "distinguished citizens."

Taken at face value, the structure seemed to accomplish two of Truman's
goals: placing atomic weapons and all atomic energy activities under civilian
control, and insulating the AEC as much as possible from political meddling.
But for a growing number of scientists and members of Congress, the bill

did not go far enough in keeping the Army from exerting undue influence over the atomic programs, because it allowed members of the military to serve on the commission. Nor, in the eyes of its critics, did it give sufficient weight to the peaceful uses of atomic energy.

Opposition from the science community coalesced around former Manhattan Project bomb builders Leo Szilard, Harold Urey, and Edward Condon and the newly formed Federation of Atomic Scientists.[41] In Congress, Arthur Vandenberg, a Republican internationalist, who was the minority leader on the Senate Military Affairs Committee, successfully blocked the headlong pace of the legislation. And within months, in the face of the growing resistance on and off Capitol Hill, Truman quietly withdrew his support of the May-Johnson bill. That opened the door for first-term Connecticut Senator, Brien McMahon, who had developed an abiding interest in atomic energy.

McMahon successfully lobbied his Senate colleagues to set up a Special Committee on Atomic Energy, and from his perch as its chairman, he crafted a legislative substitute for the May-Johnson bill. McMahon's replacement contained two key provisions that strengthened civilian control: It prohibited members of the military from serving on the commission, and it converted part-time service on the commission to full-time status. Not unexpectedly, it ran into flak from the Army and the Navy's big guns. But it successfully assuaged the concerns of both the science community and the congressional critics of the May-Johnson bill.

Unlike the protracted debate and numerous false starts that characterized the creation of the National Science Foundation, McMahon's bill quickly made it to the Senate floor on June 1, 1946. By that time, military critics had seen the handwriting on the wall and throttled back their assault on the new legislation in the Senate. After a mere three hours of debate and with a few minor changes, McMahon's bill won approval without a single dissenting vote. Within two weeks, the House began its work and, after several rounds of political jousting, passed its amended version by an overwhelming margin of 265 to 79 on July 20.

A House-Senate conference committee quickly sorted out the differences, largely adhering to McMahon's original version, and the Atomic Energy Act of 1946 received a thumbs-up from both chambers on July 26. President Truman signed it into law[42] six days later on Aug.1, 302 days a after he had sent his original message to Congress.

The Atomic Energy Commission (AEC) and National Science Foundation (NSF) stories are studies in contrast. On a grand scale, the stakes were much higher in the case of the AEC. Getting policies wrong on atomic weapons and atomic energy carried far greater risks than fumbling the outcome on basic research. Dithering might have been fine in the case of the NSF, but it was unacceptable in the case of the AEC. That is not to say that science was unimportant to the future of the nation in 1945. It certainly was: for the economy, health, environment, agriculture, transportation, and the military. But policymakers knew that that delaying science decisions for several years would not do irreparable harm.

Not so for atomic weapons. Hiroshima and Nagasaki showed how much danger such weapons posed, and both the White House and Capitol Hill understood that legislative action couldn't be put on hold. Scientists, as well, knew that developing a consensus quickly was essential. The impetus was unmistakable, and science, technology, policy, and politics converged with appropriate alacrity. The outcome created an enduring legacy, which would keep the nation safe, provide a global approach to nuclear nonproliferation, promote peaceful uses of atomic energy, and withstand the challenges of the Cold War.

"Give 'em Hell, Harry!" That was the title of the biographical play and later the 1975 movie starring James Whitmore in a one-man show about Harry Truman. The playwright, Samuel Gallu, chose it for good reason. As the stories about the National Science Foundation and Atomic Energy Commission demonstrate, Truman wasn't shy about staking out positions. As the sign on his desk loudly proclaimed, "The Buck Stops Here."

Truman's science and technology legacy extended well beyond the creation of two agencies. Perhaps he was only in the right place at the right time. But his willingness to back the policies war-time scientists and their government counterparts proposed to establish strong postwar American global leadership in science and technology was his choice. He was not the originator of the idea, but he seized the day and gave his full-throated support for the proposition.

Well in advance of the Allies' victories in Europe and the Pacific, a group of young Navy officers and reservists, known as the "Bird Dogs," had been exploring how the Navy could advance its scientific needs once hostilities had ceased.[43] Their genesis traces to an overreach by Rear Admiral Harold G. Bowen in 1941, who was then the director of the Naval Research Laboratory (NRL). Bowen attempted to gain control of all naval research under the auspices of NRL, but his consolidation plan quickly ran into opposition, ultimately leading Navy Secretary Frank Knox to establish a coordinator for research and development within his own office. Knox tapped Jerome Clarke Hunsaker, the chairman of the National Advisory Committee for Aeronautics and a graduate of the Naval Academy, for the advisory position, and Hunsaker recruited the Bird Dogs. Three years later a post-war plan for an Office of Naval Research (ONR) emerged that had the backing of the Navy higher-ups, as well as Vannevar Bush.

By September 1945, the Bird Dogs draft bill was ready to roll. It would place the Naval Research Laboratory under the control of ONR, and more importantly, provide the new office with the authority to fund basic research in the nation's universities. The White House was on board, as was an overwhelming majority of Congress. All that remained was selling university scientists on the proposition that they should be open to accepting basic research contracts from the military in a peacetime environment. In the 21st century, the idea that academic researchers would turn their backs on unrestricted federal funding seems preposterous. But at the time, there was no precedent for military support of university research outside of direct wartime needs, and it wasn't clear how scientists might react.

It took Robert D. Conrad, a navy Commander and a gifted speaker, several months of marketing to close the deal with university scientists. On August 1, 1946, by an overwhelming margin Congress passed H.R. 5911, known as the Vinson Bill. The Office of Naval Research was a reality. During the next few years, as bickering over the National Science Foundation persisted, ONR stepped in to fill the void, providing essential support for university research and allowing scientists who had returned from their wartime duties to pursue their professional goals. It was a sweet success, and illustrates once again how military demands can generate a political consensus and produce policies on a time scale that is relevant to the nation's needs.

The Navy was first out of the box. But the Air Force, which was established as a separate service on Sept. 18, 1947, moved ahead with its own research office almost immediately. Again aiming to capitalize on technical talent in the universities, the Air Force Office of Scientific Research (AFOSR) opened its doors in February 1948, and within a few years, was providing significant support for laboratories on campuses around the country. Paradoxically, the Army, which historically had led the way on military science and technology, didn't get its act together until almost a decade later.

It didn't take long for university scientists to cozy up to their military benefactors. In fact, many of the federal bureaucrats administering the ONR and AFOSR programs had been war-time colleagues of the academic scientists whom they were funding. It was, in many ways, the essence of a classic "old-boy-network." By funneling money to the scientists they knew best, the administrators might have a constricted pipeline of future researchers. But with the return of so many scientists to peacetime activities, there was no shortage of talent within the existing pool.

The hand-in-glove relationship between government program managers and academic scientists without a doubt benefitted university research programs. But the trusted connection also provided a mechanism for scientists outside government to help shape federal science and technology policies. The symbiotic link between the two communities remained largely intact well into the 1970s, when retirements and deaths began to reduce the ranks in both.

Earlier wars had highlighted the need for medical research. World War II was no exception. To address battlefield injuries and diseases endemic to far-flung theaters of combat, the Office of Scientific Research and Development (OSRD) under Vannevar Bush established a Committee on Medical Research, drawing on the expertise of civilian scientists and medical practitioners. The committee's charge was broad, but as might be expected, it placed great emphasis on the most pressing military concerns: surgery, convalescence, rehabilitation, pathology, neuropsychology, and the like.

Just as the War Department's needs for advanced military technologies required physicists, chemists, and engineers to put their fundamental research programs on hold, its specific health needs thinned the ranks of biomedical research generalists in universities and hospitals. The war pumped large amounts of money into health research, but the dollars carried with them an obligation to serve the needs of the military. From his perch atop OSRD, Vannevar Bush was in an ideal position to see how rapidly medical research was advancing and how many opportunities would exist for broader progress once the war ended.

He stated his proposition for peacetime medical research in *Science: The Endless Frontier*,[45] writing, "Notwithstanding great progress in prolonging the span of life and in relief of suffering, much illness remains for which adequate means of prevention and cure are not yet known. While additional physicians, hospitals, and health programs are needed, their full usefulness cannot be attained unless we enlarge our knowledge of the human organism and the nature of disease. Any extension of medical facilities must be accompanied by an expanded program of medical training and research." Adding further, "Progress in the war against disease results from discoveries in remote and unexpected fields of medicine and the underlying sciences." And concluding, "It is clear that if we are to maintain the progress in medicine which has marked the last 25 years, the Government should extend financial support to basic medical research in the medical schools and in the universities, through grants both for research and for fellowships."

Bush saw basic medical research as part of a grand scientific continuum, or as Nobel Laureate Harold Varmus, former director of the National Institutes of Health would write decades later, the sciences are "interdependent." To illustrate his point, in 2000 Varmus wrote the following in a *Washington Post* op-ed:[46] "Medical advances may seem like wizardry. But pull back the curtain, and sitting at the lever is a high-energy physicist, a combinational chemist or an engineer... Scientists can wage an effective war on disease only if we—as a nation and as a scientific community—harness the energies of many disciplines, not just biology and medicine. The allies must include mathematicians, physicists, engineers and computer and behavioral scientists."

That world scientific view motivated Bush to include medical research as a program within his proposed new science agency, the National Research Foundation, as he called it at the time. Had the saga of the National Science Foundation (NSF) not dragged on for half a decade, Bush's vision might have become a reality. But in the unfortunate interregnum, momentum steadily built for supporting medical research within existing government constructs. The National Institute of Health (NIH) was an obvious choice. By the time NSF opened its doors for business in 1951, NIH had become such a dominant player in medical research,[47] there was no longer a compelling rationale for Bush's proposition.

Under NIH's rubric, the federal budget for health research grew, but so too did the scope of the agency's programs. In 1948, Congress established the National Heart Institute, the National Institute of Dental Research, the National Microbiological Institute, and the National Medicine Institute—all of them within the jurisdiction of NIH. Its old name did not reflect its new expanded mandate. On June 16, 1948, recognizing the multiplicity of institutes under its roof, Congress officially changed NIH's name to the National Institutes of Health. (Note the plural in "Institutes.")

On August 15, 1950, President Truman signed the Omnibus Medical Research Act,[48] authorizing the creation of the National Institute of Neurological Diseases and Blindness and the National Institute of Arthritis and Metabolic Diseases, and granting the Surgeon General the power to establish additional institutes. By 2018, NIH would comprise 21 Institutes and 6 Centers, occupying 75 buildings on a 300-acre campus in Bethesda, Maryland, just outside Washington, D.C. and employing more than 20,000 workers.[49] It would become the largest federal research agency, dwarfing the size of the National Science Foundation by five times, and accounting for almost 50 percent of the government's research spending. The scope and prominence of the NIH reflects the attention health issues command on the political stage and the resonance they have with the general public. In the science and technology policy arena, only military affairs rise to a commensurate level.

Among Oval Office occupants of the modern era, Harry Truman's name is rarely cited as an example of a "science president." John F. Kennedy, by virtue of his commitment to the Space Program, and Barack Obama, by virtue of his engagement on climate change and his joy in hosting White House events for young scientists and innovators, are the most frequently associated with that distinction. But on the scale of accomplishments that altered the arc of science and technology policy, Truman deserves to be on the highest level of the awards podium.

Most of today's federal constructs, from agencies, bureaus, and offices within the Executive Branch, to the instruments of government that foster science research and education outside government, trace their lineage to the Truman era. The National Institutes of Health and the National Science Foundation elicit the strongest resonance with Congress and the public. Often unnoticed are the national laboratories that are extraordinary gems in the nation's science and technology enterprise. Vannevar Bush and John Steelman focused principally on the science and technology talent that could be unleashed to the nation's benefit in an America at peace. But the wartime infrastructure was just as much of a national treasure. And the Atomic Energy Commission (AEC) moved swiftly to capitalize on it.

The AEC lifted the shroud that had kept the site of the atomic bomb building facility[50] in New Mexico secret until World War II ended. The laboratory complex on a high mesa 35 miles northwest of Santa Fe quickly become synonymous with America's nuclear weapons program. Over the course of the next 70 years, Los Alamos National Laboratory grew in size and scope, expanding its activities from purely military research into basic science. By 2018, it was hosting a thousand students and employing more than 10,000 scientists, engineers, and supporting staff. It saw its annual operating budget rise to more than $2 billion.

Berkeley's Radiation Laboratory might have played second fiddle to Los Alamos during the Manhattan Project, but its role was significant nonetheless. That was true especially because it was Ernest O. Lawrence's home turf throughout the war, and the Berkeley physics department with which it was associated had been J. Robert Oppenheimer's employer[51] until he became director of Los Alamos in 1943.

Lawrence, who had left Yale University for Berkeley in 1928 after Yale turned down his promotion request,[52] established the "Rad Lab" in 1931 to house his new invention, a particle accelerator that became known as the cyclotron. Adapting his concepts to heavy ions, Lawrence proposed an electromagnetic method for enriching natural uranium to bomb-grade quality. Even though the technological challenges were immense, Leslie Groves, director of the Manhattan Project, authorized it as one of several approaches to uranium isotope separation. Racing against German scientists,[53] Groves felt compelled to pursue as many promising methods as necessary, regardless of cost.

Oak Ridge, Tennessee was the site of the Manhattan Project's isotope separation plant, and Lawrence would soon find himself commuting between the Golden State and the Volunteer State. In the end, Lawrence's concept lost out to a competing approach based on gaseous diffusion, which was more effective, cheaper, and less technologically challenging.

While Lawrence was occupied by the Oak Ridge uranium project, Edwin McMillan and Glenn Seaborg were hard at work trying to find a different path to a fission bomb. Poring over several years' worth of data from the Berkeley cyclotron, they identified a new element (94th on the periodic table) that might meet the specifications. After McMillan had departed to join the radar effort at the Massachusetts Institute of Technology, Seaborg and his colleagues who remained at Berkeley discovered that the new element they called plutonium was, indeed, fissionable. In all likelihood, it would sustain the kind of "chain reaction" a bomb required.

Plutonium was detected using the cyclotron, but its large-scale production would necessitate using reactors, similar to the one Enrico Fermi had employed when he demonstrated the first sustained chain reaction at the University of Chicago in 1942. Hanford, Washington became the site for the plutonium production plant that provided the material used in the July 16, 1945 Trinity Test of "Fat Boy" in Alamogordo, New Mexico and the bombing of Nagasaki less than two months later.

The nation had invested heavily in Los Alamos, Berkeley, Oak Ridge, and a number of other laboratories around the country. To allow them to lie fallow would have been an immense waste of resources. The Atomic Energy

Commission (AEC) was well aware of the nation's infrastructure treasure the war had left behind, and by the time Truman left office in 1952, twelve laboratories were operating under its auspices. Los Alamos and Sandia, both in New Mexico, continued the work of the Manhattan Project, but Berkeley returned to its prewar focus on basic research. In 1952, acceding to Edward Teller's persistent pressure for a new facility devoted to the hydrogen bomb,[54] the AEC opened Livermore Laboratory in northern California, about 30 miles southeast of Berkeley. Since then, Los Alamos, Sandia, and Livermore have been known simply as "The Weapons Labs."

Argonne in northern Illinois and Brookhaven on eastern Long Island—both part of the war effort—have long vied for bragging rights as the first "national" laboratory. Argonne, which focused on reactor development during World War II, redirected its mission to nondefense research after the war ended, much the way Berkeley did. Brookhaven's story is quite different, and it reveals how effective advocacy can be in achieving a science and technology policy outcome when scientists are on the same page.

Isadore Rabi and Norman Ramsey had returned from their wartime efforts to physics faculty positions at Columbia University ready to resume their basic research investigations. They were well aware of the new research tools the war effort had produced, among them, nuclear reactors. They had proven vital for producing Fat Boy's plutonium, but, as Enrico Fermi had shown, they were invaluable for studying the properties of nuclei, the core of every atom. Midwestern scientists had relatively easy access to Argonne's facilities, but scientists at Columbia, the Massachusetts Institute of Technology, and other Northeastern academic research centers would be forced to travel long distances, putting them at a great competitive disadvantage.

In March 1946, nine universities banded together to jointly make their case for a new laboratory to the nascent Atomic Energy Commission (AEC). For its site, they settled on Camp Upton, an Army installation that had served as an internment camp for a thousand Japanese Americans following the bombing of Pearl Harbor. The AEC accepted their proposal, and on March 21, 1947, the War Department transferred the 5300 underutilized acres to the new agency.

Camp Upton's name had been sullied by its use at the outset of the war, and physicists, many of whom already were questioning Truman's use of the atomic bomb, looked for an alternative. Norman Ramsey's wife, Elinor, suggested "Brookhaven,"[55] the township in which the new laboratory was to be located. And the name stuck. By the time work began on its first reactor toward the end of 1947, the AEC had officially recognized its new facility as Brookhaven National Laboratory.

Although written records are sparse, it is more than likely that Brookhaven's skids were greased by the old-boy-network, stemming from the wartime relationship Rabi, Ramsey, and others had with the post-war federal bureaucrats who had begun to populate the new science agencies. Brookhaven dabbled in some defense work over the ensuing decades, but most of its activities centered around fundamental science, dominated much of the time by nuclear and high-energy (particle) physics.

———————————

The national laboratory system continued to grow after Truman left office. Between 1962 and 1984, five more sites were added, bringing the total to today's count of 17.[56] With only one exception, the National Energy Technology Laboratory, they all function under a "GOCO" arrangement. The Government Owns them, but non-governmental Contractors Operate them. From time to time, review commissions[57,58] and select committees[59] have taken deep dives into the laboratory system, questioning whether it suffers from excessive redundancy, whether the GOCO model realizes optimal efficiencies and—especially in the aftermath of Los Alamos's Wen Ho Lee affair[60]—whether laboratory operations and practices safeguard classified research adequately. Recommendations have varied, but in the end, the national laboratory system has managed to survive for more than 70 years with only a small number of truly significant change.

Chapter 7

Growing pains 1952–1974

Harry Truman entered the White House on April 12, 1945 with the nation mourning Franklin Roosevelt's death. Expectations for him were not terribly high, but within a few months of taking office, he won over the public. His approval rating[1] soared to more than 90% following the bombing of Hiroshima and Nagasaki. Einstein, Szilard, and other like-minded physicists might have been shocked by the horror of the nuclear weapons they had created, but average American citizens did not share their fears. The war was over: that was all that mattered.

The public's peace celebration ended less than 5 years later when hostilities broke out on the Korean peninsula in June 1950. With the nation mired in another conflict so soon, Truman's popularity sank dramatically. Corruption within his administration helped drive his numbers down even further, and when he saw his public approval continuing to hover well below 30% in the summer of 1951, he decided it was time to quit. On March 29 of the following year, he made it official, announcing his intention not to run at the annual Jefferson-Jackson black-tie dinner in Washington.[2] True to his down-home Missouri manner, he spoke simply and without equivocation: "I shall not be a candidate for re-election. I have served my country long, and I think efficiently and honestly. I shall not accept a re-nomination. I do not feel that it is my duty to spend another four years in the White House." End of story—well almost.

Truman had worried what would happen to the domestic programs and the internationalist foreign policies he and Roosevelt had forged if Republicans captured the presidency. He was particularly concerned about Robert A. Taft, the isolationist Ohio senator with whom he had a particularly testy relationship.[3] In early November of 1951, he had lunched with Dwight Eisenhower, still arguably the most popular public figure in the United States, and urged him to run for president as a Democrat,[4] even offering to be his number two on the ticket if Ike wanted him.[5] Later that day, Ike reminded Truman he had been a life-long Republican and was not about to change his party affiliation simply to run for office. He was also quite content living in Paris as Supreme Commander of NATO, a position he had accepted at Truman's request in 1950. He was on an extended leave from his presidency of Columbia University and could

Navigating the Maze. https://doi.org/10.1016/B978-0-12-814710-8.00007-3

129

still return if he chose to. Running for the presidency of the United States wasn't on his bucket list.

His attitude would soon change. In the early part of 1952, he received an urgent phone call from three Republican honchos. Lucius Clay, who had been Ike's deputy when he was General of the Army and had become Commander in Chief of the U.S. forces in Europe after the war ended, was one of them. Eisenhower knew him extremely well and held him in high regard. Herbert Brownell, Jr., who had been chairman of the Republican National Committee from 1944 to 1946, and Tom Dewey, who had been the 1948 GOP standard bearer, rounded out the trio. They were globalists who supported the multinational structures the Truman Administration had championed. And they feared what an isolationist Taft presidency would do to unravel them.

Science and technology were undoubtedly as far from the trio's minds as anything could be. But the way the story eventually played out again illustrates how important serendipity is in developing and executing the policies that have shaped America and the world.

Clay, Brownell, and Dewey knew Eisenhower shared their views about internationalism and, as arguably the most popular person in America, they believed he would be a shoo-in for president. They also knew he had shown little interest in running. But they had one card to play they thought might do the trick. They were aware there was one person whom Ike feared would do the nation far greater harm than Taft—General Douglas MacArthur, the five-star general who had prosecuted the Korean War until Truman fired him for insubordination in April 1951.[6]

The public adored MacArthur, a war hero who cut an imposing figure and was a mesmerizing speaker. But Eisenhower, who had observed his extraordinary egotism close up, didn't trust him at all as a guardian of American democracy. MacArthur was scheduled to deliver the keynote address at the Republican Convention in Chicago on July 7, 1952, Brownell, Dewey, and Clay reminded Eisenhower on the phone call, warning him that MacArthur would likely be greeted with such zeal, the convention could easily turn to him as the GOP presidential nominee and upend Taft's candidacy. The only way to stop MacArthur, they said, was for Eisenhower, himself, to run.[7]

They made their point, and not long afterward, Ike agreed to toss his hat into the presidential ring. He made it official on June 4, 1952, declaring his candidacy from his home town of Abilene, Kansas. Not all Republicans were thrilled with Eisenhower's entry into the nominating fray. That became obvious following MacArthur's rousing speech,[8] when rancor over Eisenhower's nomination broke out on the convention floor.[9] It took two ballots, but when the dust from the dust-ups finally settled, Ike had soundly defeated Taft by a three to one margin.

In his short acceptance speech, Eisenhower expressed sentiments that would be almost inconceivable at a Republican Convention today. But 1952 was a different era. Although he did not mention science—and there is no reason to think he would have—his words captured his optimistic view of the future, and his subsequent policies reflected his trust in science to help achieve his goals. At the convention, he said,[10]

… it is our aim to give to our country a program of progressive policies drawn from our finest Republican traditions; to unite us wherever we have been divided; to strengthen freedom wherever among any group is has been weakened; to build a sure foundation for sound prosperity for all here at home and for a just and sure peace throughout our world…

We must use our power wisely for the good of all our people. If we do this, we will open a road into the future on which today's Americans, young and old, and the generations that come after them, can go forward—go forward to a life in which there will be far greater abundance of material, cultural, and spiritual rewards than our forefathers or we ever dreamed of…

Wherever I am, I will end each day of this coming campaign thinking of millions of American homes, large and small; of fathers and mothers' working and sacrificing to make sure that their children are well cared for, free from fear; full of good hope for the future, proud citizens of a country that will stand among the nations as the leader of a peaceful and prosperous world…

Eisenhower had little difficulty defeating Adlai Stevenson, the Democratic candidate, carrying 39 of the 48 states, securing 442 out of 531 electoral votes and capturing 55% of the popular vote. Eisenhower had a mandate. Ending hostilities on the Korean Peninsula was his first goal. Keeping the United States out of future conflicts, avoiding a nuclear confrontation with the Soviet Union, and sharing nuclear information and nuclear material for peaceful purposes were also on his list.

While he had a great respect for what science could do, he worried about what it could destroy. He also fretted about the undue influence of the military-industrial complex and its technological underpinnings, and he was concerned about the growth of a scientific elite. But none of his fears prevented him from having a close relationship with scientists and supporting their research activities. Isadore Rabi, a physics Nobel Laureate who was a member of the Columbia faculty during Eisenhower's tenure there, was especially close to him, advising him on scientific and technical matters from 1956 to 1957 and in his official capacity, also chairing the Science Advisory Committee (SAC). America's scientific enterprise grew steadily during Eisenhower's first term, and with his strong support for academic research, the National Science Foundation saw its budget grow fivefold.[11]

August 6, 1945 ushered in a new scientific era, the "Nuclear Age," the day the Enola Gay dropped Little Boy on Hiroshima. October 4, 1957 ushered in another scientific era, the "Space Age" when the Soviet Union successfully launched Sputnik, the world's first artificial satellite. By almost any scientific standard, it was extremely primitive. About the size of a beach ball, all it could do was emit two simple radio signals.

Sputnik, itself, posed no military threat, but that didn't deter the media from milking the story for all they could get out of it. In its Late City Edition, the day of the launch, under the banner headline, "*SOVIET FIRES EARTH SATEL-LITE INTO SPACE; IT IS CIRCLING THE GLOBE AT 18,000 M.P.H.; SPHERE TRACKED IN 4 CROSSINGS OVER U.S.*," *The New York Times* ran four front-page stories describing the Soviet feat.[12] There is no evidence the public panicked at all, but politicians reacted reflexively, taking the Soviet accomplishment as a dangerous challenge to American technological superiority. The result was a boon to science.

In truth, the United States barely lagged the Soviets in satellite capability. It's just that Eisenhower, in spite of his military background, had emphasized civilian, rather than defense, applications. Despite intelligence reports to the contrary—they were later proved false—American missile development for space or ballistic use was on a virtual par with the Soviet program. Eisenhower had seen to both, but he couldn't reveal how far advanced the American programs were because the information was still classified. Just over a year later, on January 31, 1958, a Juno rocket launched America's first satellite, Explorer I, into Earth orbit. And on July 28, 1959, the Atlas program, which began at the end of World War II, achieved success with the launch of an intercontinental ballistic missile (ICBM) from Cape Canaveral, Florida. Moving rapidly, the Defense Department deployed four Atlas missiles at the Vandenburg Air Force Base in California three months later.

In the first lap of the space race, the Soviet Union had bested the United States, but only barely. It could rightfully claim bragging rights to the first ICBM, as well as the first satellite, having used its intercontinental R-7 missile rocket to launch Sputnik and establishing its first operational ICBM base on February 9, 1959. But by the beginning of the 1960 election year, the two nations were on fairly equal footing. Despite evidence to the contrary, John F. Kennedy, the Democratic presidential candidate, featured the illusory "missile gap" in his winning campaign against Richard M. Nixon, who had served as Eisenhower's vice-president for 8 years.

Eisenhower understood the military exigency in challenging the Soviet Union's quest for domination in space, but he also had long harbored the dream of space exploration for research purposes. On July 29, 1958, he signed legislation creating the National Aeronautics and Space Act,[13] creating NASA as the successor to the National Advisory Committee for Aeronautics, with a mandate to reclaim American superiority in space for both military and scientific purposes.

Earlier in 1958, he had authorized the creation of the Advanced Research Projects Agency (ARPA now called DARPA)[14] in the Department of Defense to accelerate the development of emerging technologies, and on September 2, 1958, he signed the National Defense Education Act[15] to improve science education at all levels. Sputnik was unquestionably a catalyst for a ramp up in federal science spending, but its impact aligned well with Eisenhower's predisposition toward federal support of research. Whether a Taft, a MacArthur, or a Stevenson would have taken such a broad view is hard to know, but Eisenhower was the right man in the right place at the right time to assure the nation that it could deter any foreign aggression, and that it had the capability to compete technologically with the Soviet Union or any other nation in the world.

Eisenhower also understood the importance of science and technology policy within the White House, and in November 1957, as already noted, he took two steps to make sure he got the advice he needed. He gave the Science Advisory Committee a higher profile, renaming it the President's Science Advisory Committee (PSAC), and he converted the presidential science advisor's position into a full-time post, naming James Killian Special Assistant to the President and chairman of PSAC. Both remained White House constructs until President Richard Nixon abolished them in 1973. [Acted after his science advisor, Edward E. David, Jr., and PSAC, which David chaired, came out openly against his plans for an antiballistic missile defense system and a supersonic transport, basing their opposition on scientific, technical, and, in the latter case, environmental grounds.[16]]

It's a pretty good bet that few members of the general public are aware of Eisenhower's commitment to science and space exploration and the steps he took to elevate science and technology policymaking within the White House. It's probably a good bet for members of the science and technology community, as well. Apart from his devotion to golf, Ike is far better known for his enduring mark on the American landscape—the interstate highway system.

Eisenhower's interest in highways was longstanding. As a member of the 1919 Army Convoy,[17] he had traveled across the country on America's first transcontinental road, the Lincoln Highway. Several decades later, as Supreme Commander of the Allied Expeditionary Forces in Europe, he had an opportunity to see what civil engineering could achieve with sufficient resources. The contrast between the modern German autobahn network and the antiquated American highway system could not have been starker. For Eisenhower, four-lane limited-access highways would be more than a boon to commerce. As he saw it, they were a necessity for America's defense in the Cold War Era, when troops and materiel might have to be moved rapidly and at short notice.

With Eisenhower's strong backing, Congress passed the National Interstate and Defense Highways Act (officially known as the Federal-Aid Highway Act[18]) on June 26, 1956, and the president signed it into law three days later. Paid for with new taxes on gasoline and diesel fuel, which fed the coffers of the newly established Highway Trust Fund,[19] it was to become the largest public works project in the history of the country. And it would transform American life in ways that were practically unimaginable.[20]

Cars and trucks would replace long-distance trains, leading to bankruptcies and mergers of some of the most iconic railroads. Suburbs and exurbs would grow dramatically. Long-haul trucking would give containerized shipping a boost, ultimately leading to greater globalization. And carbon emissions would soar, as more vehicles took to the new highways and traveled longer distances. Of course, climate change was not yet on anybody's mind.

Eisenhower's first inaugural speech and his farewell speech bookend the way he viewed science: embodying hope for the future, but tempered with great concern about its misuse and the influence of both the "military-industrial complex" and the "scientific and technological elite." After taking the oath of office on January 20, 1953, he spoke for 20 minutes from the east portico of the Capitol. Here are some of his words:[21]

My fellow citizens:

The world and we have passed the midway point of a century of continuing challenge. We sense with all our faculties that forces of good and evil are massed and armed and opposed as rarely before in history...

Great as are the preoccupations absorbing us at home, concerned as we are with matters that deeply affect our livelihood today and our vision of the future, each of these domestic problems is dwarfed by, and often even created by, this question that involves all humankind.

This trial comes at a moment when man's power to achieve good or to inflict evil surpasses the brightest hopes and the sharpest fears of all ages. We can turn rivers in their courses, level mountains to the plains. Oceans and land and sky are avenues for our colossal commerce. Disease diminishes and life lengthens.

Yet the promise of this life is imperiled by the very genius that has made it possible. Nations amass wealth. Labor sweats to create – and turns out devices to level not only mountains but also cities. Science seems ready to confer upon us, as its final gift, the power to erase human life from this planet.

At such a time in history, we who are free must proclaim anew our faith. This faith is the abiding creed of our fathers. It is our faith in the deathless dignity of man, governed by eternal moral and natural laws...

We must be willing, individually and as a Nation, to accept whatever sacrifices may be required of us. A people that values its privileges above its principles soon loses both.

These basic precepts are not lofty abstractions, far removed from matters of daily living. They are laws of spiritual strength that generate and define our material strength. Patriotism means equipped forces and a prepared citizenry. Moral stamina means more energy and more productivity, on the farm and in the factory. Love of liberty means the guarding of every resource that makes freedom possible – from the sanctity of our families and the wealth of our soil to the genius of our scientists.

And so each citizen plays an indispensable role. The productivity of our heads, our hands and our hearts is the source of all the strength we can command, for both the enrichment of our lives and the winning of the peace…

Eight years later, on the evening of January 17, 1961, as he prepared to leave the presidency Ike expressed these thoughts from the Oval Office:[22]

My fellow Americans:

Three days from now, after half a century in the service of our country, I shall lay down the responsibilities of office as, in traditional and solemn ceremony, the authority of the Presidency is vested in my successor.

This evening I come to you with a message of leave-taking and farewell, and to share a few final thoughts with you, my countrymen…

Until the latest of our world conflicts, the United States had no armaments industry. American makers of plowshares could, with time and as required, make swords as well. But now we can no longer risk emergency improvisation of national defense; we have been compelled to create a permanent armaments industry of vast proportions. Added to this, three and a half million men and women are directly engaged in the defense establishment. We annually spend on military security more than the net income of all United States corporations.

This conjunction of an immense military establishment and a large arms industry is new in the American experience. The total influence-economic, political, even spiritual—is felt in every city, every State house, every office of the Federal government. We recognize the imperative need for this development. Yet we must not fail to comprehend its grave implications. Our toil, resources and livelihood are all involved; so is the very structure of our society.

In the councils of government, we must guard against the acquisition of unwarranted influence, whether sought or unsought, by the military-industrial complex. The potential for the disastrous rise of misplaced power exists and will persist…

Today, the solitary inventor, tinkering in his shop, has been overshadowed by task forces of scientists in laboratories and testing fields. In the same fashion, the free university, historically the fountainhead of free ideas and scientific discovery, has experienced a revolution in the conduct of research. Partly because of the huge costs involved, a government contract becomes virtually a substitute for intellectual curiosity. For every old blackboard there are now hundreds of new electronic computers.

The prospect of domination of the nation's scholars by Federal employment, project allocations, and the power of money is ever present – and is gravely to be regarded.

Yet, in holding scientific research and discovery in respect, as we should, we must also be alert to the equal and opposite danger that public policy could itself become the captive of a scientific-technological elite.

It is the task of statesmanship to mold, to balance, and to integrate these and other forces, new and old, within the principles of our democratic system – ever aiming toward the supreme goals of our free society.

There is one scientific blight on Eisenhower's 8 years in office, and it's significant. It came early on, during the "McCarthy Era," when the nation was seized by an all-consuming fear of Communism. In 1953, as Eisenhower was taking the oath of office, a second-term senator from Wisconsin gained the gavel of the Senate Committee on Government Operations.

Joseph R. McCarthy, who had used his first 6 years in the Senate to hunt for the "Red menace" wherever his leads took him, began an unrelenting search for Communist infiltration of the federal bureaucracy. Eisenhower hated him for his bullying tactics and said so, but he shared McCarthy's concern about leaks of classified information to the Soviet Union—as did the House Un-American Activities Committee (HUAC). McCarthy used his muscle without regard for the lives his investigations upended. Eventually, he overreached when he turned his fire on the military. The infamous Army-McCarthy hearings of 1954, which his intimidation provoked, riveted the nation for three months and culminated with McCarthy's censure by the Senate in December 1954. Thus defanged, McCarthy's investigative career came to an abrupt end, but not before it had tarnished the reputations of hundreds of prominent and not so prominent Americans, many scientists among them. The House committee, HUAC, might not have had the high profile of McCarthy's committee, but it was no less aggressive in its anti-Communist fervor, and probably eviscerated more Americans than its Senate counterpart.

J. Robert Oppenheimer was not a priority for McCarthy or HUAC. He didn't have to be, because the FBI had swept him up in its own search for Communist collaborators. The case, which FBI Director J. Edgar Hoover built, never demonstrated that the "Father of the Atomic Bomb" had divulged any state secrets. But Oppenheimer's association with known Communists, including his younger brother Frank—who had been a member of the party in prewar days—as well as the checkered past of his wife, Kitty, raised suspicions about his loyalty. Oppenheimer had other liabilities. He opposed the development of the hydrogen bomb—the "super," as it was called by members of the bomb-building inner circle—which Lawrence and Teller loudly touted. He was wary of peaceful uses of atomic energy, which he thought could lead to nuclear proliferation. Both of those positions clashed with Eisenhower's, and, just as significantly, with those of Lewis Strauss, chairman of the Atomic Energy Commission, whom Oppenheimer typically treated with disdain. In short, Oppenheimer gave his enemies plenty of ammunition, and they were happy to use it.[23, 24]

In November of 1953, Hoover found a letter[25] waiting for him from William Borden, who had been the congressional Joint Atomic Energy Committee's staff director from 1949 until 1953. In it, Borden laid out his case against Oppenheimer, stating that, in his "own exhaustively considered opinion, based upon years of study of the available classified evidence that more probably than not J. Robert Oppenheimer is an agent of the Soviet Union." Hoover passed the letter on to Eisenhower, who immediately cordoned off all classified information from Oppenheimer and ordered Strauss to ask for Oppenheimer's resignation from the AEC General Advisory Committee. Oppenheimer, who didn't appreciate the depth of the distrust he had engendered, labeled Borden's charges as having no merit, and following the advice of his lawyers, requested a hearing to challenge them. On May 27, 1954, after 27 hours of Oppenheimer's testimony, the three-member hearing panel by a vote of two to one, recommended that his security clearance be revoked. The AEC acted on June 29, and Oppenheimer lost his clearance that day, just 32 hours before his consulting contract with the commission was due to expire.

Oppenheimer was suddenly barred from the work he cherished, and the life he had known for a dozen years was little more than a memory. The majority of the science community condemned the AEC decision, calling Oppenheimer's treatment a sacrifice to McCarthyism. Eisenhower could have intervened, but he chose not to, leaving it to his successor, John F. Kennedy to rehabilitate the reputation of one of America's extraordinary public servants.

The Oppenheimer saga ended with several ironic twists. In 1963, Edward Teller, who had helped sink him before the AEC panel, nominated him for the prestigious Enrico Fermi Award, which the AEC supported unanimously. President Kennedy applauded the decision, committing to presenting the award personally. The White House made the announcement on the morning of November 22, 1963. Tragically, just a few hours later, Kennedy was assassinated by Lee Harvey Oswald in Dallas Texas. It was Lyndon Johnson who presented Oppenheimer with the medal and a $50,000 check in a White House ceremony on December 2, speaking these words:[26]

Members of the administration, the Senate and the House, Mr. Chairman of the Atomic Energy Commission, Dr. and Mrs. Oppenheimer, ladies and gentlemen:

One of President Kennedy's most important acts was to sign the Enrico Fermi Award for Dr. Oppenheimer for his contributions to theoretical physics and the advancement of science in the United States of America.

It is important to our Nation that we have constantly before us the example of men who set high standards of achievement. This has been the role that you have played, Dr. Oppenheimer.

During World War Two, your great scientific and administrative leadership culminated in the forging together of many diverse ideas and experiments at Los Alamos and at other places. This successful effort came to a climax with the first atomic explosion at Alamogordo on July 16, 1945.

Since the war you have continued to lead in the search for knowledge, and you have continued to build on the major breakthrough achieved by Enrico Fermi on this day in 1942. You have led in developing an outstanding school of theoretical physics in the United States of America.

For these significant contributions, I present to you on behalf of the Atomic Energy Commission and the people of the United States the Enrico Fermi Award of 1963, the Enrico Fermi Medal…

By the time Eisenhower left office, most of the modern federal science and technology policy machinery was in place. It would be up to future presidents to grease the wheels and pull the levers to get it to run smoothly and efficiently. It would also be up to scientists, technologists, business leaders, entrepreneurs, financiers and the public in general to learn how to navigate the maze that had developed over the course of two centuries. The remaining pages of the book will highlight the most significant policy changes implemented from 1960 to the present, as well as the motivations for the changes and the impact the new policies had on American science and technology.

Eleven presidents held office between 1960 and 2018, and all but one of them held science and technology in high esteem. Donald Trump, elected in 2016, is the only exception. By word and deed he showed disdain for science-based and evidenced-based policymaking during his first year in office, and there are few signs—at the time this book is being completed—that he plans to modify his thinking during his remaining years in office.

In the Epilogue, we will take a brief look at how science and technology might have affected the election outcome and how the Trump presidency might affect America's science and technology enterprise and its standing in the world. We will also see what three Washington insiders have to say about the Trump Administration's disruptive approach to policymaking and its disregard for norms. But first, a rapid journey through ten presidencies and a set of pointers on navigating the science and technology maze in the 21st century.

It had snowed heavily the night before, and a nine-inch white blanket had already covered Washington the morning of January 20, 1961, as the 35th president of the United States strode to the rostrum at the east portico of the Capitol. Despite the 22-degree temperature, John F. Kennedy disdained a coat, wearing only sweater under his suit jacket as he spoke to a shivering crowd of admirers. Theodore Sorensen, one of Kennedy's closest advisors, had collaborated with him on the speech, and the soaring rhetoric that captivated audiences around the world was a signature of his extraordinary turn of phrase.

Kennedy's soaring language might easily win an oratory contest with Eisenhower's less poetic wording, but where science is concerned, both conveyed a similar message. They captured the essence of the Cold War zeitgeist: science had given humanity the ability to destroy itself, as well as the possibility of creating a nobler, more hopeful world. The following excerpt from Kennedy's inaugural speech[27] illustrates the dueling world outlooks:

Vice President Johnson, Mr. Speaker, Mr. Chief Justice, President Eisenhower, Vice president Nixon, President Truman, Reverend Clergy, fellow citizens:

We observe today not a victory of party but a celebration of freedom – symbolizing an end as well as a beginning – signifying renewal as well as change. For I have sworn before you and Almighty God the same solemn oath our forebears prescribed nearly a century and three quarters ago.

The world is very different now. For man holds in his mortal hands the power to abolish all forms of human poverty and all forms of human life...

Finally, to those nations who would make themselves our adversary, we offer not a pledge but a request: that both sides begin anew the quest for peace, before the dark powers of destruction unleashed by science engulf all humanity in planned or accidental self-destruction.

We dare not tempt them with weakness. For only when our arms are sufficient beyond doubt can we be certain beyond doubt that they will never be employed.

But neither can two great and powerful groups of nations take comfort from our present course – both sides overburdened by the cost of modern weapons, both rightly alarmed by the steady spread of the deadly atom, yet both racing to alter that uncertain balance of terror that stays the hand of mankind's final war...

Let both sides seek to invoke the wonders of science instead of its terrors. Together let us explore the stars, conquer the deserts, eradicate disease, tap the ocean depths and encourage the arts and commerce...

And so, my fellow Americans: ask not what your country can do for you—ask what you can do for your country...

Four months after delivering his inaugural speech, Kennedy announced plans for landing a man on the Moon within a decade. His purpose was twofold: to rebuild American confidence and to preempt any military objectives the Soviet Union might harbor in space. Space science was not on his agenda. His conversation with James Webb, NASA's administrator at the time, revealed his deeply held convictions.[28] Webb argued that NASA's expanded program should have broader goals than just a Moon landing, prompting Kennedy to respond, "This is, whether we like it or not a race. Everything we do [in space] ought to be tied into getting to the moon ahead of the Russians." Winning the moon race, Kennedy continued "is the top priority of the agency and except for defense, the top priority of the United States government. Otherwise, we shouldn't be spending this kind of money, because I'm not that interested in space."

Kennedy was not in any way antiscience intellectually or philosophically. Quite the contrary. As his inaugural address illustrated, he viewed science as the embodiment of hope for the future. But, as had been the case throughout American history, military exigencies for him at that moment trumped any pure science predilections he might have had.

Kennedy surrounded himself with "whiz kids"—some of the most intellectually gifted policy wonks in the country. Jerome Wiesner, an electrical engineer, with University of Michigan and Massachusetts Institute of Technology pedigrees, was one of them. As Kennedy's science advisor and chairman of PSAC (the President's Science Advisory Committee), Wiesner advocated for arms control, fought against DDT usage, opposed manned space exploration, and supported basic research staunchly. In the early summer of 1962, to give Wiesner the policy assistance he might need, Kennedy established the Office of Science and Technology[29] (OST) within the Executive Office of the President. Having an in-house staff is one thing, influencing the presidential thinking is another. Wiesner's scientific sway is apparent in Kennedy's National Academy of Sciences policy address on Oct. 22, 1963. The following excerpts from his remarks are among the most noteworthy:[30]

> *[If] I were to name a single thing which points up the difference this century has made in the American attitude toward science, it would certainly be the whole-hearted understanding today of the importance of pure science. We realize now that progress in technology depends on progress in theory; that the most abstract investigations can lead to the most concrete results; and that the vitality of a scientific community springs from its passion to answer science's most fundamental questions. I therefore greet this body with particular pleasure, for the range and depth of scientific achievement represented in this room constitutes the seedbed of our Nation's future.*

> *The last hundred years have seen a second great change – the change in the relationship between science and public policy...*

> *As the country has had reason to note in recent weeks during the debate on the nuclear test ban treaty, scientists do not always unite themselves in their recommendations to the makers of policy. This is only partly because of scientific disagreements. It is even more because the big issues so often go beyond the possibilities of exact scientific determination...*

> *In the last hundred years, science has thus emerged from a peripheral concern of Government to an active partner. The instrumentalities devised in recent times have given this partnership continuity and force. The question in all our minds today is how science can best continue its service to the Nation, to the people, to the world, in the years to come.*

> *I would suggest that science is already moving to enlarge its influence in three general ways: in the interdisciplinary area, in the international area, and in the intercultural area. For science is the most powerful means we have for the unification of knowledge, and a main obligation of its future must be to deal with problems which cut across boundaries, whether boundaries between the sciences, boundaries between nations, or boundaries between man's scientific and his humane concerns...*

> *Scientists alone can establish the objectives of their research, but society, in extending support to science, must take account of its own needs. As a layman,*

I can suggest only with diffidence what some of the major tasks might be on your scientific agenda, but I venture to mention certain areas which, from the viewpoint of the maker of policy, might deserve your special concern.

First, I would suggest the question of the conservation and development of our natural resources...

Second, I would call your attention to a related problem; that is, the understanding and use of the resources of the sea...

Third, there is the atmosphere itself, the atmosphere in which we live and breathe and which makes life on this planet possible.

Fourth, I would mention a problem which I know has greatly concerned many of you. That is our responsibility to control the effects of our own scientific experiments...

If science is to press ahead in the four fields that I have mentioned, if it is to continue to grow in effectiveness and productivity, our society must provide scientific inquiry the necessary means of sustenance. We must, in short, support it. Military and space needs, for example, offer little justification for much work in what Joseph Henry called abstract science. Though such fundamental inquiry is essential to the future technological vitality of industry and Government alike, it is usually more difficult to comprehend than applied activity, and, as a consequence, often seems harder to justify to the Congress, to the executive branch, and to the people...

But if basic research is to be properly regarded, it must be better understood. I ask you to reflect on this problem and on the means by which, in the years to come, our society can assure continuing backing to fundamental research in the life sciences, the physical sciences, the social sciences, our natural resources, on agriculture, on protection against pollution and erosion. Together, the scientific community, the Government, industry, and education must work out the way to nourish American science in all its power and vitality. Even this year we have already seen in the first actions of the House of Representatives some failure of support for important areas of research which must depend on the National Government. I am hopeful that the Senate of the United States will restore these funds. What it needs, of course, is a wider understanding by the country as a whole of the value of this work which has been so sustained by so many of you.

I would not close, however, on a gloomy note, for ours is a century of scientific conquest and scientific triumph. If scientific discovery has not been an unalloyed blessing, if it has conferred on mankind the power not only to create, but also to annihilate, it has at the same time provided humanity with a supreme challenge and a supreme testing. If the challenge and the testing are too much for humanity, then we are all doomed. But I believe that the future can be bright, and I believe it can be certain. Man is still the master of his own fate, and I believe that the power of science and the responsibility of science have offered mankind a new opportunity not only for intellectual growth, but for moral discipline; not only for the acquisition of knowledge, but for the strengthening of our nerve and our will...

Kennedy's National Academy's Address encapsulated practically all of the lessons learned from almost two centuries of American science and technology policy. How his agenda would have unfolded, we will never know. Exactly one month to the day after he celebrated the Academy's centennial, Kennedy was assassinated.

Eisenhower began the space program, Kennedy set the nation's sights on the Moon, and Johnson was president when the Apollo 1 fire killed astronauts Roger Chaffee, Gus Grissom, and Ed White II on the Cape Kennedy launch pad on January 26, 1967. By the time Neil Armstrong and Buzz Aldrin landed the lunar module Eagle on the Moon at 20:18 UTC[31] on July 21, 1969, Richard M. Nixon was president. Four presidents, Republicans and Democrats, and more than a decade of political will: The Apollo lunar program epitomizes the enduring and nonpartisan nature of science and technology policy that characterized much of the final four decades of the 20th century.

For science, the Johnson and Nixon years were affected dramatically by the fallout from the Vietnam War, and in Nixon's case, by his tragic flaws, which the Watergate scandal exposed. The bloody Vietnam conflict, in which American military technology was on prominent display, opened fissures on university campuses. Antiwar activists waged verbal—and in some cases physical—assaults on scientists[32] and mathematicians[33] who received research support from defense agencies, such as the Office of Naval Research (ONR) and the Air Force Office of Scientific Research (AFOSR). The ONR and AFOSR money that made its way into university laboratories almost exclusively funded long-term, unclassified basic research that had no military applications. The two agencies, it should be remembered, were products of the Post-World-War-II era and policies of that time, which sought to guarantee the United States would have the necessary scientific workforce for any future military conflict, as well as the fundamental science on which future military technologies might rely.

University scientists were more than content to receive the money, so long as it came with no strings attached. They saw no hypocrisy in accepting the funding, even if they were opposed to the Vietnam War, as most of them were. Critics of the war had a different view. Scientists, they contended, were being corrupted by the very "military-industrial complex" Eisenhower had warned about in his farewell address.[34] The antiwar, antiscience contingent had two powerful allies in the United States Senate: Arkansas Democrat J. William Fulbright, who chaired the Foreign Relations Committee, and Montana Democrat Mike Mansfield, who was majority leader of the upper body.

In 1969, with Fulbright's backing, Mansfield muscled through an amendment to the Military Authorization Act prohibiting the Department of Defense (DOD)—as the War Department had been known since 1949—from funding university research that did not have any direct military application.[35] The consequences, though unintended, were profound. University research programs suffered severe disruptions. Many young scientists lost their jobs. For the first time in memory, the unemployment rate among scientists became a measurable statistic. And pessimism descended on students contemplating a scientific career, many of them opting for other professions.

Even though the Mansfield language did not appear in future legislation, the damage it did took a decade to repair. Fulbright[36] and Mansfield harbored no hostility toward science—quite the contrary, they held it in extremely high esteem—but their antiwar passions trumped their science passions and led them to formulate a policy that proved to be a destructive blow to America's science and technology enterprise. Science and technology thrive on certitude and continuity; they falter in the face of ambiguity and disruption.

Richard Nixon is remembered most for the Watergate break-in and his subsequent resignation as he was confronting certain impeachment for obstructing justice. But Watergate, in truth, was a short chapter in the annals of American history, and its impact did not prevent the nation from moving beyond it post haste. Not so for Nixon's impact on science and technology. During his second year in office, he bypassed Congress and used an executive order in December 1970 to create the Environmental Protection Agency (EPA), naming William Ruckelshaus as its first administrator. On the final day of the year, Congress passed the Clean Air Act Extension of 1970,[37] greatly expanding the federal government's mandate and authorizing the EPA to set national standards for air quality and auto emissions. It was probably Nixon's science policy high point. His final days in the White House were a mad dash to the bottom.

But in the history of environmental policy, there are few legislative actions of greater significance than the Clean Air Act. The sweeping standards for lead, sulfur dioxide, carbon monoxide, ozone, nitrogen oxides, and particulate emissions, which the EPA established under the mandate of the 1970 act and in accordance with its 1977 and 1990 amendments, have led to extraordinary improvements in public health, saving hundreds of thousands of lives and millions of dollars in medical expenditures. But its reach has been far greater than Nixon or Congress could have imagined.

Although a small group of scientists had begun to pay attention to global warming before 1972, policymakers were not only oblivious to the existential threat it posed, they undoubtedly were unaware it was even occurring. Scroll forward a quarter of a century. In 2007, the Supreme Court ruled that emissions of carbon dioxide—a critical greenhouse gas associated with global warming and climate change—constitute a pollutant under the definition of the Clean Air Act, granting the EPA the authority to regulate it. The legislation's historical arc illustrates important points: Policymaking should always be based on sound science, but sound policies should be broad enough to accommodate new scientific knowledge as it becomes available.

Health and climate impacts aside, the Clean Air Act of 1970 had two other significant consequences. It spurred innovation, quickly leading to the development of catalytic converters for cars and trucks, internal combustion engines that run on unleaded gasoline, and smokestack "scrubbers" that remove a host of pollutants on the EPA's list of undesirables. But it also led to an unforeseen deleterious result a few years later.

In 1970, American manufacturing companies competed on a fairly level domestic playing field. Regulations that affected one company largely affected all of its rivals. They were all American. But over the course of the next two decades, advances in transportation and communication technologies ushered in a globalization revolution.

The world, as Thomas Friedman described it in his 2005 best seller,[38] was becoming increasingly flat. And on the flattened globe, the playing field was becoming increasingly tilted in favor of other countries. In the 1970s and 1980s, companies in the developing world were not saddled with the costs of environmental regulations. The result, as University of Chicago economist Michael Greenstone has shown,[39] was a measurable decline in American manufacturing capacity and manufacturing jobs over a 15-year period beginning in 1972, both directly related to the costs imposed by the Clean Air Acts of 1970 and 1977.

There is little doubt the Clean Air Act has had a dramatic and enduring beneficial impact on the health of Americans, but, as in the case of the Mansfield Amendment, the policies it produced had unexpected consequences. If policymakers had known what was in store, they might have been able to plan for a future in which American manufacturing would no longer be king. But they didn't, and the rusting of American manufacturing stranded tens of thousands of workers in the following decades.

The Clean Water Act of 1972[40] was much more clear-cut. On June 22, 1969, six months after Richard Nixon took the oath of office, *The Cleveland Plain Dealer* ran the following headline: "Oil Slick Fire Ruins Flats Shipyard." The accompanying photograph said it all: Cleveland's Cuyahoga River was on fire. It was a stunning, unforgettable sight and it focused public attention on the state of the nation's waterways, lakes, and sources of drinking water in a way that nothing had previously; although twice before, once in 1956 and again in 1961, Congress had sent legislation to the White House to address the issue. The first time, President Eisenhower vetoed the bill, declaring that it was not a federal issue. The second time, President Kennedy signed the bill, but as evidence would demonstrate over the next decade, the legislation was far from all-encompassing.

Cuyahoga was a game changer. On Aug. 1, 1969, *Time* ran an article calling attention to the policy deficiencies in the starkest terms. "America's Sewage System and the Price of Optimism,"[41] began with these words:

> ALMOST *every great city has a river. The poetic notion is that the flowing water brings commerce, delights the eye, and cools the summer heat. But there is a more*

prosaic reason for the close affinity of cities and rivers. They serve as convenient, free sewers.

The Potomac reaches the nation's capital as a pleasant stream, and leaves it stinking from the 240 million gallons of wastes that are flushed into it daily. Among other horrors, while Omaha's meat packers fill the Missouri River with animal grease balls as big as oranges, St. Louis takes its drinking water from the muddy lower Missouri because the Mississippi is far filthier. Scores of U.S. rivers are severely polluted – the swift Chattahoochee, the majestic Hudson and quiet Milwaukee, plus the Buffalo, Merrimack, Monongahela, Niagara, Delaware, Rouge, Escambia and Havasupi. Among the worst of them all is the 80-mile-long Cuyahoga, which splits Cleveland as it reaches the shores of Lake Erie.

No Visible Life. Some river! Chocolate-brown, oily, bubbling with subsurface gases, it oozes rather than flows. "Anyone who falls into the Cuyahoga does not drown," Cleveland's citizens joke grimly. "He decays." The Federal Water Pollution Control Administration dryly notes: "The lower Cuyahoga has no visible life, not even low forms such as leeches and sludge worms that usually thrive on wastes…"

Congress got the message, but it took two more years and a veto override before the Clean Water Act became law.[42] The delay in drafting the legislation, followed by ten months of House-Senate haggling over dueling versions of the bill, presidential opposition to the price tag,[43] and the final congressional override illuminate the difficulties in passing legislation even when it has strong popular support. As long as it lacks the urgency of an epidemic or a war, it can be a long slog.

Three upheavals marked Nixon's final years in office. One of them—the Arab oil embargo—came from abroad. The other two were of his own making. His attempt to cover up the role he played in the Watergate break-in is common knowledge. Less well known, perhaps, were his testy reactions to science and technology assessments he didn't like.

On October 6, 1973, as Jews were attending Yom Kippur synagogue services in Israel, Egypt and Syria attacked Israeli held territories in the Sinai Peninsula and Golan Heights. So began the 1973 Arab-Israeli war, the fourth conflict in the 25 years of the Jewish state's existence. The United States, on the heels of Soviet support for the Arabs, began an airlift of supplies and weapons to Israel, along with $2.2 billion in aid. In response, 2 weeks into the war, Arab oil-producing nations followed through on their prior plans and imposed an embargo on oil exports to the United States. The war came to an end a week later, twenty days after it had begun. But the embargo continued for another five months—finally ending on March 17, 1974—and the Arab states cut oil production by 25%.

The embargo and the disruption in oil markets had a major and long-lasting impact on U.S. energy policy. The Nixon Administration took several immediate actions. It imposed price controls to limit gouging, began "odd-even" gasoline rationing to maintain supplies at the pump, and mandated a 55 mile-per-hour speed limit to improve vehicle efficiency.[44] Nixon urged Congress to codify energy efficiency and conservation standards with new legislation. Separately, he proposed reorganizing federal energy activities by creating one new department and two new agencies. A Department of Energy and Natural Resources would oversee all major energy policy programs, and in place of the Atomic Energy Commission (AEC), an Energy Research and Development Administration (ERDA) would manage all forms of energy research, while a Nuclear Regulatory Commission (NRC) would focus on nuclear power licensing and regulation.[45]

His grand plans did not materialize until after he departed the White House in August 1974 under the gathering clouds of impeachment. Two months later, Congress passed the Energy Reorganization Act of 1974.[46] The legislation abolished the AEC, replacing it with ERDA and the NRC, in line with Nixon's plan. It consolidated energy programs housed in the National Science Foundation and the Environmental Protection Agency under the ERDA umbrella. But it did not create a new energy department, as Nixon had urged, leaving such an option open to a future presidential request. The act also established the NRC as an independent regulatory and licensing body, separating government oversight of nuclear power from federal nuclear energy research and development programs.

The 1974 Energy Reorganization Act was a sweeping piece of legislation that fundamentally altered the landscape of energy policy in the federal bureaucracy. Nixon left Washington in disgrace, but his influence on energy issues continued with the passage of the Energy Policy and Conservation Act[47] in late 1975. It was the first comprehensive bill of its kind. It created a Strategic Petroleum Reserve to "diminish the vulnerability of the United States to the effects of a severe energy supply interruption, and provide limited protection from the short-term consequences of interruptions in supplies of petroleum product." It also established standards for energy efficiency, among them the first fuel economy requirements for cars and light trucks. Known today by its acronym CAFE for "Corporate Average Fuel Economy," the miles-per-gallon fleet standards reduced average gasoline consumption dramatically over the next few decades. In 1972, 14 miles per gallon typified average fleet performance. But by 1985, it had met the legislative target, almost doubling to 27.5 miles per gallon.

The result confirms an essential American science and technology proposition: set standards and let the free market of ideas generate the needed innovations. What happened in the decades after 1985, during which the CAFE standards remained unchanged, illustrates the flip side of the proposition. Absent mandated standards, an untethered market will seek to maximize corporate profits, not always in the best interest of the public. Between 1985 and 2010, for example, auto manufacturers used new technologies to build bigger and faster cars, as well as larger SUVs and light trucks. They opted to maximize their corporate bottom line rather than continue to improve the fuel efficiency of the vehicles they manufactured.[48]

By establishing the EPA in 1970, Nixon left an indelible mark on America's environmental policy legacy. A year later, he added biomedicine to his scientific footprint, when he successfully pressed Congress to pass the National Cancer Act.[49] The disease had become the second leading cause of death in the United States, and Nixon made his appeal for legislation calling for a "war on cancer."[50] His rhetoric combined the two most potent drivers of science and technology policy: disease and war. As Republican communications strategist Frank Luntz has repeatedly drummed into the political psyche, the tactical use of words is key to success.[51] And Nixon's phrasing struck a chord with the public. Congress marched ahead in lockstep, boosting funding for research in the 1971 act and creating a National Cancer Advisory Board that reported directly to the president.

Unfortunately, Nixon's interest in science was limited to the issues that had immediate social impact: energy, the environment, and cancer. His first science advisor, Lee DuBridge, a physicist and former president of Cal Tech, who had counseled both Truman and Eisenhower on science policy, resigned from his White House post in 1970, after serving only 2 years. Unable to secure Nixon's support for sustained federal research funding, facing criticism from the science community for failing to do so, and seeing much of his other science and technology advice go unheeded, he found himself in an untenable situation. An effective science advisor needs to have the trust of both the president and the science community. DuBridge found he didn't have the former and was rapidly losing the latter. Getting out was his best recourse, and he took it.

His successor, Edward David, Jr., an electrical engineer who was executive director of research at Bell Laboratories at the time Nixon tapped him as DuBridge's replacement, didn't fare any better. David joined the White House staff in the middle of September 1970, and by January 1973, he had concluded he was wasting his time. Nixon didn't seem to value his assistance or the work of PSAC—the President's Science Advisory Committee—which David, like his predecessors, chaired. The supersonic transport (SST) and anti-ballistic missile (ABM) defense system[52] underscored the disconnect. Both were high on Nixon's priority list. But some key members of PSAC found them wanting in the extreme, the SST on damage to the environment, and the ABM concept on lack of technological feasibility. And those scientific critics weren't shy about expressing their opposition publicly.[53] At the very least, Nixon found their behavior indecorous. More than likely he saw them as enemies of his administration.

It's not too surprising, then, that even before David showed himself out, Nixon had already decided he was pretty much done with having a science advisor and a science advisory committee. He had concluded that the federal science agencies had enough talent to provide him with advice if and when he might need it. Shortly after he began his second term, with David gone, he abolished PSAC and the Office of Science and Technology (OST), giving his former science advisor's staff their walking papers. H. Guyford Stever, already the director of the National Science Foundation, took on the added role of presidential science advisor on civilian issues, but there is scant evidence the president ever consulted him.[54] If there was a technological matter involving security, Nixon believed he could rely on the National Security Council for assistance, although the members of the NSC hadn't been chosen for their science and technology pedigrees.

Whether Nixon would have continued to chip away at the science policy edifice is hard to know, because the Watergate scandal[55] enveloped the White House for the next eighteen months. Facing almost certain impeachment for obstructing the investigation of the 1972 break-in at the Democratic National Committee Headquarters in the Watergate complex, he resigned his presidency on August 9, 1974.

While Nixon was banishing his science and technology gurus, Congress was bolstering its own advisory apparatus. The initiator was Emilio Daddario, a Democratic House member from Connecticut. First elected in 1958, "Mim," as he was known to his colleagues, friends, and constituents, represented his state's 1st Congressional District for 12 years. During his tenure, he chaired two subcommittees of the House Science Committee: Science Research and Development, and Patents and Science Inventions. Although his professional science credentials were minimal, his commitment to research was exceptional, so much so that after he left office, two august bodies recognized him for his unswerving devotion. The National Academy of Sciences awarded him its Public Welfare Medal in 1976, and members of the American Association for the Advancement of Science (AAAS) elected him president of the scholarly society a year later.

For more than a century, technology and Connecticut had been synonymous. Waterbury was known as the "Brass Capitol of the World in the 19th century;" the towns around Hartford were home to clockmakers; and across the state, the firearms industry, most conspicuously, had an outsized footprint—Eli Whitney, Samuel Colt, John Marlin, Richard Gatling, Oliver Winchester, and Daniel Wesson were among Connecticut's celebrated gun makers. By the end of World War II, the state had added submarine building and aircraft manufacturing to its technology portfolio. And the 1970s saw two corporate giants, Xerox and General Electric,[56] relocate their headquarters there.

Science and technology might not have been in Mim Daddario's DNA, but as a savvy politician, he knew how much they meant to his state. He also understood the importance of evidenced-based policymaking, and the combination of the two led him to conclude that Congress would be well served by having its own technology assessment capability. He made the case to his colleagues, but his inspiration didn't gain sufficient traction before he left the House for an unsuccessful gubernatorial run in 1970. Finally, 2 years later, over opposition from business leaders who worried that the new entity might generate burdensome regulation, stifling innovation,[57] Congress passed legislation establishing the Office of Technology Assessment (OTA).[58]

The act stipulated that a twelve-member bipartisan, bicameral board would oversee the office's activities, which were supposed to be apolitical and free from advocacy. Daddario became OTA's first director in 1972, but early on, some members of Congress began to question whether he could deliver on OTA's impartiality requirement. After all, he had been a Democratic member of the House, and he was a close friend of a high-profile Democratic senator, Edward "Ted" Kennedy of Massachusetts, who chaired OTA's oversight board.

In 1977, Daddario elected to step aside and was replaced by Russell Peterson, a former governor of Delaware. Peterson, unlike Daddario, had scientific bona fides, having earned a Ph.D. in physical chemistry. And as a liberal Republican outside the Washington establishment, he didn't carry around the political baggage that had weighed on Daddario. Two years later, though, Peterson left to become president of the National Audubon Society to pursue his primary interest—ecology. In 1979, John H. Gibbons, a nuclear physicist from Oak Ridge National Laboratory, took over the reins, and for 12 years, OTA was relatively stable. Under Gibbons's leadership, OTA worked hard to maintain its credibility for unbiased scientific analysis,[59] but its critics never really disappeared. Providing sound scientific assessments in the political arena, as OTA's history demonstrates, is far more difficult than students of policy generally recognize. Facts and evidence matter hugely, but politics can muddy the waters significantly, to the point where science is barely visible.

Gibbons tried to steer a steady course, but the seas almost proved to be too turbulent during Ronald Reagan's years in the White House. Trouble began in 1983 when Reagan proposed the Strategic Defense Initiative (SDI) centered on a space-based missile defense network. Until then, OTA had been largely successful in producing reports that were policy neutral, as well as politically neutral. If anything, its approach was too bland, leaving members of Congress to figure out for themselves what policy steps made sense and leading some to wonder what benefit OTA actually delivered. But Gibbons persevered with his approach, explaining in one interview, "Most congressional members welcome scientifically solid information. So that's what we offer them. We have to remember that we can never go one step further, however, and give opinions on what to do."[60] Then came "Star Wars," as SDI was known colloquially.

The Office of Technology Assessment conducted most of its work with the assistance of outside experts, because the topics it covered were so far ranging, its limited permanent staff could never provide the breadth and depth of knowledge required. In 1984, OTA released a background paper prepared by one of its contractors that took sharp issue with Reagan's missile defense proposal. In the analyst's words,[61]

The prospect that emerging "Star Wars" technologies, when further developed, will provide a perfect or near-perfect defense system, literally removing from the hands of the Soviet Union the ability to do socially mortal damage to the United States with nuclear weapons, is so remote that it should not serve as the basis of public expectation or national policy about ballistic missile defense.

Suddenly, OTA was in the cross hairs of the White House, and mostly all of Reagan's supporters in Congress. [Ironically, the author of the OTA background paper was Ashton B. Carter, a well-respected nuclear physicist, who would serve as President Barack Obama's Secretary of Defense from 2015 to 2017.] Gibbons eventually managed to right the OTA ship, but less than a decade later, the Star Wars episode would cause it to sink, although Gibbons was no longer on board when it happened.

In 1993, President Bill Clinton tapped Jack Gibbons as his science advisor, and Roger Herdman, who had been OTA's assistant director, took over. His position and OTA's place on Capitol Hill seemed secure, but its equilibrium would shortly prove unstable. The 1994 election was the turning point. After the votes were counted, Republicans found themselves in control of Congress for the first time in 40 years. "The Contract with America," the Republican policy prescriptions that framed the last six weeks of the GOP mid-term campaign, called for shrinking the size of the federal government, cutting spending, balancing the budget, reducing the number of congressional committees, and closing a raft of federal agencies and offices.

Newt Gingrich, who was the Contract's architect, became Speaker of the House. He was a science geek, but he had little use for OTA, which he saw as irrelevant and a bastion of biased liberal thinking. Curiously, however, it was not the House that turned off OTA's funding stream in the fall of 1995. There were still enough establishment Republicans around who valued its work. Not so in the Senate, which voted 54–45 on September 29 to eliminate funding for Mim Daddario's creation. [The Senate's 1995 action removed OTA's funding, but did not actually abolish the office. Several times since then, some members of Congress—among them, former Rep. Rush Holt Jr., a Democrat from New Jersey and the son of the West Virginia senator whom Harley Kilgore defeated in 1940—have tried to restore OTA's budget. But none have succeeded.]

In hindsight, it's not hard to understand why OTA drew the ire of conservative Republicans. There were three bases for their opposition. One was political, one was philosophical, and one was fundamental.

First, in line with Gingrich's thinking, they saw OTA as a Democratic think tank. Ted Kennedy's outsized role in OTA's operation at its inception fed their narrative. And when a chunk of OTA's staff accompanied Gibbons to the Clinton White House, Republicans were convinced they had proof positive.

Second, they saw OTA as the posterchild for bloated government. The House and Senate already had the Congressional Research Service (CRS) at their disposal, as well as the Government Accounting Office (now called the Government Accountability Office, but with the same acronym, GAO) and the Congressional Budget Office (CBO). Even though OTA brought a much higher level of scientific and technological sophistication to the table, in line with Daddari's vision, small government aficionados viewed the office as a luxury, not a necessity. In a crunch, they thought Congress could always turn to the National Academy of Sciences (NAS)—now the National Academies of Science, Engineering and Medicine—for technical assistance, although, as our historical tour has shown, the NAS often fell short in that capacity.

Third, although they might not have parsed it as such, they saw an inherent contradiction in OTA's operation. During its 21-year run, OTA had generated more than 750 technical analyses, each endeavoring to separate fact-based findings from the appearance of policy recommendations. But as Ash Carter's background paper on Reagan's "Star Wars" initiative showed, it's very difficult to toe that line. Scientific and technical analysis does not ipso facto produce science and technology policy, but evidence-based studies very often leave no room for debate. It was true in Carter's work for OTA, and it is the conundrum that faces science and technology policymakers who base their positions on evidence, but must function in an often-contentious political arena.

Chapter 8

A fresh start 1974–1992

Ten months before Richard Nixon resigned his presidency, Spiro Agnew left his vice-presidential office, pleading no contest to a charge of tax evasion. In accord with the Constitution's 25th Amendment, Nixon nominated a replacement. His choice was Gerald R. Ford, a congenial Republican House member from Grand Rapids, Michigan, who had risen to the post of Minority Leader in 1965. He had been a star football player at the University of Michigan and had served with distinction in the Navy during World War II. And he was immensely popular with his colleagues on Capitol Hill, as his confirmation vote demonstrated, 92-3 in the House and 398-35 in the Senate.

He had a bumbling verbal way about him, leading people who didn't know him to underestimate his intellectual capability. But Ford had an undergraduate degree in economics from Michigan and a law degree from Yale, and he valued science tremendously.[1] After taking over from Nixon in the summer of 1974, Ford selected Nelson A. Rockefeller to be his vice-president.

"Rocky" was then serving his fourth consecutive term as governor of New York where he had carved out a strong record on the environment and higher education, but on science and technology matters, by his own account, his accomplishments were less than stellar.[2] A liberal on most policy issues and a leader of the Eastern Republican Establishment, he had failed three times to secure the GOP presidential nomination. Nonetheless, the Senate and the House both confirmed him easily after Ford nominated him, by a vote of 97-10 in the Senate and 287-128 in the House. When Rockefeller took the oath of office on December 19, 1974, it marked the first time in history that neither the president nor the vice-president had been elected to their offices.

The unelected Ford-Rockefeller team commanded the Executive Branch for only 2 years. But they left office on a science and technology policy high note. While Nixon was never comfortable with elites, scientific or otherwise, and especially Ivy Leaguers, Ford was far more embracing. He was comfortable in his own skin and did not feel threatened by scientists; on the contrary, he valued their worth. It was, therefore, no accident that the National Science and Technology Policy, Organization, and Priorities Act of 1976[3] emerged from Congress on his watch. The legislation not only undid Nixon's damage to the White House instruments of science and technology policy, it enshrined

Navigating the Maze. https://doi.org/10.1016/B978-0-12-814710-8.00008-5
153

in law the relevant federal advisory and coordinating mechanisms so that no future president could eliminate them on a petulant whim. From May 11, 1976 forward, science and technology policy would have a permanent place in the White House.

The act provides one of the clearest and most compelling set of rationales for federal science and technology policy. It also illuminates the connections between science and national needs. For these two reasons, Title I of the legislation bears repeating, at least in part.

Title I—National Science, Engineering, and Technology Policy and Priorities

Findings

Section 101. (a) The Congress, recognizing the profound impact of science and technology on society, and the interrelations of scientific, technological, economic, social, political, and institutional factors, hereby finds and declares that—

(1) the general welfare, the security, the economic health and stability of the Nation, the conservation and efficient utilization of its natural and human resources, and the effective functioning of government and society require vigorous, perceptive support and employment of science and technology in achieving national objectives;

(2) the many large and complex scientific and technological factors which increasingly influence the course of national and international events require appropriate provision, involving long-range, inclusive planning as well as more immediate program development, to incorporate scientific and technological knowledge in the national decision-making process;

(3) the scientific and technological capabilities of the United States, when properly fostered, applied, and directed, can effectively assist in improving the quality of life, in anticipating and resolving critical and emerging international, national, and local problems, in strengthening the Nation's international economic position, and in furthering its foreign policy objectives;

(4) Federal funding for science and technology represents an investment in the future which is indispensable to sustained national progress and human betterment, and there should be a continuing national investment in science, engineering, and technology which is commensurate with national needs and opportunities and the prevalent economic situation;

(5) the manpower pool of scientists, engineers, and technicians, constitutes an invaluable national resource which should be utilized to the fullest extent possible; and

(6) the Nation's capabilities for technology assessment and for technological planning and policy formulation must be strengthened at both Federal and State levels.

(b) As a consequence, the Congress finds and declares that science and technology should contribute to the following priority goals without being limited thereto:

(1) fostering leadership in the quest for international peace and progress toward human freedom, dignity, and well-being by enlarging the contributions of American scientists and engineers to the knowledge of man and his universe, by making discoveries of basic science widely available at home and abroad, and by utilizing technology in support of United States national and foreign policy goals;

(2) increasing the efficient use of essential materials and products, and generally contributing to economic opportunity, stability, and appropriate growth

(3) assuring an adequate supply of food, materials, and energy for the Nation's needs

(4) contributing to the national security

(5) improving the quality of health care available to all residents of the United States

(6) preserving, fostering, and restoring a healthful and esthetic natural environment;

(7) providing for the protection of the oceans and coastal zones, and the polar regions, and the efficient utilization of their resources;

(8) strengthening the economy and promoting full employment through useful scientific and technological innovations;

(9) increasing the quality of educational opportunities available to all residents of the United States;

(10) promoting the conservation and efficient utilization of the Nation's natural and human resources;

(11) improving the Nation's housing, transportation, and communication systems, and assuring the provision of effective public services throughout urban, suburban, and rural areas;

(12) eliminating air and water pollution, and unnecessary, unhealthful, or ineffective drugs and food additives; an

(13) advancing the exploration and peaceful uses of outer space.

Declaration of Policy

Section 102. (a) PRINCIPLES.—In view of the foregoing, the Congress declares that the United States shall adhere to a national policy for science and technology which includes the following principles:

(1) The continuing development and implementation of strategies for determining and achieving the appropriate scope, level, direction, and extent of scientific and technological efforts based upon a continuous appraisal of the role of science and technology in achieving goals and formulating policies of the United States, and reflecting the views of State and local governments and representative public groups.

(2) The enlistment of science and technology to foster a healthy economy in which the directions of growth and innovation are compatible with the prudent and frugal use of resources and with the preservation of a benign environment.

(3) The conduct of science and technology operations so as to serve domestic needs while promoting foreign policy objectives.

(4) The recruitment, education, training, retraining, and beneficial use of adequate numbers of scientists, engineers, and technologists, and the promotion by the Federal Government of the effective and efficient utilization in the national interest of the Nation's human resources in science, engineering, and technology.

(5) The development and maintenance of a solid base for science and technology in the United States, including: (A) strong participation of and cooperative relationships with State and local governments and the private sector; (B) the maintenance and strengthening of diversified scientific and technological capabilities in government industry, and the universities, and the encouragement of independent initiatives based on such capabilities, together with elimination of needless barriers to scientific and technological innovation; (C) effective management and dissemination of scientific and technological information; (D) establishment of essential scientific, technical and industrial standards and measurement and test methods; and (E) promotion of increased public understanding of science and technology.

(6) The recognition that, as changing circumstances require periodic revision and adaptation of title I of this Act, the Federal Government is responsible for identifying and interpreting the changes in those circumstances as they occur, and for effecting subsequent changes in title I as appropriate.

(b) IMPLEMENTATION.—To implement the policy enunciated in subsection (a) of this section, the Congress declares that:

(1) The Federal Government should maintain central policy planning elements in the executive branch which assist Federal agencies in (A) identifying public problems and objectives, (B) mobilizing scientific and technological resources for essential national programs, (C) securing appropriate funding for programs so identified, (D) anticipating future concerns to which science and technology can contribute and devising strategies for the conduct of science and technology for such purposes, (E) reviewing systematically Federal science policy and programs and recommending legislative amendment thereof when needed. Such elements should include an advisory mechanism within the Executive Office of the President so that the Chief Executive may have available independent, expert judgment and assistance on policy matters which require accurate assessments of the complex scientific and technological features involved...

The three succeeding titles created Executive Branch machinery intended to implement the legislative goals. The following excerpts from the 1976 act capture the essence of the three components: The Office of Science and Technology Policy (OSTP), the President's Committee on Science and Technology, and the Federal Coordinating Council for Science, Engineering and Technology (FCCSET).

Title II—Office of Science and Technology Policy
Short Title

Section 201. This title may be cited as the "Presidential Science and Technology Advisory Organization Act of 1976."

Establishment

Section 202. There is established in the Executive Office of the President an Office of Science and Technology Policy (hereinafter referred to in this title as the "Office").

Director: Associate Directors

Section 203. There shall be at the head of the Office a Director who shall be appointed by the President, by and with the advice and consent of the Senate... The President is authorized to appoint not more than four Associate Directors, by and with the advice and consent of the Senate... Associate Directors shall perform such functions as the Director may prescribe.

Functions

Section 204. (a) The primary function of the Director is to provide, within the Executive Office of the President, advice on the scientific, engineering, and technological aspects of issues that require attention at the highest levels of Government...

Title III—President's Committee on Science and Technology

Establishment

Section 301. The President shall establish within the Executive Office of the President a President's Committee on Science and Technology (hereinafter referred to as the "Committee").

Membership

Section 302. (a) The Committee shall consist of—

(1) the Director of the Office of Science and Technology Policy established under title II of this Act; and
(2) not less than eight nor more than fourteen other members appointed by the President...

Federal Science, Engineering, and Technology Survey

Section 303. (a) The Committee shall survey, examine, and analyze the overall context of the Federal science, engineering, and technology effort including missions, goals, personnel, funding, organization, facilities, and activities in general, taking adequate account of the interests of individuals and groups that may be affected by Federal scientific, engineering, and technical programs, including, as appropriate, consultation with such individuals and groups...

Title IV—Federal Coordinating Council for Science, Engineering, and Technology

Establishment and Functions

Section 401. (a) There is established the Federal Coordinating Council for Science, Engineering, and Technology (hereinafter referred to as the "Council").

(b) The Council shall be composed of the Director of the Office of Science and Technology Policy and one representative of each of the following Federal agencies: Department of Agriculture, Department of Commerce, Department of Defense, Department of Health, Education, and Welfare, Department of Housing and Urban Development, Department of the Interior, Department of State, Department of Transportation, Veterans' Administration, National Aeronautics and Space Administration, National Science Foundation, Environmental Protection Agency, and Energy Research and Development Administration...

(c) The Director of the Office of Science and Technology Policy shall serve as Chairman of the Council...

(e) The Council shall consider problems and developments in the fields of science, engineering, and technology and related activities affecting more than one Federal agency, and shall recommend policies and other measures designed to—

(1) provide more effective planning and administration of Federal scientific, engineering, and technological programs,
(2) identify research needs including areas requiring additional emphasis,
(3) achieve more effective utilization of the scientific, engineering, and technological resources and facilities of Federal agencies, including the elimination of unwarranted duplication, and
(4) further international cooperation in science, engineering, and technology...

The Office of Science and Technology Policy (OSTP) has played the role envisioned for it in the 1976 legislation, essentially unchanged, although its prominence and effectiveness have waxed and waned dramatically with different administrations.[4] It had little, if any, influence during the Reagan years, and it largely flew under the radar during George W. Bush's first term. However, its lack of clout during those periods pales by comparison with its phantom profile during the beginning of Donald Trump's administration. It took more than 18 months from the time he took office for Trump to name a science advisor who would serve as OSTP's director. Its skeleton staff virtually guaranteed that OSTP would be consigned to obscurity, presumably consistent with the low esteem in which Trump held science.

It really does take a president who values science and technology, elevates its place in the policy arena, and vests his science advisor with enough authority to make OSTP a serious Washington player. That happened during George H.W. Bush's presidency and in the final 3 years of Bill Clinton's second term. But without question, Barack Obama gave OSTP its greatest visibility. In addition to valuing science and technology for the benefits they delivered to society, Obama loved science as an intellectual pursuit. He was, in the truest sense of the word, a science geek.

By contrast with OSTP, two committees specified in the 1976 act—the President's Committee on Science and Technology and the Federal Coordinating Council for Science, Engineering and Technology (FCCSET)—have undergone name changes, alterations in structure, and for periods of time, complete suspensions of activity. Carter began the mischief with an executive order[5] in 1977, abolishing the President's Committee, he said, not because he harbored a distaste for science and technology—after all, he had studied nuclear physics and reactor technology—but rather because he thought there were simply too many standing presidential committees. Four years later, George ("Jay") Keyworth, Reagan's science advisor, persuaded his boss to take a less draconian approach, recasting the committee as the White House Science Council, but with a much smaller membership than the 1976 act had authorized, and with the stipulation that it report to Keyworth rather than the president, as the original legislation had envisioned.

The Federal Coordinating Council fared no better. The 1976 legislation gave OSTP responsibility for FCCSET, enabling Congress to scrutinize its operations, and as a consequence, giving the Council much-needed teeth. But Carter's 1977 executive order removed OSTP from the picture entirely, relegating FCCSET to a White House committee chaired by Frank Press, his science advisor, thereby shielding it from congressional oversight. During Reagan's first term, FCCSET was completely missing in action. It reemerged in 1985 at the start of his second term, but it was not at all effective.

Four years later, George H.W. Bush entered the Oval Office, and appointed as his science advisor D. Allan Bromley, a Yale nuclear physicist with a personality big enough to grab anyone's attention, even in a city like Washington, where outsized egos are common calling cards. Bush gave Bromley top billing, authorizing him to raise the profile of science technology within the Administration, as well as in the nation. The President's Council of Advisors on Science and Technology (PCAST)[6,7] emerged as one of the enduring Bush-Bromley White House creations. Since its initiation in 1990, every succeeding president[8] has rechartered it. Bromley also succeeded in reinvigorating FCCSET, and it, too, has remained a White House fixture, although since 1993, under a different name—the National Science and Technology Council (NSTC).[9]

Jimmy Carter was one of the most improbable of American presidents. Although his graduate education had a technology focus, his campaign-trail persona belied any interest in science. He was the home-spun peanut farmer from Plains, Georgia.

Little known to the public outside his native state before he began his presidential run, it is quite likely Carter would not have won the 1976 race against incumbent Gerald Ford, if Ford had not pardoned Richard Nixon for his Watergate crimes just a month after Nixon resigned. The nation was tired of the Watergate scandal, tired of Washington muck, and tired of the political establishment. Ford had attempted to clear the air of the Watergate miasma with his pardon, but voters wanted to clean house entirely. It was far from a rout, but Carter's unlikely coalition of Southern and Northeastern Democrats and his reformist message carried him into the White House.

Watergate and impeachment might have vanished the from the front pages of newspapers, but not so for energy. In 1977, the Organization of Petroleum Exporting Countries (OPEC), led by Saudi Arabia, still had the world's consuming nations in a stranglehold. And the United States was one of the most vulnerable. That year, America imported nearly half the oil it used—9 million barrels out of a total of 19 million—of which OPEC nations supplied 70%.[10] Carter and the newly elected Congress were determined to do something about the threat those numbers conveyed. The result was the creation of the Department of Energy with full Cabinet status. The Findings section of the 1977 "Department of Energy Organization Act"[11] capture the sense of urgency:

Findings

Section 101. The Congress of the United States finds that—

(1) the United States faces an increasing shortage of nonrenewable energy resources;
(2) this energy shortage and our increasing dependence on foreign energy supplies present a serious threat to the national security of the United States and to the health, safety and welfare of its citizens;
(3) a strong national energy program is needed to meet the present and future energy needs of the Nation consistent with overall national economic, environmental and social goals;
(4) responsibility for energy policy, regulation, and research, development and demonstration is fragmented in many departments and agencies and thus does not allow for the comprehensive, centralized focus necessary for effective coordination of energy supply and conservation programs; and
(5) formulation and implementation of a national energy program require the integration of major Federal energy functions into a single department in the executive branch.

The legislation, which Carter signed on August 4, 1977, placed energy activities throughout the federal government under a new Department of Energy (DOE) umbrella. It subsumed the Energy Research and Development Administration (ERDA)—which had been created only 3 years before—and it transferred programs from departments as diverse as Commerce, Defense, Housing and Urban Development, Interior, and Transportation. It also established three new federal offices within the DOE: The Federal Energy Regulatory Commission (FERC), the Energy Information Administration (EIA), and the Economic Regulatory Administration (ERA). Finally, it mandated that "The President prepare and submit to Congress a proposed National Energy Policy Plan" every 2 years.

What emerged was more a messy assortment of individual federal programs and activities than a smoothly functioning energy policy engine. Years later, despite some extraordinary successes—especially in scientific research—the DOE still had the reputation in many quarters of being the most unwieldy, disjointed, and dysfunctional bureaucracy in the federal government. Without question, though, it lost that distinction in 2002 when Congress, operating in crisis mode following the 9/11 attack on the World Trade Center in New York in 2001, created the Department of Homeland Security (DHS). In both cases, Congress used a blunt legislative tool to simply marry federal programs that had very diverse rationales and cultures and hope for the best. The outcomes were similar.

Today, the public and many elected officials don't recognize the extent to which the DOE and DHS are intertwined with science and technology. The Energy Department not only manages the nation's high-tech nuclear arsenal, it is the steward of the national laboratory system and the $5-billion portfolio of fundamental research in the physical sciences. Yet, prior to being nominated for the department's top position, two Secretaries of Energy, Spencer Abraham[12] and Rick Perry,[13] publicly espoused eliminating the department. They changed their minds after learning—mostly on the job—what DOE's responsibilities actually were.

In the case of DHS, even cognoscenti are often in the dark about its science and technology nexus. But it is a department whose mission relies extraordinarily on those capabilities. Here's how it describes its "S&T" ties: "Technology and threats evolve rapidly in today's ever-changing environment. The Department of Homeland Security (DHS) Science and Technology Directorate (S&T) monitors those threats and capitalizes on technological advancements at a rapid pace, developing solutions and bridging capability gaps at a pace that mirrors the speed of life. S&T's mission is to deliver effective and innovative insight, methods, and solutions for the critical needs of the Homeland Security Enterprise."[14]

The DOE and DHS chronicles illustrate a major science and technology conundrum. Policymakers often look to science for help in responding to major security crises, but they rarely feature science in their public utterances. The names DOE and DHS, appropriately conveyed the essence of the crises— energy vulnerability in 1977 and homeland vulnerability in 2002—but did nothing to convey the crucial connection of the departments' missions to science and technology. Once a crisis passes, as it invariably must, public interest wanes, the names lose much of their emotional impact, and science, to the extent that it was ever part of the public's consciousness, is all but forgotten—until the next crisis comes along.

Jimmy Carter's term in office was plagued by rising inflation, natural gas and oil shortages, and soaring energy prices. Anyone who watched his first televised address, delivered two weeks after he was elected president, cannot help but remember it.[15] Wearing a sweater and seated next to a roaring fireplace in the White House library, he grimly laid out his vision for the future.

Large parts of the country had been in a deep freeze since the beginning of winter, and energy supplies were dangerously low when Carter entered the Oval Office. His was a somber speech delivered by a president with a demeanor to match. He called for common sacrifice and conservation, painting a decidedly pessimistic picture of the nation's condition. His assessment was bleak, as the following excerpt from his February 2, 1977 address illustrates:[16]

We must face the fact that the energy shortage is permanent. There is no way we can solve it quickly. But if we all cooperate and make modest sacrifices, if we learn to live thriftily and remember the importance of helping our neighbors, then we can find ways to adjust and to make our society more efficient and our own lives more enjoyable and productive. Utility companies must promote conservation and not consumption. Oil and natural gas companies must be honest with all of us about their reserves and profits. We will find out the difference between real short- ages and artificial ones. We will ask private companies to sacrifice, just as private citizens must do.

All of us must learn to waste less energy. Simply by keeping our thermostats, for instance, at 65 degrees in the daytime and 55 degrees at night we could save half the current shortage of natural gas.

There is no way that I, or anyone else in the Government, can solve our energy problems if you are not willing to help. I know that we can meet this energy chal- lenge if the burden is borne fairly among all our people—and if we realize that in order to solve our energy problems we need not sacrifice the quality of our lives.

Only the last sentence from the excerpt offers any hope. And his final words ended on a similarly dour note:

As President, I will not be able to provide everything that every one of you might like. I am sure to make many mistakes. But I can promise that your needs will never be ignored, nor will we forget who put us in office.

We will always be a nation of differences—business and labor, blacks and whites, men and women, people of different regions and religions and different ethnic backgrounds—but with faith and confidence in each other our differences can be a source of personal fullness and national strength, rather than a cause of weakness and division.

If we are a united nation, then I can be a good President. But I will need your help to do it. I will do my best. I know you will do yours.

Carter was genuinely concerned about energy, and he knew that science and technology held the keys to solving the problem.[17] But he had difficulty conveying such a message to a public that was, in the main, scientifically illiterate. Although he spoke about energy with great regularity, his tone and choice of words didn't help him cut through the fog of public ignorance and distrust, as his "Energy Address to the Nation"[18] on April 5, 1979 illustrates.

Good evening.

Our Nation's energy problem is very serious—and it's getting worse. We're wast- ing too much energy, we're buying far too much oil from foreign countries, and we are not producing enough oil, gas, or coal in the United States.

In order to control energy price, production, and distribution, the Federal bureau- cracy and red tape have become so complicated, it is almost unbelievable. Energy prices are high, and they're going higher, no matter what we do.

The use of coal and solar energy, which are in plentiful supply, is lagging far behind our great potential. The recent accident at the Three Mile Island nuclear power plant in Pennsylvania has demonstrated dramatically that we have other energy problems.

So, what can we do? We can solve these problems together.

Federal Government price controls now hold down our own production, and they encourage waste and increasing dependence on foreign oil. Present law requires that these Federal Government controls on oil be removed by September 1981, and the law gives me the authority at the end of next month to carry out this decontrol process.

In order to minimize sudden economic shock, I've decided that phased decontrol of oil prices will begin on June 1 and continue at a fairly uniform rate over the next 28 months. The immediate effect of this action will be to increase production of oil and gas in our own country.

As Government controls end, prices will go up on oil which has already been discovered, and unless we tax the oil companies, they will reap huge and undeserved windfall profits. We must, therefore, impose a windfall profits tax on the oil companies to capture part of this money for the American people. This tax money will go into an energy security fund and will be used to protect low income families from energy price increases, to build a more efficient mass transportation system, and to put American genius to work solving our long-range energy problems...

The balance of the address was no better in suggesting a plausibly optimistic path forward.

A poll taken not long after the speech showed how little trust the American people had in him and how poorly his message resonated with them. According to the NBC/Associated Press survey, more than half the public considered the entire energy crisis a hoax.[19] Carter's communication skills were part of the problem, but his difficulty also illustrated how hard it is to craft public policies when the public doesn't have the technical background to understand their rationale.

Carter's hope of a winning reelection to a second term was rapidly vanishing. The Iran hostage crisis, which began on November 4, 1979, pretty well killed it. Students in Teheran, who supported the Iranian Islamic revolution, stormed the American Embassy and captured 52 U.S. citizens. With negotiations for their release stalled, Carter approved a risky a helicopter rescue attempt that April. It failed spectacularly, and as voters went to the polls in November 1980, there was still no sign the crisis would soon be resolved. Between the energy and Iran fiascos, Carter's days in the White House would end in a blowout. He carried only six states plus the District of Columbia, capturing a scant 49 out of 538 electoral votes and receiving only 41% of the popular vote. [Ironically, negotiations finally led to the release of the hostages on January 20, 1981, the same day Carter departed Washington.[20]]

Ronald Reagan took the oath of office on January 20, 1981 as the 40th president of the United States. He had built his campaign around a promise to rejuvenate the American economy that had been suffering from a combination of low growth and high inflation—a condition called "stagflation"—since the 1973 Arab oil embargo. In contrast to Carter, Reagan exuded optimism, and he had built his campaign around the slogan, "Let's Make America Great Again." It harkened back to the days before Nixon, Ford, and Carter; before the Watergate scandal, the Arab oil embargo and the Iran hostage crisis; before a looming impeachment; a policy of price controls and a plea for self-sacrifice. [Unlike Donald Trump's use of the phrase, "Make America Great Again," was not rooted in racist or ethnic politics.]

During his two terms in the White House, Reagan championed policies that became the blueprint for conservative Republicans until Donald Trump's election in 2016. His prescriptions for science and technology can be summed up in relatively few words. The federal government should support research that either advances military objectives or lies outside the domain of industry. Defense research and development and long-term basic research should be in the federal portmanteau. Applied research and development should not. It's that simple.

Ronald Reagan was fond of using the radio to communicate with the American people. In the spring of 1988, he took to the airways to speak about the federal role in scientific research—not exactly a subject that would compete for ratings with his addresses on the Persian Gulf conflict, the war on drugs, or the explosion of the space shuttle Challenger. But, as his presidency was winding down, he finally affirmed his commitment to basic research and his faith in scientific progress, issues he had largely ignored during the prior 7 years.

Whether he had gotten religion on science, himself, or whether his science advisor at the time, William Graham, a physicist with a background in national security, had sparked the epiphany, is not clear, but Reagan used his bully pulpit to deliver a science sermon. His address allowed him to identify the tangible benefits of research and to express support for two extremely costly projects, the Superconducting Supercollider and the Space Station Freedom—neither of which, as the future unfolded, would make it to the finish line. But above of all, it provided him with the opportunity in his opening and closing words to frame science as a vessel of hope. In that regard, he was far ahead of his time.

Scientists, politicians, and policymakers traditionally have held science in high esteem because of its contributions to the public good—for advancing defense, medicine, economic growth, energy, and the environment, all demonstrably true. But in 2016, the first comprehensive studies[21] of public attitudes toward science revealed that Reagan's view of science as hope for the future is really how the public thinks about it in positive terms. These are his words:[22]

My fellow Americans:

Passover and Easter are festivals of hope. That's why this weekend is a good time for all of us to reflect on the enduring importance to mankind of hope and faith in the future. And nowhere do our hopes take more visible form than in the quest of science.

Science has grown, and with it, the fascination it holds for all of us. But as the pursuit of science has become ever more nationally and even multi-nationally funded, it has also become more expensive. The problem here is that science, unlike a bridge or an interstate highway or a courthouse, has no local constituency. Today, when we're witnessing some of the most exciting discoveries in the history of science, things similar to the breakthroughs associated with Einstein, Galileo, and Newton, Federal funding for science is in jeopardy because of budget constraints.

That's why it's my duty as President to draw its importance to your attention and that of Congress. America has long been the world's scientific leader. Over the years, we've secured far more patents than any other country in the world. And since World War II, we have won more Nobel prizes for science than the Europeans and Japanese combined. We also support more of what is called basic research; that is, research meant to teach us rather than to invent or develop new products. And for the past 40 years, the Government has been our leading sponsor of basic research.

The remarkable thing is that although basic research does not begin with a particular practical goal, when you look at the results over the years, it ends up being one of the most practical things government does. For example, government-sponsored basic research produced the first laser. Today, less than three decades later, lasers are used in everything from microsurgery to the transmission of immense volumes of information and may contribute to our Strategic Defense Initiative that promises to make ballistic missiles obsolete. Well, I think that over the past 50 years the Government has helped build a number of particle accelerators so scientists could study high energy physics. Major industries, including television, communications, and computer industries, couldn't be where they are today without developments that began with this basic research.

We cannot know where scientific research will lead. The consequences and spin-offs are unknown and unknowable until they happen. In research, as Albert Einstein once said, imagination is more important than knowledge. We can travel wherever the eye of our imagination can see. But one thing is certain: If we don't explore, others will, and we'll fall behind. This is why I've urged Congress to devote more money to research. After taking out inflation, today's government research expenditures are 58 percent greater than the expenditures of a decade ago. It is an indispensable investment in America's future.

Let me tell you about just a few of the many projects we'll fund this year. This year we'll begin work on the great grandchild of those particle accelerators that have meant so much to our economic growth. It's called the superconducting supercollider. And it will harness the galloping technology of superconductivity, so we can explore subatomic particles in ways we've never been able to before. We'll also continue developing the space station. When it's in orbit, the space station will let us perform once impossible experiments in the weightless and sterile environment of outer space and understand our world and universe. And we're developing new technology to allow man eventually to journey beyond Earth's orbit. Astronaut Senator Jake Garn and others in Congress have given the space program the support it needs to once again reach for the stars.

Meanwhile, back on Earth, we will be pursuing breakthroughs in biotechnology that promise to revolutionize medicine, agriculture, and protection of the environment. We're working on new ways to spread the seeds of Federal research. Working with universities across the country, we have established 14 engineering research centers devoted to basic research on emerging technologies. And we're planning 10 to 15 new science and technology centers to do the same thing in the fields of general science. All of these centers will work with industries so that what they discover can quickly lead to new and better and internationally competitive products. All of this and more is before Congress now.

Some say that we can't afford it, that we're too strapped for cash. Well, leadership means making hard choices, even in an election year. We've put our research budget under a microscope and looked for quality and cost effectiveness. We've put together the best program for the taxpayers' dollars. After all, the American tradition of hope is one we can't afford to forget.

Until next week, happy Easter and Passover. God bless you.

Reagan campaigned for election in 1980 as a hardline Cold Warrior, and during his first term, he hewed to that philosophy, rarely abandoning it and constantly stressing the need for strength against a Soviet enemy whom he characterized as "an evil empire."[23] But his anti-Soviet rhetoric masked his great fear of a nuclear mistake. He saw that risk as an unavoidable weakness in the doctrine of "Mutually Assured Destruction" (MAD), which had underpinned U.S. nuclear strategy for more than a quarter of a century.

The MAD premise was simple. Both the United States and the Soviet Union would stockpile so many nuclear weapons that a first strike by one nation could be met with an unacceptably devastating response by the other. It had worked up to that point, but it was decidedly dangerous. In the words of Caspar Weinberger, Reagan's defense secretary, it was essentially a "mutual suicide pact."[24] On March 23, 1983, in an address to the nation, Reagan unveiled his alternative with these words:[25]

...Tonight, consistent with our obligations of the ABM [Anti-Ballistic Missile] treaty and recognizing the need for closer consultation with our allies, I'm taking an important first step. I am directing a comprehensive and intensive effort to define a long-term research and development program to begin to achieve our ultimate goal of eliminating the threat posed by strategic nuclear missiles. This could

pave the way for arms control measures to eliminate the weapons themselves. We seek neither military superiority nor political advantage. Our only purpose—one all people share—is to search for ways to reduce the danger of nuclear war…

I clearly recognize that defensive systems have limitations and raise certain problems and ambiguities. If paired with offensive systems, they can be viewed as fostering an aggressive policy, and no one wants that. But with these considerations firmly in mind, I call upon the scientific community in our country, those who gave us nuclear weapons, to turn their great talents now to the cause of mankind and world peace, to give us the means of rendering these nuclear weapons impotent and obsolete…

Many foreign policy and arms control experts immediately attacked the Strategic Defense Initiative (SDI), as the program was known, asserting that it violated the ABM Treaty, which the United States and the Soviet Union had signed in 1972.[26] (The treaty originally limited each party to two ground-based missile defense sites, one near its capital and the other near an intercontinental ballistic missile launching site. Two years later, both nations agreed to an amendment that reduced the number of sites to one.) The Reagan Administration pushed back strongly, maintaining that SDI conformed to the treaty because it was not a deployment program, but rather a scientific and technological research initiative.

As research programs go, it was a huge one, receiving annual funding of $3 billion by 1987. Known derisively by its critics as "Star Wars" because a large part of its program focused on space-based systems, SDI was one of the star attractions in Reagan's defense research and development budget. It might have weathered the criticism from the arms control community, but it couldn't withstand the technological scrutiny of the scientific community.

The congressional Office of Technology Assessment (OTA) weighed in just a year after Reagan released his plan, giving it an unequivocal thumbs-down. And in July 1987, the American Physical Society, responding to a request by the Defense Department, released a comprehensive technical report, which contained the following assessment: "We estimate that even in the best of circumstances, a decade or more of intensive research would be required to provide the technical knowledge needed for an informed decision about the potential effectiveness and survivability of directed energy weapon systems. In addition, the important issues of overall system integration and effectiveness depend critically upon information that, to our knowledge, does not yet exist."[27]

The blunt wording left no doubt that SDI faced extraordinarily difficult—if not insurmountable—challenges, and that any hope of producing the required military hardware anytime soon was little more than a pipe dream. In the face of such a judgment by some of the nation's most respected physicists, SDI began to lose support in Congress. Its budget withered, and by 1993, it ceased to exist.

But in spite of its ignominious end, SDI is credited by some historians as accelerating the collapse of the Soviet Union. They contend that it showcased America's advanced technical capabilities, which was no match for the

Soviet's. Whether that stratagem was in the back of Reagan's mind when he first proposed the program remains a matter of debate. But if it was, it demonstrates how science and technology can be used to achieve national policy goals, even before they are ready for prime time.

Reagan was focused more on national security than on health issues. But 6 months after he took the oath of office, the Centers for Disease Control and Prevention (CDC) reported five cases of an unusual lung infection in five gay Los Angeles men. Their immune systems were failing, and within a week of their diagnoses, two had died. It was the beginning of the AIDS epidemic.[28]

The Reagan White House was seemingly detached from the issue. But on Capitol Hill, there were early signs of action. Henry Waxman, a Democratic representative from Los Angeles and chairman of the House Energy and Commerce Subcommittee on Health and the Environment, convened the first hearing on HIV/AIDS on April 13, 1982. Later that summer, Waxman and his fellow House member from San Francisco, Philip Burton, introduced legislation to fund AIDS research at the National Institutes of Health. The following year, Congress finally put some money on the table, appropriating $12 million for AIDS research. The bet paid off quickly: On April 23, 1984, NIH announced that Dr. Robert Gallo and his colleagues at the National Cancer Institute had definitively identified the virus that caused AIDS.

By 1985, AIDS funding had climbed to $70 million, and on September 15 that year, Ronald Reagan finally pledged to make AIDS one of his priorities. The National Institutes of Health was now fully engaged, but more importantly, the Surgeon General, C. Everett Koop, issued a report in October 1986 calling for schools and parents to counsel children about the dangers of unprotected sex.

Almost 6 years had passed since the CDC had reported the first AIDS cases, and Ronald Reagan had yet to address the epidemic publicly. That changed on May 31, 1987, when he spoke at the American Foundation for AIDS Research awards dinner. His remarks are worth reading because they reveal how important it is for a president to be fully engaged if science policy is to deliver public benefits:[29]

...I want to talk tonight about the disease that has brought us all together. It has been talked about, and I'm going to continue. The poet W.H. Auden said that true men of action in our times are not the politicians and statesmen but the scientists. I believe that's especially true when it comes to the AIDS epidemic. Those of us in government can educate our citizens about the dangers. We can encourage safe behavior. We can test to determine how widespread the virus is. We can do any number of things. But only medical science can ever truly defeat AIDS. We've made remarkable progress, as you've heard, already. To think we didn't even know we had a disease until June of 1981, when five cases appeared in California. The AIDS virus itself was discovered in 1984. The blood test became available in 1985. A treatment drug, AZT, has been brought to market in record time, and others are coming. Work on a vaccine is now underway in many laboratories, as you've been told.

In addition to all the private and corporate research underway here at home and around the world, this fiscal year the Federal Government plans to spend $317 million on AIDS research and $766 million overall. Next year we intend to spend 30 percent more on research: $413 million out of $1 billion overall. Spending on AIDS has been one of the fastest growing parts of the budget, and, ladies and gentlemen, it deserves to be. We're also tearing down the regulatory barriers so as to move AIDS from the pharmaceutical laboratory to the marketplace as quickly as possible. It makes no sense, and in fact it's cruel, to keep the hope of new drugs from dying patients. And I don't blame those who are out marching and protesting to get AIDS drugs released before the I's were—or the T's were crossed and the I's were dotted. I sympathize with them, and we'll supply help and hope as quickly as we can.

Science is clearly capable of breathtaking advances, but it's not capable of miracles. Because of AIDS long incubation period, it'll take years to know if a vaccine works. These tests require time, and this is a problem money cannot overcome. We will not have a vaccine on the market until the mid-to late 1990's, at best. Since we don't have a cure for the disease and we don't have a vaccine against it, the question is how do we deal with it in the meantime. How do we protect the citizens of this nation, and where do we start? For one thing, it's absolutely essential that the American people understand the nature and the extent of the AIDS problem. And it's important that Federal and State Governments do the same.

I recently announced my intention to create a national commission on AIDS because of the consequences of this disease on our society. We need some comprehensive answers. What can we do to defend Americans not infected with the virus? How can we best care for those who are ill and dying? How do we deal with a disease that may swamp our health care system? The commission will help crystallize America's best ideas on how to deal with the AIDS crisis...

Budgets are important, but so too is the bully pulpit. It took Reagan time to get on board, but once he did, his words were sincere, resonant, and compelling. Although AIDS would continue to be a national health problem long after he left office, his 1987 speech must be regarded as a turning point. The HIV/AIDS story drives home a significant point: Putting teeth into federal science and technology policies often requires getting a presidential imprimatur. Without it, policies can languish for years with few tangible outcomes.

George H.W. Bush entered the White House on January 20, 1989 as the 41st president of the United States, having served as Ronald Reagan's vice president for 8 years. In late April, he nominated D. Allan Bromley to be the director of the Office of Science and Technology Policy (OSTP). Bromley, a Yale nuclear physicist, had first met Bush while he was serving on Reagan's White House Science Council, and in the summer of 1987, he had invited the vice president to speak at the dedication of Yale's new 20-million-volt electrostatic accelerator.[30]

It was a somewhat unusual request: It's not often that a university, even a prominent one, invites the vice-president of the United States to the dedication of a physics research facility—let alone with any expectation of an acceptance. But Bush had been a Yalie, and Bromley was a dogged political animal, who exuded confidence and could turn on the charm when he had to. Bush came, and during his brief visit, as Bromley recounts it, "I volunteered to be of as much assistance as I could during his continuing election campaign" for president.

Bromley was able to deliver on his promise 18 months later. While the nation was consumed with the 1988 election, rumors had begun to circulate in the science community that two electrochemists, Stanley Pons and Martin Fleischmann, had made a preliminary observation of a nuclear reaction in a test tube. They were seeking major support from the Department of Energy for their "cold fusion" studies, which, if successful, would transform the world. There would be cheap, abundant, clean energy essentially forever.

By the time Bush had taken his oath of office, cold fusion had become an incredibly hot topic—for science, as well as public policy. Was it real, or was it a hoax? The Bush team needed to know. A lot was riding on it. Billions of dollars were at stake. And Bromley, who had Bush's trust, was in the right place to provide the answer.

From the perspective of nuclear physicists, the Pons-Fleischmann work was almost certainly flawed. But it was not so much a hoax as really bad science. Bromley's scientific pedigree gave him the credibility to render that judgment. He was, quintessentially, the right person at the right time to secure his future in the Bush Administration.

In late April 1989, the president nominated him to be Director of the Office of Science and Technology Policy (OSTP), and on August 4, 1989, the Senate unanimously confirmed him. Bush also conferred on him an additional title that did not require Senate confirmation, but would prove invaluable to his influence within the White House.

Bromley had acquired considerable Washington experience as a member of Reagan's White House Science Council and as president of the American Association for the Advancement of Science, which is based in the nation's capital. He knew that power in the Executive Branch depended on two things: budget authority and proximity to the president. The first would require negotiations with the Office of Management and Budget; the second would hinge on his persuasiveness and his rapport with Bush.

Bromley made his case for presidential access and visibility within the administration, and Bush agreed to his request, naming him Assistant to the President for Science, and giving him a seat at the Cabinet table, as well as a small but highly prized office in the West Wing of the White House. Even with such unusual access and a powerful perch, Bromley could see that he would face major challenges. Ford had worked hard to reinvigorate the science and technology policy apparatus in the White House, after Nixon had eviscerated it.

But Carter had failed to pursue Ford's course with any enthusiasm, and Reagan had ignored it entirely.

By the time Bromley arrived, he found OSTP atrophied, exiled by the former administration to a location away from the White House complex, pilloried by Congress for failing to respond to committee requests, and extraordinarily ill-equipped to tackle any significant problems. It needed to be restructured, relocated, reinvigorated, and expanded to meet the needs of the modern era. Bromley moved swiftly to accomplish those goals, and by the time Bush left office in 1993, OSTP had become the policy hub Congress intended in the 1975 legislation.

In most nations around the world, science and technology activities are concentrated in just a few ministries or departments. In the United States, there are more than twenty major players. Harmonizing science and technology portfolios across federal departments and agencies is a daunting task, one that the 1976 policy legislation recognized when it established the Federal Coordinating Council for Science, Engineering and Technology (FCCSET). But neither Carter nor Reagan empowered FCCSET to carry out the mission Congress had envisioned.

Bromley saw things differently, and with Bush's backing, he reprised FCCSET's original role. To address the interdisciplinary nature of science in the modern era, he created a robust framework for integrating science and technology activities across the federal government at the highest levels. In so doing, he laid the foundation for future administrations, at least until the Trump presidency.

Bromley worked hard to make science a prominent feature in the Bush White House, and found a willing adherent in the president, who frequently attended the meetings of the reconstituted President's Council of Advisors on Science and Technology (PCAST). Bromley was the antithesis of Jay Keyworth and Bill Graham, who had maintained relatively low profiles during the years they had served as Reagan's science advisors. Bromley was assertive and unafraid of breaking new ground. Technology policy and international scientific affairs offered him the opportunity to do so.

Despite his academic background, Bromley had long believed that industry's role in research and development was crucial to American leadership in science and technology, and that his White House predecessors had focused too strongly—almost exclusively, in his eyes—on the "S" (science) and far too little, if at all, on the "T" (technology) in crafting federal S&T policy. He was determined to remedy the imbalance, at the risk of incurring the wrath of Republicans who viscerally saw—and continue to see—technology policy as encroaching on the prerogatives of the free market. He developed a set of guiding principles that gained Bush's support and remain largely accepted today. (More about them in a later discussion of Pasteur's Quadrant in Chapter 12.)

Bromley was a true science globalist, demonstrably exceeding all of his predecessors in that regard. As he wrote in his reminiscences about his time in the

White House, "Science and technology have always been among the most inter-national of human activities, with individuals frequently having closer ties to colleagues on the other side of the world than with those on the other side of the hall. I have long been convinced, however, that we have an enormous amount to gain by internationalizing both science and technology to a much greater degree. During my years in Washington I devoted substantial effort to building bridges between the United States and the rest of the world."

Bromley orchestrated bilateral science agreements with a number of nations during his tenure in the Bush Administration, but more significantly, with Bush's strong backing, he brought a bold proposal to the 1992 ministerial meet-ing of the Organization of Economic Cooperation and Development (OECD) to deal with large science projects. Shaped by his experience with the Supercon-ducting Supercollider (SSC)—the massive and hugely expensive high-energy physics accelerator project in Waxahachie, Texas—Bromley became con-vinced that such major science ventures were becoming too expensive for any one nation to handle on its own. He proposed establishing an international Megascience Forum, under OECD auspices, which would convene meetings of leading scientists from around the world to plan future big projects. Unlike the SSC, which had American fingerprints all over it, and had only sought interna-tional financial support late in the game, approved Megascience projects would have an international imprimatur and an international buy-in from the outset.

Bromley had an exceptionally strong motivation to sell his plan to the OECD ministers at their March 1992 Paris meeting. Two months earlier, he had been in Tokyo with President Bush for a meeting with Japanese Prime Minister Kiichi Miyazawa. Japanese support for the SSC was one of the items on the agenda.

Unlike the Reagan years, when physical science support had benefited from the spillover effects of strong military spending, the early 1990s were marked by severe constraints on discretionary federal budgets. The cost of the SSC had continued to spiral upward, and other projects were competing for scarce dollars in the Energy Department's spending plan. By the end of 1991, it was becoming increasingly clear that the SSC needed a financial lifeline from another country to avoid drowning. The amount was not trivial: at the minimum, a billion dollars was needed. With Europeans pursuing their own high-energy mega-project at the CERN laboratory in Geneva, Switzerland, Japan was the only hope.

The story that unfolded illustrates, once more, the serendipitous nature of science and technology policy. In this case it didn't end well.

As Bush's third year in White House was coming to a close, all SSC eyes were on Bromley. He had a personal bond with the president, and an excellent rapport with John Sununu, a mechanical engineer with a Ph.D. from MIT, who was White House chief of staff when Bromley arrived on the scene in 1989. Despite his sometimes fraught relationship with Richard Darman, the prickly director of the Office of Management and Budget, Bromley had generally nav-igated the West Wing with great success. To get Japan on board with a $1.5

billion commitment to the SSC, the Science Council of Japan had told Bromley in October 1991, all Bush had to do was make a formal request of Miyazawa during the upcoming summit between the two nations.[31] Bromley delivered the message to Sununu, who was a strong supporter of research and had Bromley's back. Sununu promised that the SSC would be high on the priority list for the planned December meeting in Tokyo.

And then, the unraveling began. With Bush's approval rating tanking, Sununu submitted his resignation in early December, and Bush replaced him with Samuel Skinner, who had been serving as Secretary of Transportation. Skinner had little interest in science, and even less in the SSC. His background was law, accounting, and business. Commerce and trade policy were the issues that got his juices flowing.

With Skinner in the driver's seat, the agenda for the summit was thoroughly reworked to focus almost exclusively on trade, especially quotas on auto parts. But the date of the summit slipped from December to January, buying Bromley a little time for securing Bush's agreement to keep the SSC on the table, even if it came at the end of the Tokyo meeting. The president finally gave his approval and invited Bromley to accompany him on Air Force One in anticipation of successfully securing Japan's billion-dollar commitment to the SSC.

Everything seemed in order until the fateful evening of January 8, 1992. Bush's Japanese hosts had arranged a lavish state banquet at the prime minister's residence to celebrate the close bonds between the two countries, and, following protocol, seated the president next to the prime minister on the dais. Bush was in the middle of a 12-day swing through Asia and was feeling the effects of the grueling and stressful trip. This is how Newsweek reported on the unfortunate events that unfolded:[32]

> *[Bush] had traveled through 16 time zones in 10 days and had just been creamed by the Emperor of Japan at tennis. As he stood in the receiving line before a state dinner in Tokyo last week, President Bush had to excuse himself to go into the bathroom and throw up. Most ordinary men would have called it a night and headed for bed. But Bush, ignoring the advice of his doctor, doggedly returned to his duties. Still, the Secret Service was quietly warned that he might not make it through the meal.*

> *He didn't. Between the second course (raw salmon with caviar) and the third (grilled beef with peppery sauce), the president pushed back in his chair and fainted. His chin slumped to his chest, his body reeled to his left, and he vomited onto the pants of his host, Prime Minister Kiichi Miyazawa. Horrified, Barbara Bush leaped to her feet and held a napkin to her husband's mouth, and a Secret Service agent vaulted over the table to catch the president before he tumbled. As Prime Minister Miyazawa cradled the head of his guest, Bush's entourage gently lowered him to the floor. The president's eyes fluttered open, and he quipped to his personal physician, Dr. Burton Lee, "Roll me under the table until the dinner's over." The panicky moment passed; within a few minutes, Bush was on his feet, white as a sheet, but gamely smiling.*

Bush might have been smiling, but the awkward episode left SSC supporters weeping. The president never managed to get to the remaining items on the summit's agenda, and without the commitment of Japanese funds, the accelerator project began a slow slide to its demise. There was a glimmer of hope in 1993 that Japan might opt in, but discussions between the Clinton and Miyazawa administrations failed to materialize,[33] and on June 24 of that year, as the cost of the SSC was on track to exceed $10 billion,[34] the House of Representatives voted 280 to 150 to kill the project.[35] The Senate gave it a reprieve, as it had the previous year, voting 57 to 42 on September 30 to keep the dollars flowing.[36] But less than a month later, after considerable rancor and parliamentary maneuvering, SSC House opponents succeeded in flexing their muscle and eliminated continued SSC funding from the appropriations conference report. On October 26, the House voted 332 to 81 to terminate the SSC, and a day later, the Senate followed suit by a vote of 89 to 11.[37]

Shortly before the House prevailed on the conference report, I received a phone call from Sen. Joseph Lieberman (D-Conn.), a longtime friend, who had been carrying water for Yale high-energy physicists since 1988, when he had won his first Connecticut Senate election. Seeking to avoid an embarrassing negative vote on the SSC, Lieberman offered a plan—which he said had the backing of the Senate leadership—to add $150 million to the high-energy physics annual operating budget if high-energy physicists would cease lobbying for the SSC, which he saw as totally futile at that juncture.

I passed the information on to leaders of the high-energy community, who rejected the proposal, hoping the Senate could still find a way to block the House action. But Lieberman's prediction of how the Senate would vote was right on the mark. In the end, the high-energy physics community lost not only the SSC, but also the consolation prize of an additional $150 million in annual federal research funding.

The Lieberman-SSC episode illustrates a failing among scientists. They often get so caught up in their own world, they cannot see the political landscape shifting. Whether it's arrogance, lack of sophistication, or simply having blinders on, matters little. The takeaway is the same: Knowledge isn't the same as political savvy. To be successful in the policy arena, both are necessary.

Allan Bromley had the good fortune to serve a president with whom he had a personal relationship, someone who respected and trusted him, and consequently supported his efforts to inject new life into an atrophied White House science and technology policy apparatus. Historians lionize Vannevar Bush, who served Franklin Roosevelt during World War II, as the architect of the of the American science and technology policy edifice we see today. His vision was inspirational, but as we have seen, the policy path the nation traveled in the subsequent years had far more bumps and twists than Bush could have imagined.

Bromley might not have been in the same class with Vannevar Bush as a visionary, but his scientific stature, keen political savvy, and tenacious nature produced a policy legacy that has endured for more than a quarter of a century. White House science advisors, who followed him—Jack Gibbons and Neal Lane during the Clinton years, Jack Marburger during the George W. Bush Administration, and John Holdren during Obama's tenure—all benefited tremendously from the policy structures he revived, expanded, or created. The George H.W. Bush-D. Allan Bromley era, even though it lasted only one presidential term,[38] was an appropriate capstone to the two-hundred-year history of science and technology policy that preceded it. It set the stage for Part Two of our narrative: "Science and Technology Policy in the Modern Age."

Part II

Science and technology policies in the modern age

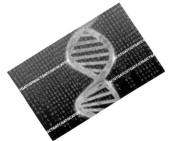

Human Genome
Science nature
February 2001

Apple iPhone
First Release
June 29, 2007

Bromley

Gibbons
Photo by
Scott Bauer,
USDA Agricultural
Research Service

Lane

Marburger
Photo by
Tina Hager,
White House
Photo Office

Holdren

Presidential science advisors

Chapter 9

Crossing new intersections 1992–2000

Bill Clinton's successful 1992 presidential run marked the end of 12 consecutive years of Republican White House control. It also marked the bicentennial of the first American large cent, the coin that carried the motto, Liberty, Parent of Science and Industry. In the intervening 200 years, science protected America's liberty, winning wars, conquering disease, and keeping the nation safe. And for the most part, America's science and technology policy focused on those imperatives.

During the Cold War era, from 1945, when World War II ended, until 1991, when the Soviet Union collapsed, American policymakers saw the physical sciences and engineering as protectors of Western democracy. Physicists, in particular, benefited from Washington's flattering attention. For decades, research dollars flowed freely, driven not only by near-term defense needs, but also by a hedge against the possibility of a hot war whose outcome again would be determined by technological superiority. The way elected officials looked at it, keeping the physics bench filled with all-star researchers was a worthwhile national investment

Scientists are generally not the venal sort, but they are more than willing to receive support when proffered, especially if it comes without any quid pro quos. As federal dollars flowed to universities and high-tech industry, the Cold War era became a golden age for scientific research. It featured discoveries of esoterica, such as neutrinos, quarks, and parity non-conservation, none of which had—and still do not have—any evident practical application, as well as the development of practical devices, such as the transistor, the laser, and large-scale integrated circuits, which revolutionized telecommunications and commerce. It was also the golden age of antibiotic development: What had begun as the serendipitous discovery of penicillin by Alexander Fleming in 1928 became a disease-fighting juggernaut.

The Cold War era was also a belle époque for vertically-integrated technology industries and their central research and development laboratories. Bell

Navigating the Maze. https://doi.org/10.1016/B978-0-12-814710-8.00009-7

Labs in New Jersey, Xerox PARC in California, and IBM in New York were powerhouses of innovation. They benefited tremendously from monopolistic or near-monopolistic holds on their sectors. General Electric, RCA, and Westinghouse might not have had such strangleholds, but they, too, were technological forces to be reckoned with.

America seemingly had it all: robust state support of public universities; generous federal spending on academic research and construction of large national facilities; and industrial laboratories that were unequaled in the world. In America, liberty had truly become the parent of science and industry. War and disease might have been the prime drivers of science and technology policymaking for more than 200 years, but in economic circles, the social return on investment in research was slowly becoming recognized as an extraordinarily beneficial outcome.

Robert M. Solow, a Massachusetts Institute of Technology professor of economics, received the 1987 Nobel Prize in Economic Sciences for "his contributions to the theory of economic growth." Solow had begun his work three decades earlier, and in a series of articles[1-3] published between 1956 and 1960, he made the case that labor and capital investments only accounted for a small contribution to changes in economic growth. According to his models, technical progress, which he defined as improvements in production technology, was principally responsible for economic growth. But outside the economics arena, and especially among science and technology policymakers, Solow's work went largely ignored.

In 1991, another economist made a foray into the arcane science and technology forest, and this time a few people heard a tree fall. Edwin Mansfield, who had been studying the economics of technology[4] since the early 1960s at the University of Pennsylvania, published an analysis[5] that caught the attention of several policy wonks, D. Allan Bromley, George H.W. Bush's science advisor, among them.[6] Mansfield had examined the impact of academic research on innovation, surveying the products and industrial processes of 76 major companies across seven areas of manufacturing during the decade ending in 1985. He found that across sectors, more than a tenth of all *new* products depended *critically* on *recent* academic research. In the case of the pharmaceutical industry, he found that the impact was even greater. There, more than a quarter of new drugs would not have been developed in the absence of research performed in universities.

As striking as those conclusions were, it was Mansfield's macroeconomic findings that had an even more profound impact on Bromley. Although the study was fraught with complexity and suffered from inherent uncertainties, Mansfield's conclusion was profound. The social rate of return—the payback, not just to the innovator, but to society in general—on research conducted in universities, he concluded, was 28%. A year later, after revising his analysis,[7] Mansfield estimated the rate could be as high as 40%, a truly remarkable number.

From a pro-science policy perspective, the timing could not have been worse. Mansfield's modification appeared in June, just as the country was becoming consumed with the 1992 election, and politics, not economic theory, occupied center stage. In such an atmosphere, it's more than likely Bromley was unaware of the significant upward revision.[8] In any case, 6 months later, following Bush's November loss to Bill Clinton, he would find himself headed back to Yale and untethered from daily senior-staff White House meetings.

Bromley was not unique in overlooking Mansfield's 40% revision. To this day, policymakers invariably cite the 28% social rate of return on investment. And most of them neglect to mention one of Mansfield's other significant findings, published in 1980: dollar for dollar spent, basic research has a larger impact on industrial productivity than applied research and development.[9]

Mansfield's 1992 revision was unfortunate in its timing, but it only reinforced his earlier emphasis on the economic benefits of science. And that focus would prove to be a game changer just a few years later.

Had social media been around on November 8, 1989, images of the fall of the Berlin Wall would have gone viral. They made the end of the Cold War a pictorial reality, and they presaged changes in American foreign and domestic policy that could barely have been imagined just a few years before. The breakup of the Soviet Union, which followed in short order, left the United States as the world's only superpower, and ushered in an era in which domestic issues would quickly take precedence over foreign affairs and military interests. It also weakened the rationale for federal support of scientific research, which had been tied to defense needs for half a century.

As the nation turned inward, the flow of federal research dollars to universities and national laboratories was suddenly in danger of being throttled back. Scientists, especially physicists, chemists, and engineers, who had taken their special status for granted, found themselves at risk. Bill Clinton's election and his choice of Jack Gibbons, the director of the congressional Office of Technology Assessment (OTA), as his science advisor did little to allay their fears. Gibbons had both science and science policy credentials, but Clinton had made it clear that his vice-president, Al Gore, was going to be in charge of all things technological.

Albert Arnold Gore, Jr. was no stranger to Washington. He had been a member of Congress for 16 years, serving 8 years in the House and eight in the Senate, before being elected vice-president in 1992. The son of Albert Gore, Sr., an iconic Tennessean who had represented his constituents in Congress for almost three decades, Al Jr. spent his formative years living in the Fairfax Hotel on Embassy Row, as the posh stretch of Massachusetts Avenue is known. A son of privilege, he attended prep school before enrolling at Harvard in 1965. Although he majored in government, he developed a passion for science, math, and philosophy; and after serving in the military, journalism, religion, and law, as well. His wide range of interests could easily have earned him the label 'dilettante,' but a conversation with him would quickly have revealed him more a renaissance man.

Gore's breadth of knowledge and exceptional memory were on display at a meeting of the American Physical Society in Arlington, Virginia in 1991. He had already carved out a reputation in policy circles on two subjects: the dangers of anthropogenic (human-induced) climate change, which gained him popular notoriety, and the impact of technology on American life at the close of the 20th century, which had attracted the attention of wonkish techies. He was scheduled to address about a thousand physicists on the latter subject that April afternoon, but his interest in the former had taken him to the Arctic, where bad weather had interfered with his timely return to Washington. Arriving in Arlington an hour late and having forgotten to bring his prepared speech with him, he gave a 45-minute extemporaneous lecture—using no notes—on the role of quantum theory in 20th century philosophy.[10]

To address a crowd of physicists on such a subject took a lot of self-assurance and more than a small dose of egotism. Gore had both, and they were often on display during his tenure as Clinton's science and technology (S&T) policy guru. Although Clinton generally preserved the S&T policy structures he had inherited from the Bush Administration—actually expanding the scope of FCCSET and renaming it the National Science and Technology Council (NSTC)—he was quite disengaged from them. Unlike Bush, who often attended meetings of the President's Council of Advisors on Science and Technology (PCAST), Clinton never once during his first 2 years in office joined gatherings of the group, which his science advisor, Jack Gibbons, co-chaired. He did so for the first time in the late spring of 1995, and how his attendance at that meeting came to pass once again illustrates the importance of personal contacts in developing science and technology policy.

The story begins with Gore's role as the Administration's point man on anything scientific and his management of the science and technology portfolio. Despite his intellectual interest in the subject, his focus was almost solely on technology and climate change.[11] By the summer of 1994, leaders of the science community had become acutely aware of that tilt, and began to express concern that federal support for research—basic research, especially—was rapidly dropping off the Clinton Administration's radar screen. The apprehension caused a number of prominent scientists to rethink their long-held aversion to advocating for their profession.

After more than 50 years of special treatment during the Cold War, physicists, chemists, and biologists had grown accustomed to seeing lobbying in their own self-interest as demeaning—although high-energy physicists had shown no qualms about making the case for the Superconducting Super Collider (SSC). They viewed such activity as beneath their elite status in American society.

Many scientists went even further. They argued that weighing in on any policy matter, even those that fell outside the realm of self-interest, could damage science's reputation as an impartial arbiter. But rising concerns about research funding began to chip away at those arguments, and by 1994, several organizations had decided to test the lobbying waters.

The Federation of American Societies for Experimental Biology, the Joint Committee for Biomedical Research, and Research!America waded in on behalf of the life sciences, enlisting the support of Rep. John Edward Porter, the new chairman of the House Appropriations Subcommittee on Labor, Health, and Human Services responsible for funding the National Institutes of Health. But as great as the anxiety was that biologists were feeling, it paled by comparison with the worries that began to intrude on the physics community.

Among scientists, physicists had benefited most from the federal largesse during the Cold War era, given their role in nuclear weaponry and other instruments of combat. If they had any doubts about the end of their "chosen" status following the collapse of the Soviet Union, the cancellation of the SSC in 1993 surely dispelled them. In the summer of 1994, the American Physical Society decided it needed to beef up its Washington presence and start to lobby for research. It tapped me for the role.

Several months after I took up residence inside the capital Beltway, Republicans gained control of Congress for the first time in 40 years. It was more than a warning shot across the bow of the Clinton ship of state, and it prompted the president to bring on board several new policy advisors in early 1995. One of them was William E. Curry, Jr., who had lost his bid for governor of Connecticut the previous November. Bill, whom I had gotten to know well through my political work in our home state, was charged with helping Clinton reinvent his domestic policy.

After Bill arrived in January 1995, we began to meet for dinner regularly to talk policy, politics, and gossip, continuing our schmoozing as we browsed the shelves of the Georgetown Barnes and Noble. Two months into the new year, I was sitting in Bill's office in what was then called the Old Executive Office Building (now the Eisenhower Executive Office Building), when he popped a question: What could Clinton do for science? Other than boost funding for research, which we both recognized would be difficult with conservative Republicans now in charge of the House of Representatives. I told him, for starters, Clinton ought to begin to attend PCAST meetings. Bill pulled up the White House calendar on his computer, checked the date of the next PCAST gathering and, as I sat with him, called the president's scheduler requesting her to add it to the president calendar. "It's done," he said, "and I'll put you down as a guest. You know it wouldn't have happened if you hadn't asked."

On the morning of the PCAST meeting weeks later, the White House called to tell me Jack Gibbons had taken me off the list of attendees. No guests would be allowed, according to the protocol he had established. The meeting would take place without me, but with the president attending, and that's all that mattered, I noted mentally.

Late that afternoon, Curry called me. "How did the PCAST meeting go? How did the president do?" he asked

"Jack took me off the list of attendees," I recall saying.

After a few moments of silence, Bill replied, "I guess he didn't know you and I were the ones responsible for getting Clinton there."

"Apparently not," I agreed.

Some months later, I saw Jack at a Washington function, and he told me how happy he was that the president had finally made time to sit down with PCAST. I never let on how it had happened. Jack was a person of extraordinary character whom I counted as a friend, and I was content to allow him to take credit for Clinton's presence at the meeting.

I've decided to recount the story now, only because it so clearly illustrates how much science policy can depend on timing and friendships. Jack died in 2015, but I have a feeling he would agree with my decision to reveal what happened were he still alive today.

By the end of the 1990s the science and technology policy landscape in Washington had changed dramatically. Economic growth had become a prime rationale for research funding; dramatic medical advances had provided a strong argument for more investment in medical research; advocacy groups had proliferated; for the first time in decades, Congress had begun to dig into the science policy weeds; the boundaries between scientific disciplines had begun to blur; Europe and Asia were nipping at the heels of America's science and technology supremacy; and, responding in part to changes in the tax code and in part to the revolution in information technology, industry had virtually abandoned its support of long-term research. Almost overnight, the policy maze had become extraordinarily more difficult to navigate.

About the same time Edwin Mansfield had published his analysis of the rate of social return on investment in basic research, Michael Boskin and Lawrence Lau, both Stanford University economists at the time, had completed their study[12] of the economic growth of nations. Analyzing the role of the three largest drivers of growth in the United States, the United Kingdom, France, West Germany, and Japan, they concluded that technical progress far outpaced capital and labor in spurring economic growth. More specifically, they found that technical progress is by far the most important direct source of economic growth for the industrialized countries in the sample, accounting for more than 50% of the growth in real aggregate output (more than 80% for the European countries).[13]

You might think that science and technology policymakers and advocates for federal support of scientific research would have immediately seized on the result, but it actually took 6 years for the Boskin-Lau report to make a significant splash in Washington. It came following a news conference in the National Press Building on March 4, 1997, at which Allan Bromley, president of the American Physical Society, and Paul Anderson, president of the American Chemical Society, released a "Joint Statement on Scientific Research" on behalf of a coalition of 22 science and engineering societies. The statement, which Bromley and Anderson had discussed earlier that day on C-SPAN's "Washington Journal," read:[14]

As the federal government develops its spending plans for Fiscal Year 1998, we call upon the President and Members of Congress to renew the nation's historical commitment to scientific research and education by providing the requisite funding for the federal agencies charged with these responsibilities. Our call is based upon two fundamental principles that are well accepted by policy makers in both political parties.

- *The federal investment in scientific research is vital to four national goals: our economic competitiveness, our medical health, our national security and our quality of life.*
- *Scientific disciplines are interdependent; therefore, a comprehensive approach to science funding provides the greatest opportunity for reaching these goals.*

We strongly believe that for our nation to meet the challenges of the next century, agencies charged with carrying out scientific research and education require increases in their respective research budgets of 7 percent for Fiscal Year 1998. These agencies include, among others, the NSF, NIH, DOE, DOD, and NASA. The increases we call for strike a balance between the current fiscal pressures and the need to invest in activities that enable long-term economic growth and productivity. Such increases would only partially restore the inflationary losses that most of these agencies suffered during the last few years.

Prudent planning argues for strengthening the respective activities of major research agencies, as already recognized in pending legislation. To constrain still further federal spending on their scientific programs would jeopardize the future well-being of our nation.

Bromley met with reporters after the news conference and stressed several points. With technology underpinning economic growth, as Boskin and Lau had concluded, and with academic research providing the high rate of social return on investment, as Mansfield had found, he argued that federal support of research should be pegged to the gross domestic product (GDP). The proposed increase for the coming fiscal year, he said, should be the first installment of a 10-year plan to double federal research support. That would roughly restore funding of science as a percentage of the GDP to what it had been 30 years earlier. We knew what economic bang research had provided in the intervening years, Bromley maintained, and he was confident it would produce the same kind of return in the future. He also pointed out that industrial support of long-term research had practically vanished—except in the pharmaceutical sector—making the federal role even more important than it had been several decades before.

The Joint Statement quickly garnered champions on Capitol Hill, and by the fall, a chorus calling for doubling federal science support over 10 years had grown to more than one hundred science and engineering societies. On October 22, 1997, a throng of the society leaders and a bevy of reporters packed the Mansfield Room in the Capitol to release a "Unified Statement" containing the proposal Bromley had alluded to the previous March. The advocacy effort, unprecedented for scientists and engineers, had also yielded legislative fruit.

The gathering included three members of the Senate, Phil Gramm, a Texas Republican, Joseph I. Lieberman, a Connecticut Democrat, and Pete V. Domenici, a New Mexico Republican. Domenici wielded considerable clout as chairman of the Senate Energy and Water Development Appropriations Subcommittee, and Gramm and Lieberman were two original co-sponsors of the National Research Investment Act of 1998 (S. 1305), which authorized doubling civilian basic research over a decade,[15] as the "Unified Statement" had proposed. That Domenici would lend his weight to the effort was not surprising, since New Mexico was home to two Department of Energy laboratories, Los Alamos and Sandia. He had been instrumental in securing funding for the human genome project in the 1980s, and for years he had been a prime go-to person for science in the Senate.

At first blush, Gramm and Lieberman would not have been obvious crusaders for scientific research. Gramm, a conservative Democrat turned Republican was a fiscal hawk, an unabashed free marketer with a doctorate in economics. But he had also majored in physics as an undergraduate at the University of Georgia. Lieberman was a moderate Democrat with a Yale undergraduate degree in economics and political science and a Yale law degree. Neither claimed to have professional experience in science. But as we have seen, policymaking can produce strange bedfellows, although generally there's a backstory. The Gramm-Lieberman sponsorship of science certainly had one. It involved personal relationships and key staffers.

Allan Bromley had developed a friendship with Phil Gramm during his years as George H.W. Bush's science advisor. Gramm, prior to being elected to the Senate in 1984, had served in the House of Representatives for three terms. His district was Texas's 6th, home to Waxahachie, the site of the SSC, and it was a given that Gramm would be one of its biggest cheerleaders. The SSC had also been one of Bromley's priorities, and it was natural that he and Gramm would bond over it. Although their bond didn't save the SSC, it proved vital to sustaining Gramm's interest in science and his willingness to promote a 10-year doubling of civilian research in 1997, first with his Republican bill,[16] S. 124, in January 1997, and 9 months later with the bipartisan legislation S. 1305.

Getting a piece of legislation drafted often requires getting the attention of a dedicated staffer. In Gramm's office, that proved to be Mike Champness. An engineer by training, he bird-dogged the issue, and it was no accident the bill called for doubling civilian research, which encompassed engineering, rather than civilian scientific research, which could have excluded it.

The dossier on Lieberman was slightly different. As a Democrat, he was not reflexively opposed to federal spending as much as Gramm. But he was far enough away from the liberal brand that he could strike a deal with a conservative Republican if the time and the issue were ripe for action. The National Research Investment Act provided such an opportunity. Bromley and I divided the responsibilities: Gramm was his; Lieberman was mine.

I first met Joe at a dinner party in Woodbridge, Connecticut in 1971, and we remained political allies for many years. I had helped him secure Democratic support for his successful Attorney General runs in 1982 and 1986, and when he ran for the United States Senate in 1988, I assisted him with his defense policy. Calling on him to consider co-sponsoring S. 1305 wasn't a heavy lift. His legislative director, William Bonvillian, made it even easier because he was an avid supporter of science who would later move on to a private sector position representing the Massachusetts Institute of Technology in Washington.

Personal relationships and legislative staff engagement are important for achieving policy goals, but often they are not sufficient. Having a strong buy-in from the chairmen of the committees of jurisdiction—in this case the three authorization committees, Senate Commerce, Science and Transportation (CST), Energy and Natural Resources and Health, Education, Labor and Pensions (HELP)—can tip the scales in a dramatic way. And the Gramm-Lieberman bill had none of them, although Domenici, who was one of its co-sponsors, was a senior member of the Energy Committee, and as chairman of the Senate Energy and Water Development Appropriations Subcommittee, he controlled the purse strings of the Energy Department's science programs. It's also very likely the link between science and economic growth was still too novel to attract widespread endorsement. And the argument about future impacts was not as compelling as the case for immediate social program needs. In the end, S. 1305 never made it across the finish line, but it would gain renewed traction soon enough.

A month after Gramm-Lieberman met its demise in the late spring of 1998, like the proverbial Phoenix, a bill predicated on many of the same economic rationales arose from the legislative ashes. It contained a few significant sweeteners to attract conservative members, authorizing doubling civilian research funding over twelve rather than 10 years; requiring the president as part of his annual budget request to provide assessments—including possible terminations—of federal research programs, as well as a prioritized list of his requests for research; and directing the White House to commission a National Academy of Sciences study of the research evaluation methodologies.

The new bill,[17] the "Federal Research Investment Act," carrying the label S. 2217, also had the backing of Bill Frist, a Tennessee Republican who chaired the Science, Technology, and Space Subcommittee of the CST, and John D. ("Jay") Rockefeller, a West Virginia Democrat, who was its ranking member. Just as significantly, the bill had the support of Sen. John McCain, the Arizona Republican who chaired the full CST Committee and had assigned one of his staffers, Elizabeth Prostic, the task of shepherding it through the legislative sausage-making process. The changes in leadership and authorization language paid off. The Senate passed S. 2217 by unanimous consent on October 8, 1998, just as the 105th Congress was closing out its second session. There wasn't enough time left for the House to act. But with the unanimous vote in their pocket, the bill's sponsors vowed to reintroduce it once the new Congress convened in January 1999.

The Findings section of the bill put Congress firmly on record in recognizing economic growth as a prime impetus for federal support of science and technology. It also signaled a new reality: other nations were catching up rapidly with American technological superiority, and on a globalized landscape, economic competition was the new normal. The language is concise and worth repeating because its long-term impact was substantial:

SEC. 2. GENERAL FINDINGS REGARDING FEDERAL INVESTMENT IN RESEARCH.

(a) VALUE OF RESEARCH AND DEVELOPMENT—The Congress makes the following findings with respect to the value of research and development to the United States:

(1) Federal investment in research has resulted in the development of technology that saved lives in the United States and around the world.

(2) Research and development investment across all Federal agencies has been effective in creating technology that has enhanced the American quality of life.

(3) The Federal investment in research and development conducted or underwritten by both military and civilian agencies has produced benefits that have been felt in both the private and public sector.

(4) Discoveries across the spectrum of scientific inquiry have the potential to raise the standard of living and the quality of life for all Americans.

(5) Science, engineering, and technology play a critical role in shaping the modern world.

(6) Studies show that about half of all United States post-World War II economic growth is a direct result of technical innovation; and science, engineering, and technology contribute to the creation of new goods and services, new jobs and new capital.

(7) Technical innovation is the principal driving force behind the long-term economic growth and increased standards of living of the world's modern industrial societies. Other nations are well aware of the pivotal role

of science, engineering, and technology, and they are seeking to exploit it wherever possible to advance their own global competitiveness.

(8) Federal programs for investment in research, which lead to technological innovation and result in economic growth, should be structured to address current funding disparities and develop enhanced capability in States and regions that currently underparticipate in the national science and technology enterprise.

True to their word, the sponsors of the "Federal Research Investment Act" reintroduced the bill at the start of the 106th Congress,[18] and the Senate passed it (now with the label, S. 296) on July 26, 1999, again by unanimous consent. There were several major hurdles still to be overcome, and they illustrate how hard it is to implement science and technology policy, especially when money is involved.

Congress carries out most of its work through its committees and subcommittees, and where broad science issues are concerned, there are many that may claim jurisdiction on a single piece of legislation. Bills that are so legislatively challenged can proceed with either sequential referrals—one committee completes its work on the bill before the next one considers it—or concurrent referrals—multiple committees carry out their work on the bill at the same time—neither of which guarantees a smooth or efficient process.

Subtle but significant differences in the House and Senate jurisdictional structures make the legislative maze even more difficult to navigate. The House Science, Space, and Technology Committee (its current name) oversees the operations of the National Institute of Standards and Technology (formerly the National Bureau of Standards), the National Science Foundation (NSF), NASA, the National Oceanographic and Atmospheric Administration (NOAA), and the basic science programs in the Department of Energy (DOE). But it has no responsibility for DOE's management structure or the functions of the National Institutes of Health (NIH) and the Environmental Protection Agency (EPA). Those fall under the jurisdiction of the House Energy and Commerce Committee.

The Senate juggles the authorization and oversight tasks of science agencies just enough to create confusion. It assigns all energy issues to the Energy and Natural Resources Committee, EPA to the Environment and Public Works Committee, NIH to the Health, Education, Labor, and Pensions (HELP) Committee, and three of the remaining "N" alphabet agencies—NASA, NIST, and NOAA—to the Commerce, Science, and Transportation committee. It gives that committee and the HELP Committee joint responsibility for NSF.

In both chambers, defense research falls within the province of the Armed Services Committee, and agricultural research is part of the portfolio of the Agriculture Committee (called Agriculture, Nutrition and Forestry in the Senate). The fragmentation of congressional oversight and policymaking for science and technology mimics the fragmentation of science and technology across all federal agencies or departments: Transportation, State, Justice, Education, Intelligence, Homeland Security and Treasury all have a toehold in the science enterprise, and each falls under the jurisdiction of a different committee or subcommittee.

The balkanization of science within the federal bureaucracy has been a feature of American life for almost a century and a half. It led the Allison Commission[19] to consider establishing a Department of Science in the mid-1880s. And although the commission ultimately rejected the plan, the concept gained new currency in 1995, after Republicans won control of the House of Representatives. Having wandered the halls as the minority for 40 years, they began reshaping the Capitol landscape to match their guiding principles contained in the "Contract with America"—much as the Israelites did when they entered the Land of Canaan after 40 years in the desert.

A Department of Science was one concept that captured the imagination of Newt Gingrich, the new outspoken Speaker of the House and architect of the "Contract." He put it on the plate of the House Science Committee and its chairman, Robert "Bob" Walker of Pennsylvania, a fiscal conservative with a bent for rabble rousing, but, like Gingrich, in many ways a science geek. Walker served one 2-year term as Science Committee chair, and by the time he retired from the House in 1997, a Department of Science was no longer an agenda item. Changing the way the federal government organized the science and technology functions would have been a monumental task, and many policymakers believed the consolidation would carry substantial risks to a system that, for all its occasional inefficiencies and redundancies, actually worked quite well.

In spite of the thrashing about, balkanization of science and technology policy remained—and still remains—a feature of the federal bureaucracy. Following S. 296's initial referral to the House Science Committee on July 27, 1999, it fell victim to the jurisdictional strains. Even if multiple referrals could have been avoided, however, the bill would have had a difficult time in the House, because the legislation authorized new spending without any "offsets"—reductions in spending on other federal programs—a violation of the statutory rules known as "PAYGO," to which House conservatives insisted on strict adherence.

Setting aside the authorization obstacles facing the legislation, the objective of the "Federal Research Investment Act"—increasing government support for research—ultimately would have required Congress to appropriate the funds. And given PAYGO rules, achieving that goal would have been even more difficult than getting agreement on general policies and proposed spending targets. Realizing a budgetary success, history has shown, is not for novices.

However difficult the budgetary maze might have been in 1999, at least there was a road lawmakers and bureaucrats traveled, and there were rules of the road they generally followed. By 2018, the road had been washed out, and the rules, ditched. On March 22 that year, the House of Representatives voted 256 to 167 to pass H.R. 1625. Its first page read:

H. R. 1625

One Hundred Fifteenth Congress
of the
United States of America

AT THE SECOND SESSION

Begun and held at the City of Washington on Wednesday, the third day of January, two thousand and eighteen

An Act

To amend the State Department Basic Authorities Act of 1956 to include severe forms of trafficking in persons within the definition of transnational organized crime for purposes of the rewards program of the Department of State, and for other purposes.

If you proceeded no farther, you would have no idea the bill provided $1.3 trillion dollars in federal discretionary spending for the fiscal year that had begun the on the first day of the previous October.

The bill was actually the "Consolidated Appropriations Act of 2018,"[20] an "Omnibus" spending package 878 pages long. The State Department title, a carry-over from pending legislation, was merely a parliamentary vehicle for getting money out the door before the government would have to shut down. The Senate passed it 65 to 32 a day after the House had acted, and the president signed it into law a few hours later. Everyone was acting under the gun.

Hardly any members of Congress had read all 878 pages of the bill, and there is little doubt President Trump had barely thumbed through it. That was not what Congress had in mind when it passed the "Congressional Budget and Impoundment Control Act of 1974," which prescribed the sausage-making process of funding the federal government and its programs.[21]

The 1974 act stipulated that the federal fiscal year would begin on October 1, rather than July 1, as had been the case until then. But the legislation did much more. It prescribed a complex budgetary process, illustrated by the following chart, which uses fiscal year (FY) 2018 as an example.

Calendar year		Activity	Activity
2015	Oct	OMB FY 2018 presidential budget planning	OMB coordination with OSTP on FY 2018 science budgets
	Nov	OMB FY 2018 presidential budget planning	OMB coordination with OSTP on FY 2018 science budgets
	Dec	OMB FY 2018 presidential budget planning	OMB coordination with OSTP on FY 2018 science budgets
2016	Jan	OMB guidance to agencies for FY 2018 budget requests	
	Feb	Agency FY 2018 budget analysis and preparation	
	Mar	Agency FY 2018 budget analysis and preparation	
	Apr	Agency FY 2018 budget analysis and preparation	
	May	Agency FY 2018 budget analysis and preparation	
	Jun	Agency FY 2018 budget analysis and preparation	
	Jul	Agency FY 2018 requests submitted to OMB	
	Aug	OMB FY 2018 analysis and presidential budget preparation	
	Sep	OMB FY 2018 analysis and presidential budget preparation	
	Oct	OMB FY 2018 analysis and presidential budget preparation	
	Nov	OMB FY 2018 analysis and presidential budget preparation	OMB "pass backs" to agencies of proposed FY 2018 presidential budget (thanksgiving)
	Dec	Agency responses to proposed FY 2018 presidential budget	
2017	Jan	Completion of FY 2018 presidential budget	
	Feb	Release of FY 2018 presidential budget (first Monday in February)	Start of congressional budget hearings
	Mar	Congressional FY 2018 budget hearings	Preparation of congressional budget resolution
	Apr	Passage of congressional 302(a) budget resolution (April 15)	Authorization and appropriations hearings
	May	302(b) allocations of funds to appropriations subcommittees	Authorization and appropriations hearings
	Jun	House appropriations subcommittee "mark-ups" of spending bills	Authorization hearings
	Jul	House passage of appropriations bills	Senate appropriations subcommittee "mark-ups" of spending bills
	Aug	Summer recess	
	Sep	Senate passage of appropriations bills	House-senate reconciliation and final passage of appropriations bills
	Oct	Start of FY 2018 spending (October 1) after presidential signing	

As the timeline illustrates, October 1, the last line in the chart, is the start of a new fiscal year, but the budget process leading up to it actually begins 24 months earlier with budget planning by the Office of Management and Budget (OMB) with science oversight by the Office of Science and Technology Policy (OSTP).[22] Usually, shortly after New Year's Day, OMB, reflecting the priorities of the president, gives federal agencies guidance for preparing their budgetary requests for the fiscal year not yet on the horizon. In the case of the 2018 fiscal year, that would have happened early in 2016. By the end of the summer, agencies and departments submit their budget requests to OMB, and just before Thanksgiving, the budget office provides them with their "pass backs," which reflect how much the president intends to propose for their activities. They have about a month to appeal the White House decisions before the presidential budget plan is set in stone. For the 2018 fiscal year beginning on October 11, 2017, that would have occurred just prior to Christmas in 2016.

According to the 1974 statute, the president is expected to send his budget request to Congress on the first Monday in February. Once Congress receives the "President's Budget," the House and Senate Budget Committees begin work on a resolution that establishes an aggregate bottom line for federal spending known as the 302 (a). It also provides a proposed budgetary breakdown by "function" for all the activities of the federal government. By April 15, after sorting out their differences, the House and Senate are expected to adopt a joint Budget Resolution, setting the stage for the twelve appropriation subcommittees in each chamber to begin their work.

As their first order of business, the subcommittee chairs, known as "cardinals," negotiate how much money each will have at its disposal. Once the haggling ends and the so-called 302 (b) allocations are set, each subcommittee begins the "mark-up" of its spending bill. A positive vote from the subcommittee sends the appropriations bill to the full committee for approval, and eventually to the chamber floor for passage.

Constitutionally, the House must act first and deliver its bill to the Senate for further action. If the Senate amends the bill, as it often does, a conference between appropriators of both chambers takes place to resolve the differences, and both chambers vote the compromise either up or down without further amendments. If the outcome is positive, the bill winds up on the president's desk for his signature. If it is negative, the House begins anew. Once signed, the bill becomes law, and money flows to the federal agencies covered at the beginning of the new fiscal year. But if the president declines to sign the bill, both chambers have the option to override his veto with a two-thirds vote. If they can't, they have to try to craft a new bill that can gain presidential approval.

Each stage of the budgetary process draws different people to the table. Scientists and engineers populate agency advisory committees and put their intellectual heft behind programs and initiatives. Members of the science and technology community and leaders of companies that have skin in the game often testify at congressional hearings, using those opportunities to influence policy and spending decisions.

The President's Council of Advisors on Science and Technology (PCAST) and the White House Office of Science and Technology Policy, if they are operational, supply the president with policy recommendations and budget advice. The National Academy of Sciences, through the National Research Council, periodically assesses disciplinary needs and opportunities, hoping the White House and Capitol Hill will consider its decadal studies as reliable forecasts of the technological future.

Think tanks also throw their weight around, including the Heritage Foundation and the Cato Institute, for example, on the right; the Center for American Progress on the left; the liberal-leaning Brookings Institute and the conservative-leaning American Enterprise Institute in the center; and organizations such as the Bipartisan Policy Center and the Aspen Institute, which try to steer clear of partisan politics. Science and technology advocacy groups and lobbyists of every stripe knock on any available door to plead their case, basing their petitions on whatever issues might gain traction: defense and homeland security, cures for disease, economic growth, education and jobs, environmental challenges, transportation and, with increasing frequency in recent years, innovation, entrepreneurship, global competitiveness, and artificial intelligence (AI).

The buzz is incessant, a cacophony produced by economists pushing their hottest analyses, pollsters peddling their juiciest numbers, academics pontificating on their freshest ideas, scientists trumpeting their coolest discoveries, medical researchers promoting their latest advances, titans of industry and entrepreneurs touting their greatest innovations, military contractors marketing their newest capabilities—in truth, lobbyists all, using every tool of the trade from grass-tops to grass-roots, from op-eds to ads, from marches to money, from assurances to threats, always assuming that outcomes are never certain. It's amazing that elected officials and bureaucrats can keep their cool in the face of the bombardment, and in the end, often—but not always—make smart decisions. But that is the essence of American democracy.

It's worth repeating what Winston Churchill said when he addressed the British House of Commons in 1947:[23]

> *Many forms of Government have been tried, and will be tried in this world of sin and woe. No one pretends that democracy is perfect or all-wise. Indeed it has been said that democracy is the worst form of Government except for all those other forms that have been tried from time to time.*

Nonetheless, the sponsors of the 1974 Budget Act could not have imagined the hyper-partisanship that has plagued Washington since the beginning of the 21st century, the havoc it has wreaked with regular order in Congress, and in the aftermath of the 2016 election with its subsequent chaos, and the threat to American democracy, itself.

In 1999, the House and Senate were still making the effort to pass all twelve discretionary appropriations bills on time. In the end, they managed only a third of them. But, as the succeeding years would prove, 1999 was the pinnacle of success. From 2000 until 2017, Congress failed to pass even one bill by the October 1 deadline. The consequences for scientific research, which appears in nine of those bills—although significantly in only six (Agriculture; Commerce, Justice and Science (NASA, NIST, NOAA, NSF and OSTP); Defense; Energy and Water Development; Interior and the Environment (EPA) and Labor, Health and Human Services (NIH)—have been substantial.

In the absence of regular order—that is, reporting out the individual bills from the appropriations, passing them on the floors of both chambers and sending them to the president's desk for signature—Congress has resorted to two remedies: a mammoth "Omnibus" bill that contains all the spending goodies in one gift box, often, with special interest sweeteners dropped in at the last minute; or a series of "Continuing Resolutions" that maintain spending at the previous year's level, or lower, and typically prohibit any new initiatives. Both legislative prescriptions poison the careful planning and follow through that research and development require. When political practicalities and immediacy trump thoughtful policy, the outcome is far from beneficial.

Economic growth as a rationale for federal support of science retained its cachet, but it never gained enough currency during Clinton's presidency to mimic the successful campaign for doubling funding of biomedical research. The difference is not hard to understand. As the history of science and technology policy demonstrated, mitigating epidemics and curing disease had been winning issues among lawmakers almost from the founding of the republic. Major advances in medicine during the last quarter of the 20th century only strengthened the historical predisposition of Congress to commit funds to those causes. It also did not hurt that members of Congress were living longer and increasingly serving in office well into their dotage. Undoubtedly, many of them saw themselves as prime beneficiaries of National Institutes of Health breakthroughs.

Still, activating congressional support for any policy almost always requires two key elements: Hill champions and public backing. Medical research was no exception, and Research!America, had been at work tackling both, since its founding in 1989. With Mary Wooley at its helm, the organization and its partners took their case to the public and by 1997, they had also found three members of Congress willing to spearhead the doubling effort: Tom Harkin, a Democratic senator from Iowa; Arlen Specter, then a Republican senator from Pennsylvania; and John Porter, a Republican representative from Illinois. The three were no ordinary members. They controlled the NIH purse strings, and during the five-year period from 1998 to 2003, the NIH budget doubled. Finding champions for a cause is important; finding the right champions, even more so.

As the 1990s were winding down, biomedical research was on a roll. Money was one metric, but revolutionary scientific advances were just as indicative. Nowhere was that more evident than the Human Genome Project.[24]

It had gotten off to a slow start in 1985, when Robert Singleton's DNA sequencing proposal elicited a tepid response from the NIH. If the nation's premiere biomedical agency had been the sole Washington player, the project might well have vanished from sight, at least for a time. But science is not the province of a sole federal agency—as it is in a ministry of science in most other countries—and the Department of Energy's (DOE) Office of Health and Environmental Research (later renamed the Office of Biological and Environmental Research) stepped into the breach.

Within two years, funding for the nascent genome project appeared as a line item in Ronald Reagan's presidential request for DOE's fiscal year 1988 budget. But it still needed a Capitol Hill champion, and it found an enthusiastic one in a key Republican senator, Pete V. Domenici from New Mexico. He chaired two influential Senate committees—Budget and Energy and Natural Resources—and shaped by the presence of two national laboratories in his home state, he had an abiding interest in all DOE matters.

As the project worked its way through the appropriations process, NIH got genome religion, and Congress provided planning funds for a joint NIH-DOE effort. By 1990, the genome project had become a reality, and was targeted for completion fifteen years down the road. James Watson, who had shared the Nobel prize with Francis Crick in 1962 for unraveling the mystery of DNA's double helical structure, took the project lead at NIH. And David Galas assumed the reins at DOE.

It was painstaking work, but almost from the outset, two things were clear. If it succeeded, the result would stand as one of the crowning achievements in the annals of science. Cracking the code of human life was the biological equivalent of deciphering the telltale cosmological signs of the big bang that began the universe. The scientific splendor of the Human Genome Project was without question.

But there was more at stake than the grandeur of accomplishment and the fame that accompanies it. There was the possibility of fortune at the end of the genome rainbow, and such a prospect did not escape the attention of the growing ranks of bio-pharmaceutical companies. It also did not escape the attention of NIH director Bernadine Healy, who saw the project as a perfect fit for the Bayh-Dole Act.

Also known as the Government Patent Policy Act of 1980,[25] the goal of the bipartisan legislation was to bridge the chasm between the discoveries resulting from basic research—by that time, mostly carried out by university scientists using federal funds—and the potential commercial applications the discoveries might spawn. Two senators, Indiana Democrat Birch Bayh and Kansas Republican Robert Dole, teamed up to co-sponsor legislation they hoped would give the flagging economy a boost. Until that time, the federal government had retained all rights to the intellectual property of federally sponsored research. The Bayh-Dole act allowed universities and other non-profit institutions, as well as small businesses, to patent any inventions emanating from such research, so long as the government had unfettered access to those inventions at no additional cost. How universities shared the profits of the inventions with its faculty innovators was up to each university. In the ensuing years, especially in the biomedical arena, many top universities reaped major financial benefits from patents and licensing agreements attached to the research outcomes.

Healy judged that the Human Genome Project might generate major profits for NIH researchers, and she put her considerable public weight behind a proposal to allow the project's intellectual property to be patented. She also put herself on a collision course with James Watson, the highly respected but prickly Nobelist with a big reputation and an ego to match, who believed that every outcome of the Human Genome Project should be available to all comers free of charge. But Healy held all the high cards, and Watson found himself fighting a losing battle. Facing allegations of financial impropriety, which he claimed were trumped-up charges designed to force him out, he resigned his position as head of NIH's genome program on April 10, 1992.[26] Of course, Healy disputed Watson's version of the episode, but whatever the truth, Watson was out, and Healy was in.

The following year, NIH became the lead agency, with Francis Collins, a guitar-playing University of Michigan geneticist, recruited to run the operation. Although DOE no longer occupied center stage, its new project director, Ari Patrinos, would be thrust into a crucial policy role seven years later.

Fame and the possibility of the pot of gold at the end of the genome rainbow began to attract a number of international participants—prime among them, the Wellcome Trust of London—and the pace of the ambitious research program picked up. George H.W. Bush's science advisor, D. Allan Bromley, had already recognized that science was on a path toward globalization, and had convinced the Organization for Economic Co-operation and Development (OECD) to create the Megascience Forum in 1992. But in making his case, he had drawn heavily from his experience with big physics facilities, such as the Superconducting Super Collider (SSC). If the Human Genome Project was on his radar screen, he never let on.[27] Although it escaped Bromley's attention in 1992, the global nature of the project would become an undeniable truth by the end of the decade. It would provide a fitting coda: science at the close of the 20th century was a truly international enterprise.

The genome project attracted more than international partners. Its profit-making potential caught the attention of Tony White, a hard-charging, no-nonsense corporate executive, who had taken over the reins of Perkin-Elmer in 1995. By the time White arrived on the scene at its Connecticut headquarters, the analytical and optical instrumentation company, founded in 1937, had become an unfocused conglomerate in almost total disarray. The company's technological image had been badly tarnished in 1990, when it botched the fabrication of the Hubble Space Telescope mirror, nearly crippling the $2-billion NASA instrument. And during the next few years, its Wall Street image fared no better. The price of a share of its stock remained mired in the single digits.

In 1993, Perkin-Elmer had purchased Applied Biosystems (ABI), a California company specializing in sequencing proteins, amino acids, and DNA, but the acquisition had little impact on Perkin-Elmer's bottom line or its stock price. As the outside world saw it, the company was still a lumbering dinosaur. But Tony White saw it differently. He recognized that a biotechnology revolution was on the horizon, and ABI could give Perkin-Elmer an opportunity to become a major player. In short order, he split the company in two, pumping assets into a cutting-edge life sciences piece and sucking resources out of the stodgy instrumentation business he left behind on the cutting room floor. He went on a buying spree, bolstering Perkin-Elmer's life sciences enterprise with a bevy of small biotech acquisitions. One of his purchases would soon spark a titanic battle between the public and private sector for bragging rights to the human genome—and any riches it might return to private investors.

In 1997, White gobbled up PerSeptive Biosystems, a protein analysis company, which a freshly minted MIT Ph.D., Noubar Afeyan, had started seven years earlier. With the acquisition documents not yet finalized, Afeyan, still a relative newbie, made an audacious proposal. Using ABI's technology, he argued, Perkin-Elmer could upstage the Human Genome Project and complete the entire sequencing in just three years, five years ahead of the public research program's timetable.

Riches beckoned, and White jumped at the idea. He had the financial resources and the technology, but he needed a scientist with the knowledge, charisma, drive, and daring to turn Afeyan's wild dream into a reality. Timing is everything, and perhaps the only man who had those attributes had just become available. What transpired would alter the landscape of science and technology policy in ways that no one could have imagined.

Craig Venter was born in Salt Lake City, Utah in 1946 and grew up on the wrong side of the tracks—literally—in the scruffy bay-side neighborhood of Milbrae, California. He had been the quintessential bad boy, a surf bum, a rule breaker, far more interested in girls and fast cars than passing his classes and getting into college. But he was not without talent. He had been a high school swimming champion, and he had proved himself adept at building things. Most of all, he had shown himself to be a risk taker with a huge ego.

Unable to get a student deferment, because he hadn't made it into a senior college, Venter enlisted in the Navy. He soon found himself headed to Vietnam at the age of twenty following a court martial for failing to obey a command. Ever resourceful and always living on the edge of an ethical cliff, he finagled a posting to the field hospital in Da Nang where he worked first in the emergency room, then in the infectious disease clinic, and finally in the relative safety of the operating room. Those medical experiences would prove to be life changing.

When his stint in Vietnam ended, Venter, whose IQ cracked 140, enrolled in San Mateo Community College, and three semesters later, transferred to the

University of California at San Diego (UCSD). Now in a hurry to establish himself in biomedicine, he raced through his Bachelor of Science degree, and physiology and pharmacology Ph.D. in just five years. Skipping the postdoctoral waypoint on the traditional career path of an academic scientist, he landed a junior faculty position at the State University of New York at Buffalo. Despite his rude and crude manner and his outlandish garb, he had no trouble getting grants and building a significant research group. But Buffalo was not in the same league as MIT, Berkeley, or even UCSD. And Venter had set his sights higher. In 1984, he and his second wife—Claire Fraser, his former student-paramour—left western New York state for the NIH campus in Bethesda, Maryland.

There, he got hooked on genomics. He knew nothing about DNA sequencing, but an article in *Genomics* about automated sequencers captured his attention. Venter was not one for bureaucratic process, and he found a way to pay for two of the sequencing machines ABI had developed. By 1989, two of them were up and running in his lab. It was at just the point James Watson was launching the Human Genome Project at NIH.

Never diffident about self-promotion and always willing to take risks, Venter was also a superb salesman. He had little trouble convincing Watson the sequencing machines could give the Human Genome Project a big boost and were worth an investment of $5 million. But neither he nor Watson had counted on the federal government's bureaucratic procedures. A single line in an agency manual could slow down a speeding train to a glacial pace with little anyone could do about it.

Venter, who considered bureaucracy a hindrance to scientific research, nonetheless had to justify his request in a written proposal subject to peer review. He had to behave as if he were an ordinary NIH scientist—which, of course, he was—even though he thought himself superior to everyone else around him, with the exception of Watson, whom he revered. Venter's proposal failed to garner support from the science community on two successive attempts. According to James Shreeve, author of *The Genome War*, Venter's NIH lab manager, Richard McCombie, explained the rejections, this way: "There were two reasons why the grants were rejected. First, we were way ahead of everybody else, and nobody realized it. And second, Craig was an asshole, and everybody realized it."

After that episode, there was little doubt Venter would leave NIH if the right opportunity came along. It happened less than two years later, when HealthCare Investment Corporation offered him an astounding $70 million over seven years to set up a new non-profit basic research institute devoted to genomics. Venter accepted, although the proposition came with a stipulation. The Institute for Genomic Research (TIGR), as Venter named his new laboratory, would be required relinquish the commercial rights associated with any of its discoveries to a new for-profit company, Human Genome Sciences (HGS). As an incentive, HealthCare gave him a ten percent interest in HGS. To keep him honest, it brought in William Haseltine as head of HGS.

The seeds of an eventual parting of the ways were sown at the outset. Haseltine, like Venter, arrived at the joint enterprise with a reputation as a smart, opportunistic scientist. But he also brought with him a business-world notoriety as a cutthroat executive, bent on maximizing profits at almost all costs. To anyone who knew the two men, it was clear that eventually Haseltine, not Venter, would be calling the shots, and Venter would have to either bend or bail.

It took nearly five years for Venter to conclude he needed to exercise the second option and concede he had made a mistake. His reputation as a serious scientist was in jeopardy. Academics were pillorying him for refraining to publish his research findings quickly and openly, and at the same time Haseltine, with his eye on HGS's bottom line, was berating him for releasing too much information publicly too quickly. Venter and Haseltine's unhappy marriage ended in July 1997 when TIGR and HGS officially parted ways.

Venter should have learned a lesson about commercial support of research, but the lure of big bucks for his big scientific genome quests would remain too enticing. Less than 6 months passed before Tony White made him an offer he couldn't refuse. Perkin Elmer intended to establish a new company to sequence the human genome, White told him, using ABI's technology. He wanted Venter to run it and work with Michael Hunkapiller, the brains behind ABI's automatic sequencers, to beat the Human Genome Project to the finish line.

Venter's competitive ego took over. He loved racing his yacht for sailing trophies. Now he would be racing Francis Collins for scientific glory. But having been stung by Haseltine's venality, he wanted to make sure White would not put him in another scientific straitjacket for the sake of corporate profits. The human genome, Venter insisted, had to be made publicly available, and at no cost to anyone who wanted it. White and Hunkapiller went along with his request, but reserved the right to seek patents on a limited number of genes that held promise for profitable medical applications. They struck the deal, and in May 1998, Perkin Elmer established a new company with Venter as president.

Celeritas is the Latin word for speed, and Venter, for an obvious reason, settled on the name Celera for his new enterprise. The Human Genome Project had an eight-year lead. But Venter, the consummate risk-taker, was convinced that by adopting an unconventional—as yet unproven—approach to sequencing, and by taking advantage of Hunkapiller's technical genius and the largest and fastest bank of computers in the private sector, he would sail to victory. It would soon become clear that operating without government funding also tilted the playing field heavily in Celera's direction.

Although Venter had insisted that the final human genome sequence should be in the public domain, he never committed make Celera's intermediate steps publicly available until the project was complete. By contrast, under the terms of the global public collaboration, the Human Genome Project (HGP) partners had agreed to deposit their vetted results every step of the way in an open database, known as GenBank. In short, Celera had access to all of HGP's data, while Collins and his partners had access only to what Venter would allow them.

In the annals of science and technology policy, it's hard to find a comparable example of such a public-private competitive asymmetry.

For HGP and especially NIH, the stakes were enormous. If Celera won the race, the nuances of how it pulled off the victory would be lost on policymakers, and most significantly, congressional appropriators. In the eyes of elected officials, it would feed a simple narrative: the private sector was far more adept at research—even basic research—than the government. The consequence for future NIH budgets and activities was obvious, and Collins understood it well.

Moving forward, there were three possibilities. Celera could form a collaboration with the HGP global partnership for the duration of the project. Celera could compete head to head with NIH, DOE, the Wellcome Trust, and the other HGP partners around the world. Or all of the participants could try to find some middle ground, however elusive it might be.

Collins was in no mood to compromise, and he took every opportunity to denigrate Celera's unconventional "shotgun" genomic approach. But as the leader of the multinational effort, he was in a position to wage more than just a scientific battle. He had the political access Venter lacked. He was prepared to use it at the right time, and Venter knew it.

Collins's boss was the director of NIH, Harold Varmus. A well-respected Nobel Laureate, he wielded plenty of clout of his own. But of even greater importance, he was close to Bill Clinton, having co-chaired Scientists and Engineers for Clinton-Gore during the 1992 presidential campaign. Venter had little doubt the president would weigh in strongly on HGP's behalf if Collins wanted him to.

The same was true across the pond. John Sulston, who was leading the Sangster Center's part of the collaboration supported by the Wellcome Trust in England, was similarly well positioned with British Prime Minister Tony Blair. Venter might have had a congressional card to play, but Collins and Sulston had the White House and 10 Downing Street ready to trump it.

The Department of Energy had much less to lose by collaborating with Celera. Even though it had been the originator of the Human Genome Project, it had long become NIH's poor, almost forgotten cousin. Ari Patrinos, who had been leading DOE's effort for half a dozen years, saw Celera as a promising partner. And in the fall of 1998, less than 6 months into the company's rookie year, he approached Venter with a proposal for a public-private partnership. Venter was interested, but there were several major issues that needed to be resolved. The other HGP partners—particularly NIH and the Sangster Center—would have to agree. And the gap between Celera's profit motive and HGP's open science ethic would have to be bridged.

The debate churned for several months before the proposition foundered. Celera had to abide by the ground rules Tony White had laid down when Perkin Elmer established the company: Celera would have first dibs on any intellectual property the project generated—other than the genome itself. As a result,

Venter's team couldn't submit its step-by-step gene results to the GenBank until White and his team had used their 3-month proprietary time window to file for patents on them. That proviso ultimately proved a bridge too far for the HGP partners to cross. And after months of intrigue and accusations by each side that the other was not acting in good faith, the public and private ventures reverted to what they had been at the outset, adversaries and competitors.

In retrospect, even if Collins and Venter had struck an acceptable agreement, it is hard to see how a key policy issue could have been resolved to the satisfaction of both parties. The private sector needed financial deliverables, and the public sector needed societal benefits. The federal government was spending billions to achieve the latter. Would it be fair to taxpayers if Perkin Elmer and its shareholders became the prime beneficiaries of the government program by snapping up patent rights that flowed from the genome project? Or did PE, as it had renamed itself by then, rightly deserve the spoils because it had shown the moxie to gamble on an unproven scientific approach—even though it was piggybacking on an existing government program?

The decision not to pursue a partnership might have begged those questions. But it raised another one. If Celera had the ability to sequence the human genome, and agreed to make the raw genome publicly available free of charge to anyone who wanted it, why should taxpayers continue to pick up the tab for HGP's research program? That question had gnawed at Collins from the time Venter had entered the fray.

By the time the saga ended a few years later, all three of the questions remained largely unanswered. They remain so, even today. It's possible the human genome saga is a one-off, but given the rapid pace of technological innovation in the 21st century, it's more than likely similar policy conundrums will surface again.

Before it ended, the race for sequencing the human genome produced more than scientific euphoria. It initiated a shockwave that began at the White House on March 14, 2000 and traveled at light speed to Wall Street in New York and "The City" in London, eventually reaching Silicon Valley in California months later. The story bears repeating because it illustrates how important it is for government officials, especially those at the highest levels, to remain in the loop, get their facts straight, and make certain their communications are accurate.

Patrinos's failed attempt to set up a collaboration between HGP and Celera produced several immediate fallouts. Collins revved up the engine of the public project, committing to produce a "genome draft" by 2000. Venter went all in, asserting that Celera would complete the entire sequencing before HGP had published its draft. White's intellectual property team submitted filings for individual genes at an astonishing rate. Each side accused the other of producing flawed science, making impossible promises and engaging in shameful, if not libelous, activities. The unrelenting salvos sent Celera's stock price on a wild ride, soaring upward on every glowing parry by Venter and crashing on every negative riposte by Collins. As the two traded barbs, a number of lawmakers began to wonder whether the government was wasting money on HGP and should simply let Celera finish the work on its own. Others suspected Celera was blowing smoke when it promised to release the completed genome publicly. Otherwise why would investors keep throwing money at the company?

That was the state of affairs in January 2000, when Bill Clinton decided to intervene. He gave his science advisor, Neal F. Lane, a simple command:[28] "Neal, you need to fix this." Lane had replaced Jack Gibbons at the White House in 1998 following a five-year stint as director of the National Science Foundation (NSF).

When Clinton began staffing his administration in the winter of 1992, he made it clear that he wanted the White House to reflect the diversity of America. Lane did not help the roster on the gender or ethnic front, but unlike many past high-level science officials, his pedigree bespoke Oklahoma and Texas, rather than Massachusetts, New York, or California. He was a respected theoretical atomic physicist with experience in university administration, a commitment to civic responsibility, and more than a passing knowledge of Washington. His calm, gentlemanly demeanor stood in stark contrast to the hubbub that swirled around the Clinton White House, especially in its early days. But as NSF director, he was exiled to Virginia.

Unlike the Office of Science and Technology Policy (OSTP), which was located in the Old (Eisenhower) Executive Office Building adjacent to the White House, NSF was situated in the suburb of Arlington during much of Lane's tenure. Being close to the levers of power is important for gaining political advantage, but it's also important for keeping current with major issues. Five miles—the distance from 1600 Pennsylvania Avenue to NSF's headquarters—might not seem far, but during Washington's rush hour, it can be a hellish bumper-to-bumper drive.

As Lane has admitted privately, when he moved back to the District of Columbia as OSTP director and presidential science advisor, he knew almost nothing about the Human Genome Project.[28] He had simply been out of the loop. Opponents of NSF's relocation to Arlington had warned of such a hazard in 1992, when the agency was ordered to leave town to save money.

Fortunately, Lane was a quick study, and by the time the president had given him the "fix this" order, he was up to speed. His plan for ending the genome free-for-all followed a plan he and his British counterpart, Sir Robert May, had discussed a year earlier. It would involve two White House events. In March, on behalf of the entire HGP collaboration, Bill Clinton and Tony Blair would reiterate that the raw sequence of the genome would be available publicly, and would not be subject to patent protection, although individual genes could be. Several months later, Clinton and Blair would announce the completion of the first draft of the human genome, with Collins and, hopefully, Venter both attending.

Lane was trying to paint a smile on the genome baby, but it almost turned into a scowl. After negotiations with Celera had broken down over openness and the

204 PART | II Science and technology policies in the modern age

GenBank issue, word had spread that Collins, through a third party, had begun exploratory discussions with the biopharmaceutical company Incyte, a potential commercial competitor of Celera. Would Incyte consider a partnership with HGP on the final phases of the genome project, subject, of course, to HGP's requirements for openness? That was the gist of the rumor that reached Venter. If it was true, he concluded bitterly, the government was out to destroy him.

Stoicism was not in Venter's DNA, and he had no qualms about using the media to publicize what he saw as an unethical and even illegal assault on Celera. On March 13, *USA Today* broke the Incyte story[29] under the headline, "Feds May Have Tried to Bend Law for Genome Map." It was Collins's turn to be furious. He took Venter's accusations personally. In truth, each side saw the other as the greedy party, and itself as the aggrieved party. But as dicey as the situation was, it was about to become even worse less than twenty-four hours later.

Lane had suggested the National Medals of Science and Technology ceremony, scheduled for March 14, as the venue for the release of the joint statement by Clinton and Blair. The upbeat atmosphere in the White House East Room, he thought, and the ceremony's focus on scientific discovery might temper the nastiness of the genome rhetoric. It was a good idea, and it might well have succeeded, had it not been for a damaging communications gaffe.

Flashback to Chapter 1—Before the president entered the East Room, Lane's senior advisor, Jeffrey M. Smith, handed him one of America's first coins to use as a prop. The 1792 large cent, which bore the inscription, "Liberty, Parent of Science and Industry," would provide a perfect segue to the medal ceremony and the genome statement. The orchestration was perfect, and it's doubtful any members of the audience knew what had happened at a White House press briefing several hours earlier.

It's the job of the White House press secretary to keep the media informed and cast the president in a favorable light. Joe Lockhart, who had been on the job since 1998, was good at both. He was a loyal White House team player, a conscientious worker who made sure he was on top of the issues of the day. He could be funny, but he could also be biting in his criticism of journalists who made the president look bad.

The morning of March 14 is likely one he will never forget. In giving the media a heads up on the genome statement to which Clinton and Blair had agreed, he had given the impression they would be calling for a ban on *all* intellectual property rights associated with the human genome. The implication was that individual genes fell under the same ban, and would have no commercial value. In the East Room, a few hours later, Clinton sought to clarify the matter, giving a big boost to the government-funded international collaboration in the process. His words made it clear that only the raw genome was off limits to patent filings:[30]

Perhaps no science today is more compelling than the effort to decipher the human genome, the string of 3 billion letters that make up our genes. In my lifetime, we'll go

from knowing almost nothing about how our genes work to enlisting genes in the struggle to prevent and cure illness. This will be the scientific breakthrough of the century, perhaps of all time. We have a profound responsibility to ensure that the life-saving benefits of any cutting-edge research are available to all human beings.

Today, we take a major step in that direction by pledging to lead a global effort to make the raw data from DNA sequencing available to scientists everywhere to benefit people everywhere. To this end, I am pleased to announce a groundbreaking agreement between the United States and the United Kingdom, one which I reconfirmed just a few hours ago in a conversation with Prime Minister Blair and one which brings the distinguished British Ambassador here today.

This agreement says in the strongest possible terms our genome, the book in which all human life is written, belongs to every member of the human race. Already the human genome project, funded by the United States and the United Kingdom, requires its grant recipients to make the sequences they discover publicly available within 24 hours. I urge all other nations, scientists, and corporations to adopt this policy and honor its spirit. We must ensure that the profits of human genome research are measured not in dollars but in the betterment of human life. [Applause] *Thank you.*

Already, we can isolate genes that cause Parkinson's disease and some forms of cancer, as well as a genetic variation that seems to protect its carriers from AIDS. Next month the Department of Energy's joint genome project will complete DNA sequences for 3 more chromosomes whose genes play roles in more than 150 diseases, from leukemia to kidney disease to schizophrenia. And those are just the ones we know about.

What we don't know is how these genes affect the process of disease and how they might be used to prevent or to cure it. Right now, we are Benjamin Franklin with electricity and a kite, not Thomas Edison with a usable light bulb.

As we take the next step and use this information to develop therapies and medicines, private companies have a major role. By making the raw data publicly available, companies can promote competition and innovation and spur the pace of scientific advance. They need incentives to throw their top minds into expensive research ahead. They need patent protection for their discoveries and the prospect of marketing them successfully, and it is in the Government's interest to see that they get it.

But as scientists race to decipher our genetic alphabet, we need to think now about the future and see clearly that in science and technology, the future lies in openness. We should recognize that access to the raw data and responsible use of patents and licensing is the most sensible way to build a sustainable market for genetic medicine. Above all, we should recognize that this is a fundamental challenge to our common humanity and that keeping our genetic code accessible is the right thing to do.

We should also remember that, like the Internet, supercomputers, and so many other scientific advances, our ability to read our genetic alphabet grew from decades of research that began with Government funding. Every American has an investment in unlocking the human genome, and all Americans should be proud of their investment in this and other frontiers of science.

But his words did little to calm the stock market. Celera's shares plunged fifteen percent that morning, taking the entire biotech sector down with it. Following the ceremony, Lane and Collins met with reporters in the White House Briefing Room to try to stem the bleeding, but their efforts failed. The Nasdaq was cratering.

The following morning, at the White House senior staff meeting, Lane was the object of some black humor. "I recall" Lane told students at a Rice University symposium in 2018, "that Clinton's economic advisor Gene Sperling said—in jest—'I just want to congratulate Neal Lane for doing what Alan Greenspan (then Chairman of the Federal Reserve) has been trying to do for months." (He meant, of course, deflate a stock bubble.) I don't think I was enjoying the moment that much. In the end, it all worked out. Once investors understood the confusion, the markets recovered."[31]

Well, not quite. The tech-heavy Nasdaq Composite Index hit its high of 5,133 on March 10. Within 30 days, it had lost 20% of its value, and by the end of 4 years, it was down almost 75%.[32] Lockhart's botched media advisory probably did not cause the Nasdaq crash. But the knee-jerk reaction of biotech traders to the misinformation he provided was a clear sign that the tech bubble was about to burst. The dot-com explosion in Silicon Valley, which had propelled the euphoria on Wall Street and in The City, was over.

At the Rice University symposium Lane closed with these words: "My message from this story is that facts and truth matter, but they are not enough. They have to be translated and communicated effectively to be of use. And that process can get muddled, as it did in this case."

He might well have added that in an age when the economy—and for that matter, much of life, in general—is so heavily dependent on science and technology, policymakers must be extremely careful what they say and how they say it. A few words accidentally misplaced or a few actualities unintentionally mischaracterized can wreak havoc.

The genome controversy ended on a positive note a few months later. The rapprochement occurred only after Collins and Venter had clashed at a congressional hearing. Throughout the rancorous rock throwing, DOE's Ari Patrinos, the supporting American actor in the genome drama, had been able to maintain his friendship with both Collins and Venter. And after several private cheese, pizza, and beer meetings at his home, he arranged a truce and a fitting end to the rivalry. Celera and HGP would declare a tie in the rivalry. Venter and Collins would both participate in the June White House event Lane had suggested, and pictures of both of them would appear on the cover of *Time* and the front page of *The New York Times*. They would also publish their scientific results at the same time.

On June 26, 2000, President Clinton again spoke from the podium in the East Room, beginning his remarks with these words of gratitude:[33]

> *Good morning. I want to, first of all, acknowledge Prime Minister Blair, who will join us by satellite in just a moment, from London. I want to welcome here the Ambassadors from the United Kingdom, Japan, Germany, France. And I'd also like to acknowledge the contributions not only that their scientists, but also scientists from China, made to the vast international consortium that is the Human Genome Project.*
>
> *I thank Secretary Shalala, who could not be here today; and Secretary Richardson, who is here. Dr. Ruth Kirschstein, Dr. Ari Patrinos, scientists of the Department of Health and Human Services and the Department of Energy, who have played an important role in the Human Genome Project.*
>
> *I want to say a special word of thanks to my Science Advisor, Dr. Neal Lane. And of course, to Dr. Francis Collins, the director of the International Human Genome Project; and to the Celera president, Craig Venter. I thank Senator Harkin and Senator Sarbanes for being here, and the other distinguished guests...*

Tony Blair added these thoughts:

> *...Scientists from Japan and Germany, France, China, and around the world have been involved, as well as the United Kingdom and the United States. And this undertaking, therefore, has brought together the public, private and non-profit sectors in an unprecedented international partnership. In particular, I would like to single out the Wellcome Trust, without whose vision and foresight, Britain's 30-percent contribution to the overall result would not have been possible. And I would like, too, to mention the imaginative work of Celera and Dr. Craig Venter, who in the best spirit of scientific competition, has helped accelerate today's achievement.*
>
> *For let us be in no doubt about what we are witnessing today—a revolution in medical science whose implications far surpass even the discovery of antibiotics, the first great technological triumph of the 21st century. And every so often in the history of human endeavor there comes a breakthrough that takes humankind across a frontier and into a new era. And like President Clinton, I believe that today's announcement is such a breakthrough—a breakthrough that opens the way for massive advances in the treatment of cancer and hereditary diseases, and that is only the beginning.*

The well-orchestrated event brought some tranquility to the human genome dispute. A week later, following Lane's script, *Time's* cover[34] featured a picture of Venter and Collins with the headline "Cracking the Code: The Inside Story of How these Bitter Rivals Mapped Our DNA, Life Historic Feat that Challenges

Medicine Forever," and in mid-February 2001, *Nature* published HGP's results[35] while *Science* published Celera's.[36]

But the final chapter of the saga was not written until more than a decade later. It fell to the American Civil Liberties Union and its science adviser, Tania Simoncelli, to close the circle. In 2009, the ACLU filed suit against Myriad Genetics, a Utah-based company, which had exercised its right to patent individual genes, in this case the rights to the BRCA genes associated with breast cancer. The Association for Molecular Pathology was the plaintiff, but Simoncelli was the protagonist. She argued compellingly that genetic patent protections harmed patients financially, stifled medical research, and damaged healthcare.

The entire biopharmaceutical industry watched as the case made its way through the federal court system. It reached the United States Supreme Court in the fall of 2012, and on June 13, 2013 the justices ruled unanimously that naturally occurring genes could not be patented.[37] Unlike the stock market meltdown that occurred on March 14, 2000, Wall Street barely hiccupped. But *Nature* took notice, naming Simoncelli one of "Nature's 10 People Who Mattered in 2013."[38]

The Executive and Legislative Branches might have been the battlegrounds of the genome war. But in the end, the Judicial Branch provided the final word in one of the most significant science and technology policy chapters in American history.

Craig Venter's character was on full display during his successful quest for the human genome: brash, boastful, smart, focused, and capable. He accepted that controversy came with the turf, and he wasn't shy about expressing his opinions. He hasn't changed much since.

In 2004, Venter and Daniel Cohen, the principal scientist at the Parisian company GENSET, made a bold forecast. "If the 20th century was the century of physics," they wrote,[39] "the 21st century will be the century of biology." It hasn't quite worked out that way.

Biotech has blossomed, but the physical sciences, computation, engineering, and mathematics haven't been slouches. Physicists discovered the Higgs boson, and together with astronomers and computer scientists, they confirmed the existence of the gravitational waves Einstein had predicted a century earlier. Streaming videos, smart phones, tablets, LED lighting, autonomous vehicles, robots, drones, high-speed trading, and augmented reality have come to characterize life in the 21st century. Even in medicine, biology hasn't been flying solo.

Harold Varmus, NIH's director from 1993 until 1999, was on target in his 2000 *Washington Post* essay, "Squeeze on Science," referenced previously. Commenting on the disparity in federal support for the sciences, he opined:[40]

... The NIH does a magnificent job, but it does not hold all the keys to success. The work of several science agencies is required for advances in medical science...

But Congress is not addressing with sufficient vigor the compelling needs of the other science agencies, especially the National Science Foundation and the Office of Science at the Department of Energy. This disparity in treatment undermines the balance of the sciences that is essential to progress in all spheres, including medicine.

I first observed the interdependence of the sciences as a boy when my father—a general practitioner with an office connected to our house—showed me an X-ray. I marveled at a technology that could reveal the bones of his patients or the guts of our pets. And I learned that it was something that doctors, no matter how expert with a stethoscope or suture, wouldn't have been likely to develop on their own.

Effective medicines are among the most prominent products of medical research, and drug development also relies heavily on contributions from a variety of sciences. The traditional method of random prospecting for a few promising chemicals has been supplemented and even superseded by more rational methods based on molecular structures, computer-based images and chemical theory. Synthesis of promising compounds is guided by new chemical methods that can generate either pure preparations of a single molecule or collections of literally millions of subtle variants. To exploit these new possibilities fully, we need strength in many disciplines, not just pharmacology.

… Perhaps the last century's greatest advance in diagnosis, MRI is the product of atomic, nuclear and high-energy physics, quantum chemistry, computer science, cryogenics, solid state physics and applied medicine.

In other words, the various sciences together constitute the vanguard of medical research…

Although the op-ed's focus was medicine, it captured the essence of science and technology in the 21st century.

If anything, the sciences are more interdependent today than they were when Varmus addressed the matter in 2000. Some of the institutions that govern science and technology have recognized the new reality and taken steps to address it, but some have either turned a blind eye or have been unable to surmount structural impediments. Nine months before the Washington Post published Varmus's essay, and 2 months before Joe Lockhart's media release unintentionally tanked the Nasdaq, Bill Clinton delivered a speech about science at the California Institute of Technology. In it, he highlighted the value of science and technology to the nation; stressed the importance of the federal role in sustaining it over the long term; called attention to the interdependence of different fields of research; and, underscoring the last point, introduced one of his signature policy achievements—the National Nanotechnology Initiative. The following excerpt summarizes those four key points:[41]

…As all of you know, Albert Einstein spent a lot of time here at Caltech in the 1930's.[42] And 3 weeks ago, Time magazine crowned him the Person of the Century. The fact that he won this honor over people like Franklin Roosevelt and

Mohandas Gandhi is not only an incredible testament to the quantum leaps in knowledge that he achieved for all humanity but also for the 20th century's earth-shaking advances in science and technology.

Just as an aside, I'd like to say because we're here at Caltech, Einstein's contributions remind us of how greatly American science and technology and, therefore, American society have benefited and continue to benefit from the extraordinary gifts of scientists and engineers who are born in other countries, and we should continue to welcome them to our shores.

But the reason so many of you live, work, and study here is that there are so many more questions yet to be answered: How does the brain actually produce the phenomenon of consciousness? How do we translate insights from neuroscience into more productive learning environments for all our children? Why do we age— the question that I ponder more and more these days. [Laughter] I looked at a picture of myself when I was inaugurated the first time the other day, and it scared me to death. [Laughter] And so I wonder, is this preprogrammed, or wear and tear? Are we alone in the universe? What causes gamma ray bursts? What makes up the missing mass of the universe? What's in those black holes, anyway? And maybe the biggest question of all: How in the wide world can you add $3 billion in market capitalization simply by adding ".com" to the end of a name? [Laughter]

You will find the answers to the serious questions I posed and to many others. It was this brilliant Caltech community that first located genes on chromosomes and unlocked the secrets of chemical bonds and quarks. You were the propulsive force behind jet flight and built America's first satellites. You made it possible for us to manufacture microchips of ever-increasing complexity and gave us our first guided tour on the surface of Mars. With your new gravitational wave observatory, you will open an entirely new window on the mysteries of the universe, observing the propagating ripples which Einstein predicted 84 years ago.

Today I came here to thank you for all you're doing to advance the march of human knowledge and to announce what we intend to do to accelerate that march by greatly increasing our national investments in science and technology.

The budget I will submit to Congress in just a few days will include a $2.8 billion increase in our 21st century research fund. This will support a $1 billion increase in biomedical research for the National Institutes of Health; $675 million, which is double the previous largest dollar increase for the National Science Foundation in its entire 50-year history; and major funding increases in areas from information technology to space exploration to the development of cleaner sources of energy.

This budget makes research at our Nation's universities a top priority, with an increase in funding of more than $1 billion. University-based research provides the kind of fundamental insights that are most important in any new technology or treatment. It helps to produce the next generation of scientists, engineers, entrepreneurs. And we intend to give university based research a major lift.

The budget supports increases not only in biomedical research but also in all scientific and engineering fields. As you know, advances in one field are often dependent on breakthroughs in other disciplines. For example, advances in computer science are helping us to develop drugs more rapidly and to move from sequencing the human genome to better understanding the functions of individual genes.

My budget supports a major new national nanotechnology initiative worth $500 million. Caltech is no stranger to the idea of nanotechnology, the ability to manipulate matter at the atomic and molecular level. Over 40 years ago, Caltech's own Richard Symonds asked, "What would happen if we could arrange the atoms one by one the way we want them?" Well, you can see one example of this in this sign behind me, that Dr. Lane furnished for Caltech to hang as the backdrop for this speech. It's the Western hemisphere in gold atoms. But I think you will find more enduring uses for nanotechnology.

Just imagine, materials with 10 times the strength of steel and only a fraction of the weight; shrinking all the information at the Library of Congress into a device the size of a sugar cube; detecting cancerous tumors that are only a few cells in size. Some of these research goals will take 20 or more years to achieve. But that is why—precisely why—as Dr. Baltimore said, there is such a critical role for the Federal Government...

President Clinton and his science advisor, Neal Lane, had clearly understood the nature of science and technology at the close of the 20th century—their centrality in American life and the degree to which the various threads of the science and technology fabric are interwoven. One of Clinton's last acts before he left office was signing legislation[43] on December 29, 2000 that established the National Institute of Biomedical Imaging and Bioengineering (NIBIB) at NIH. It integrated the physical sciences into NIH's portfolio, formally validating Varmus's proposition.

It's worth a brief digression to reflect on how that legislation materialized. The story illustrates the role that turf and connections play in the policy arena. Two years before the bill reached Clinton's desk, the biophysics community had successfully enlisted the support of several key members of Congress[44] to promote the NIBIB concept. But the legislative effort had stalled because the White House was not backing it.

I had been working in Washington for four years, when the biophysicists[45] approached me for advice. In that short time, I had learned enough to suspect that turf more than policy was the problem. The biophysicists had begun their discussions on Capitol Hill before they checked the NIH box. Varmus, who was still the NIH director, could have been an ally. Instead, he became an obstacle—a least temporarily.

Most federal administrators abhor an end run around them. And I felt confident that Varmus, however brilliant and renowned, was not an exception. The rationale for an NIBIB seemed extremely strong, and the solution seemed extremely straightforward. Clinton's Office of Science and Technology Policy (OSTP) needed to make it an Administration initiative, and allow NIH to take credit for it. A phone call to a science colleague at OSTP and a one-page description of the proposal began the process.[46] A year later, NIBIB became a reality, although by that time Varmus had moved on to become president and CEO of Memorial Sloan Kettering Hospital in New York.

_____

Shortly after 10:00 o'clock in the morning on April 2, 2013, Francis Collins, now the director of NIH, walked up to the lectern in the White House East Room and introduced Barack Obama to the expectant crowd that had gathered. Obama, whom many long-time observers of the Washington scene have called the "Science President," was about to announce a bold new inter-agency, inter-disciplinary, public-private partnership that kicked Varmus's characterization of 21st century science up a notch. With the usual twinkle in his eyes and his impeccable delivery, the president began to speak:[47]

Thank you so much. Thank you, everybody. Please have a seat. Well, first of all, let me thank Dr. Collins not just for the introduction, but for his incredible leadership at NIH. Those of you who know Francis also know that he's quite a gifted singer and musician. So I was asking whether he was going to be willing to sing the intro-duction—[laughter]—and he declined.

But his leadership has been extraordinary. And I'm glad I've been promoted Sci-entist in Chief. [Laughter] Given my grades in physics, I'm not sure it's deserving, but I hold science in proper esteem, so maybe that gives me a little credit.

Today I've invited some of the smartest people in the country, some of the most imaginative and effective researchers in the country—some very smart peo-ple—to talk about the challenge that I issued in my State of the Union Address: to grow our economy, to create new jobs, to reignite a rising, thriving middle class by investing in one of our core strengths, and that's American innovation.

Ideas are what power our economy. It's what sets us apart. It's what America has been all about. We have been a nation of dreamers and risk takers, people who see what nobody else sees, sooner than anybody else sees it. We do innovation better than anybody else, and that makes our economy stronger. When we invest in the best ideas before anybody else does, our businesses and our workers can make the best products and deliver the best services before anybody else…

…[T]he investments don't always pay off. But when they do, they change our lives in ways that we could never have imagined. Computer chips and GPS technology, the Internet—all these things grew out of Government investments in basic research. And sometimes, in fact, some of the best products and services spin off completely from unintended research that nobody expected to have certain applications. Businesses then use that technology to create countless new jobs.

So the founders of Google got their early support from the National Science Foun-dation. The Apollo project that put a man on the Moon also gave us, eventually,

CAT scans. And every dollar we spent to map the human genome has returned $140 to our economy—$1 of investment, $140 in return. Dr. Collins helped lead that genome effort, and that's why we thought it was appropriate to have him here to announce the next great American project, and that's what we're calling the BRAIN Initiative.

As humans, we can identify galaxies light years away, we can study particles smaller than an atom. But we still haven't unlocked the mystery of the three pounds of matter that sits between our ears. [Laughter] But today, scientists possess the capability to study individual neurons and figure out the main functions of certain areas of the brain. But a human brain contains almost 100 billion neurons making trillions of connections. So Dr. Collins says it's like listening to the strings section and trying to figure out what the whole orchestra sounds like. So as a result, we're still unable to cure diseases like Alzheimer's or autism or fully reverse the effects of a stroke. And the most powerful computer in the world isn't nearly as intuitive as the one we're born with.

So there is this enormous mystery waiting to be unlocked, and the BRAIN Initiative will change that by giving scientists the tools they need to get a dynamic picture of the brain in action and better understand how we think and how we learn and how we remember. And that knowledge could be—will be—transformative.

In the budget I will send to Congress next week, I will propose a significant investment by the National Institutes of Health, DARPA, and the National Science Foundation to help get this project off the ground. I'm directing my bioethics commission to make sure all of the research is being done in a responsible way. And we're also partnering with the private sector, including leading companies and foundations and research institutions, to tap the Nation's brightest minds to help us reach our goal…

The BRAIN (Brain Research through Advancing Innovative Neurotechnologies) Initiative was unique in many ways. Developed and led by Philip Rubin and Tom Kalil at the Office of Science and Technology Policy, it was a grand scientific challenge that mined the resources and expertise of multiple federal agencies, most prominently the Defense Advanced Projects Agency (DARPA), NIH and NSF. But even more significantly, it tapped into the rapidly growing world of science philanthropy. It brought on board at the very outset the Allen Institute for Brain Science, the Howard Hughes Medical Institute, and the Kavli Foundation. On a grander scale, it cast a strong spotlight on the central role technology can play in addressing complex scientific problems.

With a few succinct sentences, the president also used the opportunity to underscore the enduring impact of federally funded scientific research and the serendipitous outcomes that characterize basic research. The audience that morning comprised mostly scientists, Washington insiders, banks of TV cameras, and a battery of reporters. Despite the coverage and the promise the BRAIN Initiative held, it's likely the announcement created little more than a transitory blip on the general public's radar screen. It's the reality promoters of science have come to expect: Most of the public loves the innovations science delivers, but has little interest in what led to them and what is necessary to sustain the delivery pipeline.[48]

The Executive Branch of the federal government is far better equipped to orchestrate cross-disciplinary initiatives than the Legislative Branch. As the Nanotechnology and BRAIN initiatives demonstrate, a well-functioning OSTP can use interagency mechanisms, such as the President's Council of Advisors on Science and Technology (PCAST) and the National Science and Technology Council (NSTC), to plan and execute such efforts. And a president who fully appreciates the potentialities of PCAST and NSTC can have an outsized impact on America's science and technology enterprise. Barack Obama was a rare White House occupant who understood that instinctively. He met with his council of advisers regularly and used the NSTC levers effectively, creating an enviable science legacy.

Congress has no comparable capabilities. Biomedicine and the physical sciences fall under separate authorizing committees and separate appropriations subcommittees. Likewise, different committees and subcommittees have responsibility for science-related energy, environment, homeland-security, and defense matters.

One story illustrates an unintended consequence of the jurisdictional division of responsibilities among congressional committees. The narrative involves one of the Department of Energy's (DOE) premier research facilities located on the campus of the Stanford Linear Accelerator Center, now known as SLAC National Laboratory. Originally built for high-energy physics research, SPEAR, the acronym for Stanford Positron Electron Accelerating Ring, was one of the sites of the "November 1974 Revolution"—the discovery of the charm quark. For leading the SPEAR research effort, Burton Richter received the Nobel Prize in Physics in 1976,[49] sharing it with Samuel Ting of MIT, who had carried out complementary work at Brookhaven National Laboratory. That same year, Martin Perl, who received a Nobel Prize in 1995[50] for his work at SPEAR, led a research team to make another dramatic discovery: the tau lepton, a fundamental particle similar to the electron but about 3,500 times more massive.

Almost from the beginning, scientists outside the high-energy community had recognized that the "synchrotron radiation," which SPEAR generated as an unwanted by-product of its operation, could be harnessed to study matter at the atomic scale. For chemists, material scientists, condensed matter physicists, and biologists, it was a boon. They piggy-backed their research on the high-energy physics operation, using the intense X-ray light to study everything from semiconductor surfaces to protein molecules with a precision that had previously been unthinkable.

By 1990, high-energy physicists had milked SPEAR for all they could, and SLAC converted SPEAR into a dedicated synchrotron X-ray light source. But repurposing an old accelerator designed for a different use is not the same as building a new one from scratch and optimizing it for the new objective. It was clear that SPEAR soon would be unable to compete with a new generation of machines that were specifically designed to produce the intense X-ray beams scientists needed. One had actually been operating since 1983,[51] and two new ones[52] were scheduled to join the club within four years. The Stanford Synchrotron Radiation Laboratory (SSRL), as the research center was officially known, would either have to build a new accelerator or close its doors.

Arthur Bienenstock had been SSRL's director for nearly twenty years when he left Stanford to become OSTP's Associate Director for Science in 1998. He began to hear complaints from NIH, which was supporting biomedical research at SSRL, that its programs were subject to the whims of the DOE, which funded SLAC, and the National Science Foundation, which shared responsibility for supporting SSRL. NIH demanded to have a voice in SPEAR operations and any plans for its future. Officials at DOE were equally vocal. They complained that a third of the users of the synchrotron facility received support from NIH for their research projects, but paid little, if any, of SPEAR's operating costs. They also noted that the DOE's science budget was stagnant, while NIH's was on a course to double over the next five years. In short, DOE was poor, and NIH was rich.

Bienenstock did what a smart policymaker should do. Instead of relying on his own instincts and knowledge—which was considerable—he set up an interagency committee to answer two questions. First, was a new machine warranted? And second, if so, should DOE fly solo, following past project practices, or should it break new ground by pursuing a joint effort with NIH? The memorandum of understanding signed on May 27, 1999 by Harold Varmus, on behalf of NIH, and Martha Krebs, director of the DOE Office of Science, on behalf of the Energy Department, contained the answers to both questions. Upgrade the facility, and pursue it as a joint DOE-NIH project.

Keith Hodgson, who had been named SSRL's director after Bienenstock left, underscored the novelty of the cooperative approach, writing at the time[53] in *Synchrotron Radiation News*, "This is the first time an outside agency has agreed to assist the funding of a major DOE basic science research facility...." He further noted that the arrangement was "sensible," because structural biologists, by that time, had grown to about half of the 1,600 researchers using the SSRL facility. The approach exemplified the interdependence of the sciences.

Less than four years later, on March 8, 2004, SPEAR3, as the new facility was called, delivered its first new beam of intense X-ray light. The policies Bienenstock had set in motion had worked out, and the marriage of the physical and biological sciences had been accomplished—although not without a glitch on Capitol Hill, which, might have been anticipated with even less than perfect hindsight.

The joint project, budgeted at a total of $53 million, needed two separate appropriations subcommittees in each congressional chamber to act on the proposed spending plan. The House Labor, Health, and Human Services

Subcommittee, which had responsibility for NIH, did its part, passing a funding bill that was in line with the project's budget profile. But when the Energy and Water Development Subcommittee, which had purview over DOE, "marked up"[54] its funding measure, the SPEAR3 upgrade was absent. In the "stove-piped"[55] world of divided congressional responsibilities, the Energy and Water Subcommittee, it seemed, had been unaware of the Labor and Health Subcommittee's action.

The error was rectified before the appropriations bill reached the House floor for a final vote. And in the end, no damage was done. But the hiccup highlights what can happen when legislative machinery fails to keep up with science and technology's changing complexion.

Congress has more than a machinery problem. In an era when science and technology intersect nearly every facet of American life, members of Congress have little personal experience to draw on. If you walked the corridors of the House and Senate office building over the course of the last half century, the odds are you never encountered a single member of either chamber who had an advanced degree in engineering or the natural sciences. Since 1950, there have been only six House members with doctorates in any of those fields: three physicists,[56] one chemist,[57] one mathematician,[58] and one electrical engineer.[59] The Senate lays claim to only one, a geologist and former astronaut.[60]

In 1973, the American Association for the Advancement of Science (AAAS) initiated a program to help address the paucity of scientific and technological expertise on Capitol Hill. If scientists and engineers were not running for office and serving in Congress, AAAS hoped the community could at least provide technical assistance to House and Senate offices and committees. The AAAS Congressional Science and Engineering Fellowship program[61] was a bold experiment. Would professional scientists and engineers be willing to give up a year of their careers to spend 12 months toiling away in a warren of small cubicles to help inform the legislative process? Would members of Congress and their staff welcome scientists and engineers into their fold? And finally, would AAAS fellows be able to adapt to a culture and an environment that differed dramatically from the ones to which they were accustomed?

Science demands precision, perseverance, and evidence-based objectivity. Congress, when it is functioning properly, depends on compromise, timeliness, and political exigency. When AAAS began the fellowship program, it was not obvious the gap between the two sets of principles could be bridged. But with more than four decades behind it, the AAAS experiment, which drew the financial support of many scholarly science and engineering societies, has proved to be an extraordinary policy winner. Over the course of that period, the program sponsored more than a thousand fellows, each on a one-year grant. At the conclusion of their stint, some of the fellows returned to their home institutions; some of them elected to remain on Capitol Hill, becoming regular staff members; some of them took positions in federal agencies; and one of them—Rush Holt—took the political plunge, running for Congress and serving for a dozen years.

For much of the AAAS program's history, science and technology policy remained above the political fray, although there were a few occasions when partisanship intervened. Reagan's Strategic Defense Initiative (SDI) or "Star Wars," as liberals derisively called it, and stem cell research, which religious conservatives, mostly Republican, opposed, are two examples. But, other than disagreements over where federal responsibility for research ends and private responsibility begins, Republicans and Democrats tended to work together, especially on big picture items.

Chapter 10

Years of anxiety 2001–2008

As science and technology's impact on economic growth became a more widely accepted paradigm, lawmakers began to focus on the relationship between federal support of research and America's ability to compete globally. The issue was not really new. Economic challenges from Japan had captured headlines during the Reagan era and spurred the creation of a commission to propose an American response. Its report led the private sector to create the Council on Competitiveness[1] in 1986, but its impact on the public sector produced little more than a yawn. It would take two decades for lawmakers to enact legislation addressing the assessments and recommendations contained in the 1985 report produced by the President's Commission on Industrial Competitiveness—once again proving that science and technology policy often moves at a glacial pace, unless wars or epidemics intervene.

Chaired by Hewlett-Packard's CEO John Young, the commission defined competitiveness as "the degree to which [a nation] can, under free and fair market conditions, produce goods and services that meet the test of international markets while simultaneously expanding the real income of its citizens." After examining America's standing according to four standard indicators—labor productivity, real wage growth, real returns on capital employed in industry, and position in world trade—it found U.S. competitiveness declining in relation to its major trading partners on all four counts. Its commentary on technology is worth noting, because it both reprises some old propositions—a Department of Science, for example—and emphasizes issues—a permanent research and development tax credit—that would remain on the policy agenda for several decades. The following summary is taken from a memorandum regarding a March 29, 1985 Senate Finance Committee hearing on the Industrial Competitiveness Report:[2]

Assessment

In order for technology to be a continuing and greater competitive advantage, the U.S. must

- Create a solid foundation of science and technology that is relevant to commercial uses;

Navigating the Maze. https://doi.org/10.1016/B978-0-12-814710-8.00010-3

- Apply advances in knowledge to commercial products and processes; and
- Protect intellectual property by strengthening patent, copyright, trademark and trade secret protection.

Although the U.S. spends a greater share of its GNP on research and development (R&D) than its international competitors, much of the R&D is for defense and space programs in which commercial application is an incidental objective. Furthermore, government needs better management of R&D funds. In any event, private R&D incentives are needed to fuel advances in commercially useful new technologies. Reversing inadequate support for university research is the starting point. Greater attention to manufacturing technology is essential to translating new product technologies into commercial success. Finally, greater protection must be given intellectual property to enhance incentives for investments in innovation.

Recommendations

- Create a Cabinet-level Department of Science and Technology to coordinate and integrate fragmented government efforts and highlight the importance of science and technology.
- Make the R&D tax credit permanent and make it available for total R&D spending (instead of just incremental spending), for accounting expenses, and for development of equipment and processes involved in prototype development.
- Increase and manage better government support for basic research at universities.
- Improve manufacturing technology and manufacturing-related university curriculum.
- Improve international protection for intellectual property rights.

By 2005, it was no longer Japan—by then, in the midst of a 10-year economic funk—that got the juices flowing on Capitol Hill, but rather China. The 1985 threat had come from an ally; the 2005 threat came from an adversary. At least that's how a number of members of Congress saw it. Frank Wolf, a Republican who represented a northern Virginia district in suburban Washington was one of them.

On a sunny afternoon in early March of 2005, four of us were gathered outside Frank Wolf's office in the Cannon House Office Building waiting to be ushered in. Norman Augustine, the retired CEO of Lockheed-Martin, Charles "Chuck" Vest, who had recently stepped down as president of MIT, Jack Crowley, MIT's head of government affairs, and I, representing the American Physical Society (APS) and the newly created Task Force on American Innovation. We had secured a 15-minute meeting with Wolf, who was chairman of the House Appropriations Subcommittee that handled some of the key federal science and technology accounts, including NASA, the National Science Foundation (NSF), and the National Institute of Standards and Technology (NIST).

For more than a year, Steve Pierson, a staff member in the APS Washington Office, and Tobin "Toby" Smith, a senior government relations officer at the Association of American Universities, had been assessing the global stature of the United States in research and development (R&D). We wanted more than a snapshot of our nation's present-day standing. We wanted to know where we had stood decades back, and what the current trends predicted for the future. The graphs Steve and Toby generated forecast troubling times for American competitiveness. By a number of measures—patents, publications, R&D spending relative to the size of the economy, and advanced technical degrees—we were on track to lose our international leadership in science, technology, and innovation sometime between 2015 and 2020.

The challenges came from both Europe and Asia, especially China at the end of the decade. We had captured the breadth and depth of the problem, and we believed Wolf would find the issue compelling, if not extremely alarming. Norm Augustine, who had strong Republican credentials, having served as Acting Secretary of the Army during the Ford Administration, took the lead in the discussion. Chuck Vest and I waited, trying to gauge Wolf's interest and reaction. Finally, Chuck, whose 15 years as MIT's president gave him unusual standing on Capitol Hill, began to paint the picture of the future. As he was speaking, I spread out on the table in front of us the graphs Steve Pierson and Toby Smith had created. They provided a compelling visual counterpoint to Vest's words, and they grabbed Wolf's attention immediately.

At that moment, a knock at the door broke the policy spell.

"Excuse me Mr. Wolf," the staffer said, poking his head into the office, "your next meeting has arrived."

"Tell them to wait," Wolf replied. And then turning to me, he asked, "Where did you get the data?"

"It's all publicly available, from the National Science Board's *Science and Engineering Indicators*, the Bureau of Economic Analysis, the Bureau of Labor Statistics, the Patent Office, and various scientific publishers," I replied, handing him a copy of a report, *The Knowledge Economy: Is the United States Losing Its Competitive Edge*, which the Task Force on American Innovation[3] had just released.

Norm Augustine picked up the thread of the discussion. And then came another knock at the door. We were 15 minutes over our allotted time.

"Mr. Wolf, you now have two groups waiting to see you. What should I tell them," the staffer asked.

"Tell them to wait."

We spent the next 15 minutes discussing the next steps. Wolf laid out his ideas: First, an Innovation Summit run by the Commerce Department to highlight the problems. Second, a rapid study and a report by the National Academies of Science (NAS) and Engineering (NAE) and the Institute of Medicine

(IOM)—now the National Academy of Medicine—to recommend solutions. And, third, legislation to implement the National Academies' recommendations. It was an ambitious plan.

The first step was the easiest. Wolf controlled the purse strings of the Commerce Department in the House, and he could simply write the Summit into the department's budget. The other two, he wisely noted, should have bipartisan buy-in. The timing was propitious.

A month earlier in February, at their annual joint meeting, the councils of the National Academy of Sciences and the National Academy of Engineering had taken a close look at the standing of the United States in what they termed "today's global knowledge-discovery enterprise."[4] What they found concerned them deeply. The American science and technology enterprise was flagging, and they warned that if the trend were to continue, it "would inevitably degrade [the nation's] social and economic conditions, and in particular, erode the ability of its citizens to compete for high-quality jobs."

In mid-May, the National Academies' Committee on Science, Engineering, and Public Policy (COSEPUP) held a meeting to develop a course of action. Lamar Alexander, former president of the University of Tennessee and former secretary of education was one of the speakers. He was now the junior senator from Tennessee, and as a Republican, he chaired the Energy Subcommittee of the Senate Energy and Natural Resources Committee, working closely with the chairman of the full committee, Pete V. Domenici, the New Mexico senator who had jump-started the Human Genome Project almost two decades earlier.

Competitiveness was now one of the top items on Domenici's agenda, and he had asked Alexander to help orchestrate a bipartisan strategy to address it. Alexander came to the COSEPUP meeting to give the attendees a heads-up: Within 2 weeks, the Academy would be getting a request to study the problem and provide recommendations to reinvigorate American science and the nation's global competitiveness. The letter arrived on May 27, 2005, signed by Alexander and New Mexico Senator Jeff Bingaman, the Democratic ranking member of the Energy Subcommittee.

Alexander's role was baked in by virtue of the 55 to 45 majority Republicans maintained in the Senate that year. But he was relatively new to the science and technology table, winning his Senate seat only 2 years earlier. Bingaman, his co-signer, had the real bona fides. For the better part of two decades, along with Pete Domenici, his fellow New Mexican, he had been a prime go-to legislator on all things science.

Bingaman valued science, but he also valued scientists. Bypassing lawyers and politicos, he had named Robert M. Simon, a chemist with a doctoral degree from MIT, as his committee staff director in 1999. And in 2004, with Simon and Adam Rosenberg, an AAAS congressional fellow who worked for the Energy Committee, he co-authored a hard-hitting article for *Issues in Science and Technology*.[5] In it, he warned that the lack of federal investment in research was putting America's future economic growth at risk.

A year later, shortly after he and Alexander sent their letter to the Academy, Bingaman offered a set of policy prescriptions designed to shore up the nation's technological competencies. Expressing his views to the nation's physicists[6]— who were mostly academics—he led with a proposition he knew would appeal to their self-interests: increased funding of long-term research. But he also advanced ideas for strengthening America's near-term competitiveness: expanding R&D tax credits, developing more and better science parks, and providing incentives for investments in high-tech manufacturing on American soil.

As Bingaman, Alexander, and Domenici were beating the Senate drums, Wolf was moving ahead rapidly in the House. He enlisted the support of two Republican members of the Science Committee, Sherwood "Sherry" Boehlert, its chairman from New York, and Vern Ehlers from Michigan, chairman of its Research and Education subcommittee, and the first research physicist to serve in the House. Wolf also reached out to Bart Gordon, a Tennessee Democrat, who was the Science Committee's ranking member. Shortly after the Academy received the request from Alexander and Bingaman, an even more urgent plea arrived from Boehlert and Gordon.

When Congress petitions the National Academies for help, it often provides funding for the work. But in this case, the gang of four making the request believed the issue needed a rapid response. And they had little expectation appropriators could move their bills fast enough to meet their timeline. With the competitiveness buzz on Capitol Hill getting louder by the day, the Academies decided to take up the challenge using their own resources. Norm Augustine chaired the hastily assembled ad hoc team, which COSEPUP gave the unwieldy name the Committee on Prospering in the Global Economy of the 21st century an Agenda for American Science and Technology. Marketing had never been a strength of the Academies, and the new committee's title broke no new ground in that regard. But the committee's membership was truly impressive: major university presidents, Nobelists, Fortune 500 CEOs, and former federal officials, 20 in all.

The charge to the committee was specific. Determine "the top ten actions, in priority order, that federal policymakers could undertake to enhance the science and technology enterprise so that the United States can successfully compete, prosper, and be secure in the global community of the 21st century." And identify a "strategy, with several concrete steps, [that] could be used to implement each of those actions." The committee was free to identify the thematic principles it would use in its deliberations. It settled on two: job creation and clean, affordable, and reliable energy. The first was noncontroversial. The second was phrased broadly enough so that it would not trigger a debate over climate change, an issue that was already so politically fraught that it could potentially marginalize the committee's findings and recommendations.

But even at the risk of running afoul of the House and Senate petitioners, the Augustine Committee rejected the idea of identifying ten prioritized actions, opting instead for four major recommendations, arguing further that all of them should be viewed as one "coordinated set of policy actions." Ignoring one, the report warned, would substantially weaken the other three, noting, for example, that "there is little benefit in producing more researchers if there are no funds to support their research." The interdependence of science and technology policies intriguingly mirrored the interdependence of the sciences that Harold Varmus identified in his *Washington Post* op-ed of 2000.

The National Academies and its operational arm, the National Research Council, were notorious for their thorough but laborious approach to all issues that made it over the transom. But Congress made it clear it was in no mood to wait, and the Augustine Committee's marching orders had a 10-week term limit. Remarkably, it met its goal and produced a truly formative and informative report articulating some of the most compelling policy imperatives for 21st century American science and technology. The report remains as relevant today as it was in 2005. And its Findings and Recommendations in the Executive Summary are worth reading:[7]

Findings

Having reviewed trends in the United States and abroad, the committee is deeply concerned that the scientific and technological building blocks critical to our economic leadership are eroding at a time when many other nations are gathering strength. We strongly believe that a worldwide strengthening will benefit the world's economy—particularly in the creation of jobs in countries that are far less well-off than the United States. But we are worried about the future prosperity of the United States. Although many people assume that the United States will always be a world leader in science and technology, this may not continue to be the case inasmuch as great minds and ideas exist throughout the world. We fear the abruptness with which a lead in science and technology can be lost—and the difficulty of recovering a lead once lost, if indeed it can be regained at all.

The committee found that multinational companies use such criteria as the following in determining where to locate their facilities and the jobs that result:

- Cost of labor (professional and general workforce).
- Availability and cost of capital.
- Availability and quality of research and innovation talent.
- Availability of qualified workforce.
- Taxation environment.
- Indirect costs (litigation, employee benefits such as healthcare, pensions, vacations).
- Quality of research universities.
- Convenience of transportation and communication (including language).
- Fraction of national research and development supported by government.

- Legal-judicial system (business integrity, property rights, contract sanctity, patent protection).
- Current and potential growth of domestic market.
- Attractiveness as place to live for employees.
- Effectiveness of national economic system.

Although the US economy is doing well today, current trends in each of those criteria indicate that the United States may not fare as well in the future without government intervention. This nation must prepare with great urgency to preserve its strategic and economic security. Because other nations have, and probably will continue to have, the competitive advantage of a low wage structure, the United States must compete by optimizing its knowledge-based resources, particularly in science and technology, and by sustaining the most fertile environment for new and revitalized industries and the well-paying jobs they bring. We have already seen that capital, factories, and laboratories readily move wherever they are thought to have the greatest promise of return to investors.

Recommendations

The committee reviewed hundreds of detailed suggestions—including various calls for novel and untested mechanisms—from other committees, from its focus groups, and from its own members. The challenge is immense, and the actions needed to respond are immense as well.

The committee identified two key challenges that are tightly coupled to scientific and engineering prowess: creating high-quality jobs for Americans, and responding to the nation's need for clean, affordable, and reliable energy. To address those challenges, the committee structured its ideas according to four basic recommendations that focus on the human, financial, and knowledge capital necessary for US prosperity.

The four recommendations focus on actions in K–12 education (*10,000 Teachers, 10 Million Minds*), research (*Sowing the Seeds*), higher education (*Best and Brightest*), and economic policy (*Incentives for Innovation*) that are set forth in the following sections. Also provided are a total of 20 implementation steps for reaching the goals set forth in the recommendations.

Some actions involve changes in the law. Others require financial support that would come from reallocation of existing funds or, if necessary, from new funds. Overall, the committee believes that the investments are modest relative to the magnitude of the return the nation can expect in the creation of new high-quality jobs and in responding to its energy needs.

The committee notes that the nation is unlikely to receive some sudden "wakeup" call; rather, the problem is one that is likely to evidence itself gradually over a surprisingly short period.

10,000 TEACHERS, 10 MILLION MINDS, AND K–12 SCIENCE AND MATHEMATICS EDUCATION

Recommendation A: *Increase America's talent pool by vastly improving K–12 science and mathematics education.*

SOWING THE SEEDS THROUGH SCIENCE AND ENGINEERING RESEARCH

Recommendation B: *Sustain and strengthen the nation's traditional commitment to long-term basic research that has the potential to be transformational to maintain the flow of new ideas that fuel the economy, provide security, and enhance the quality of life.*

BEST AND BRIGHTEST IN SCIENCE AND ENGINEERING HIGHER EDUCATION

Recommendation C: *Make the United States the most attractive setting in which to study and perform research so that we can develop, recruit, and retain the best and brightest students, scientists, and engineers from within the United States and throughout the world.*

INCENTIVES FOR INNOVATION

Recommendation D: *Ensure that the United States is the premier place in the world to innovate; invest in downstream activities such as manufacturing and marketing; and create high-paying jobs based on innovation by such actions as modernizing the patent system, realigning tax policies to encourage innovation, and ensuring affordable broadband access.*

The committee's first inclination was to name its report, *The Gathering Storm*, but a number of members thought the title should convey a more optimistic message[8] and amended it to *Rising Above the Gathering Storm*, which has the unfortunate acronym, *RAGS*. But its content was far richer than its acronym implied, and an expectant Capitol Hill was poised to accept its call for engagement, reacting to it with a fervor no other National Academies' work had ever elicited. The report contained twenty separate action items, and over the course of the next 2 years, Congress and the president embraced a majority of them— although in the case of the White House, not without some unexpected twists caused by a comparatively weak Office of Science and Technology Policy.

The National Academies released the *RAGS* report on October 12, 2005, meeting its 10-week deadline. Lamar Alexander moved swiftly following its publication, arranging a dinner for his colleagues so that Norm Augustine could present its results. Amazingly, one third of the Senate showed up.

Science was suddenly roaring down the track, and Augustine was driving the train. He had the reputation, credentials, and connections to keep it running. It was no accident, for example, that Shirley Tilghman, president of Princeton University, convened a meeting on *RAGS*, which Democratic House Leader Nancy Pelosi and other senior political leaders attended. Norm, after all, was a Princeton grad and a former member of its board of trustees. Of course, it didn't hurt that he was a winner of the National Medal of Technology, retired CEO of Lockheed-Martin, former Acting Secretary of the Airforce, former Under Secretary of the Army and a member of the President's Council of Advisors on Science and Technology (PCAST). It also didn't hurt that physicist Rush Holt, Jr. was the local congressman.

On December 6, 2 months after the National Academies released the Augustine report, Wolf's National Summit on Competitiveness took place at the Reagan Center in downtown Washington. The half-day convocation, which drew 63 senior officials from the high-tech industry, universities, and government, issued a six-page statement[9] that began with a stark warning. "The National Summit on Competitiveness has one fundamental message; if trends in U.S. research and education continue, our nation will squander its economic leadership, and the result will be a lower standard of living for the American people."

Echoing the recommendations of the *RAGS* report, the Summit called for action on three fronts: (1) Revitalizing fundamental research by doubling federal investments focused on science, engineering, and mathematics over 10 years, and allocating at least eight percent to high-risk, high-payoff programs; (2) Expanding the STEM (science, technology, engineering, and math) talent pool by doubling the number of science, math, and engineering bachelor's degrees over 10 years, expanding the number of K-12 science and math teachers, reforming immigration policies to attract talented STEM workers from around the world, and creating more public-private partnerships to support STEM careers; and (3) Stimulating deployment of advanced technologies—especially in nanotechnology, high-performance computing, and energy technologies—to ensure national security and sustain American global economic leadership.

A presidential initiative is a well-worn method of advancing any new policy agenda. But at the close of the Competitiveness summit, there was little indication that the White House had any intention of creating one. It's occupant, George W. Bush, except for several speeches he delivered during a 3-day speaking tour following the release of the *RAGS* report, had rarely indicated any interest in science and technology. He was mostly known for his controversial opposition to embryonic stem-cell research and his reluctance to accept the scientific consensus on climate change.

Shortly before he took office in January 2001, Bush indicated that he intended to abolish the Office of Science and Technology Policy (OSTP). He backed away from his plan, only after he learned its elimination would require an act of Congress. But he had so poisoned the well with his apparent disdain for the office that he ran into early trouble finding anyone who would agree to be its director and simultaneously serve as his science advisor, as all directors had since OSTP was established in 1976. Finally, in June, he found a taker in John H. Marburger, III, a physicist with a Princeton and Stanford University pedigree, who had burnished his credentials as president of the State University of New York at Stony Brook and director of Brookhaven National Laboratory. Marburger, a lifelong Democrat, would be serving a Republican president, whose election many Democrats refused to accept, because it had effectively been decided by a five to four vote of the United States Supreme Court, smacking of partisanship.

Marburger didn't need the job, but he felt a responsibility to do what he could to inject science into White House policy-making, as well as to provide guidance on administration appointments. He did not see his party affiliation posing any difficulties. As he told *The New York Times*, "If there's any subject that should be bipartisan, it's science."[10] It was probably a naïve view.

The contrast between D. Allan Bromley's relationship with George H.W. Bush and Marburger's relationship with his son, "W," could not have been greater. Bromley had an office in the West Wing of the White House and had convinced Bush "41" that OSTP should be located in the Old Executive Office Building (OEOB) across the West Executive Drive on the west side of the White House. He also had a seat at the Cabinet table.

Marburger had none of those trappings. He reported to Bush "43" through the White House Chief of Staff, Andrew "Andy" Card, and when he showed up for work on October 23, 2006, he found that his office was on Pennsylvania Avenue near 18th Street, more than a block away from the White House. But Jack, as all of his friends called him, never complained publicly. A congenial person, he put his head down and made the best of his situation. He would provide the president with science and technology policy advice whenever the occasion arose.

If Jack had any failing, it was painting a rosier picture of science under Bush than was really the case. That flaw was in evidence on May 20, 2005, when he gave an after dinner speech at a meeting of the American Physical Society's Division of Atomic, Molecular, and Optical Physics in Lincoln, Nebraska.[11] That he traveled halfway across the country to give a 30-minute talk to a group of physicists on a corn-belt campus suggested that either he had something monumental to share with a Midwestern, largely agricultural community, or he had little on his plate and could afford to be away from Washington. By the time he wrapped up, he had lent credence to the latter and reinforced the perception that he was burnishing Bush's support of research more than it deserved.

Every presidential science advisor is obligated to provide the occupant of the Oval Office with the highest level of technical advice he or she can. But once White House policies are set, as a member of the Administration, the science advisor must support them publicly. The other option is to resign if the policies are too odious. At the University of Nebraska, Jack Marburger tried to walk a tightrope. He touted the Administration's commitment to research, but did not own up to the parsimony of its research budget requests. In reality, he had few options, but in the minds of many physicists in the audience, his parsing of the policies was a bridge too far.

I had known and considered Jack a friend since the late 1970s, when he was dean of the University of Southern California College of Letters, Arts, and Science and had tried to entice me to move there from Yale. I knew he wouldn't take my criticism personally, so after his talk ended and as the banquet hall was emptying, I presented him with the budget reality, as I saw it: The Bush Administration was allowing federal support of science to stagnate or decline at the same time the economy, overwhelmingly driven by science and technology, was growing.

I drew the analogy of the United States to a flourishing high-tech company and argued that such a company would eventually cease to exist if it didn't continuously renew its technology portfolio. It could do so by investing in research and development or acquiring intellectual property from another company that had made the required investments. In the case of our nation, the second option does not exist. The only debatable issue is whether government support of long-term, basic research is indispensable, or whether the industry has the ability to pick up the tab. The next chapter will reprise the narrative I used with Marburger, emphasizing that in 21st century America, industry not only has few incentives to pay for long-term research, basic or applied—especially if it carries high risks—it has huge disincentives to do so. In that case, the federal government's role is vital.

The proper metric for a high-tech company's research and development spending, I suggested, was a percentage of the company's bottom line, defined by either its profits or its revenues. For the nation, by analogy, I argued, it was a percentage of the total economy, defined by the gross domestic product (GDP) that mattered—as Allan Bromley and Paul Anderson had at their Washington press conference 8 years earlier. Developing a universally accepted model to determine a GDP target percentage is almost impossible based on macroeconomic theory,[12] but using the results of past performance as a predicate for current policy is more than a reasonable proposition.

Jack readily acknowledged that federal support of long-term research as far back as 1950, but especially between 1960 and 1980, had transformed American life in the succeeding decades. It generated all the technologies Steve Jobs used in Apple's iPhone: the large-scale integrated circuit, the global positioning system—usually abbreviated as GPS—and the touch screen. It led to laser-enabled technologies in telecommunications, manufacturing, retail, entertainment, and medicine. Today, those technologies account for more than one third of the American economy.[13,14] It delivered medical diagnostic tools, such as CT and PET scanning, magnetic resonance imaging (MRI), and proton therapy. It produced the Internet, and through a many decades-long commitment to high-energy physics, the World Wide Web and the browser.[15]

Jack and I also spoke briefly about the synergy between defense and non-defense research: The impact on national security of programs at the Departments of Agriculture, Commerce and Energy, and agencies, such as NASA, the National Institutes of Health, and the National Science Foundation. As well as the flip side, the spill-over of Defense Department research and development projects into the civilian domain.

In policy circles, those synergies are often called dual-use technologies. And over the years they have been significant. That is especially true of the work sponsored by the Advanced Research Projects Agency. Established during the Eisenhower era, ARPA—which later acquired the prefix "Defense" and the acronym DARPA—developed the mystique early on of a technological magician. The Transit satellite network, which was the precursor of GPS, carried the ARPA label. Tiros, the first weather satellite, was an ARPA creation. So, too, was the computer mouse and the first computer networking system, ARPANET.[16]

The defense research agency also gave the field of materials science a jump start, eventually delivering carbon and polymer-matrix composites that made "stealthy" aircraft difficult to detect. They helped the military, but they also revolutionized sports equipment, wind turbines, cars, and boats. Historically, (D)ARPA's directors recognized the important role risk-taking plays in delivering blockbuster outcomes, and the agency was fortunate to be led over the years by talented technologists and managers. If it had any failing, it was its lack of diversity in its leadership. From its inception in 1958 until 2009, every (D)ARPA director was a white male, almost always over the age of forty.[17] The testosterone streak ended with Barack Obama's appointment of Regina E. Dugan in 2009 and Arati Prabhakar in 2012, a hopeful sign perhaps, of the growing influence of women in defense science policy.

I am convinced Jack recognized the shortfalls in the Bush science budgets. He knew fully well that federally-funded research and development, as a percentage of U.S. GDP, had peaked at 1.86 percent[18] in 1963 at the height of the space program and had been declining since, reaching 0.74 percent in 2005.[19] Competitiveness aside, he surely recognized that such a trend did not bode well for the future of a nation whose economic growth depends overwhelmingly on science and technology.

Following our conversation at the University of Nebraska, Jack suggested we meet again in Washington after Labor Day. The calendar might have read September, but Washington was still hot and sultry. We met for lunch at a restaurant not far from the Eisenhower Executive Office Building, as OEOB was now known, and where OSTP was again more appropriately located. By that time, the Augustine Committee was well underway, and Jack wanted to wait for the *RAGS* report to be released before settling on a strategy.

A month later, the report was out, and Administration budget-making was in full swing. But when Jack and I spoke by phone, he held out little hope the president would take any action. Science, it seemed, was still a low-priority item for the White House. Still, he said he would try to get some face time with President Bush and his chief of staff, Andy Card. Three more weeks passed without any movement. At about that time, Norm Augustine called me. He knew I had been talking to Marburger, and asked whether Jack had been able to use the Academies' report to his advantage.

"Apparently not," I told him. "He's made no progress."

Norm has a great deal of patience, but with Thanksgiving approaching and the presidential budget request close to being put to bed, his patience was wearing thin. "Why," he asked, "do we have to go through Marburger? I know he's a good guy, but he doesn't seem to have the president's ear."

Clearly that was the case.

"I know Card and Bolten pretty well," Augustine continued, referring to Josh Bolten, the director of the Office of Management and Budget. "I also know Cheney from his days as defense secretary. Do you see any reason why I shouldn't go to them directly?"

"No," I said, "I think Jack gave it his best shot, and he'll be fine if you get the president to act." I was also thinking that Jack would take credit for anything Norm achieved, but I kept my thoughts to myself. One Washington lesson I learned early was to let good people in high positions take credit for worthy science and technology policies that could improve their chances of becoming more influential. Jack Marburger was certainly someone who could use a leg up.

Andy Card, Josh Bolten, and Vice President Dick Cheney were the Administration's big three when it came to science and technology budgets. I accompanied Norm to the meeting with Bolten, which took place in early December in one of the small offices on the second floor of the West Wing. Bolten seemed persuaded by Norm's arguments, but made no commitment to follow through. Norm was guardedly optimistic, and said if Card and Cheney got on board, he believed the president would take action. The other two meetings, which I did not attend, took place before Washington emptied out for the Christmas and New Year holidays.

The stars seemed to be aligning, but we didn't know with any certainty how Bush would respond. We found out a month later, when he delivered his State of the Union Address.[20] One section of the speech focused on competitiveness, and for science, the message could not have been better. In Bush's words,

...to keep America competitive, one commitment is necessary above all: We must continue to lead the world in human talent and creativity. Our greatest advantage in the world has always been our educated, hard-working, ambitious people. And we're going to keep that edge. Tonight I announce an American Competitiveness Initiative to encourage innovation throughout our economy and to give our Nation's children a firm grounding in math and science.

First, I propose to double the Federal commitment to the most critical basic research programs in the physical sciences over the next 10 years. This funding will support the work of America's most creative minds as they explore promising areas such as nanotechnology, supercomputing, and alternative energy sources.

Second, I propose to make permanent the research and development tax credit to encourage bolder private sector initiative in technology. With more research in both the public and private sectors, we will improve our quality of life and ensure that America will lead the world in opportunity and innovation for decades to come.

Third, we need to encourage children to take more math and science, and to make sure those courses are rigorous enough to compete with other nations. We've made a good start in the early grades with the No Child Left Behind Act, which is raising standards and lifting test scores across our country. Tonight I propose to train 70,000 high school teachers to lead Advanced Placement courses in math and science, bring 30,000 math and science professionals to teach in classrooms, and give early help to students who struggle with math, so they have a better chance at good, high-wage jobs. If we ensure that America's children succeed in life, they will ensure that America succeeds in the world.

Preparing our Nation to compete in the world is a goal that all of us can share. I urge you to support the American Competitiveness Initiative, and together we will show the world what the American people can achieve.

All that remained were the dollars needed to support the initiative, and 2 days later, on February 2, 2006, the president put his monetary stamp on ACI, as the American Competitiveness Initiative was quickly dubbed in Washington jargon. Two paragraphs in the presidential letter accompanying the ACI budget release summed it all up:[21]

To build on our successes and remain a leader in science and technology, I am pleased to announce the American Competitiveness Initiative. The American Competitiveness Initiative commits $5.9 billion in FY 2007 to increase investments in research and development, strengthen education, and encourage entrepreneurship. Over 10 years, the Initiative commits $50 billion to increase funding for research and $86 billion for research and development tax incentives. Federal investment in research and development has proved critical to keeping America's economy strong by generating knowledge and tools upon which new technologies are developed. My 2007 Budget requests $137 billion for Federal research and development, an increase of more than 50 percent over 2001 levels. Much of this increased Federal funding has gone toward biomedical research and advanced security technologies, enabling us to improve the health of our citizens and enhance national security. We know that as other countries build their economies and become more technologically advanced, America will face a new set of challenges. To ensure our continued leadership in the world, I am committed to building on our record of results with new investments—especially in the fields of physical sciences and engineering. Advances in these areas will generate scientific and technological discoveries for decades to come.

The bedrock of America's competitiveness is a well-educated and skilled workforce. Education has always been a fundamental part of achieving the American Dream, and the No Child Left Behind Act is helping to ensure that every student receives a high-quality education. Accountability and high standards are producing positive results in the classroom, and we can do more to provide American students and workers with the skills and training needed to compete with the best and brightest around the world. Building on our successes, the American Competitiveness Initiative funds increased professional development for teachers, attracts

new teachers to the classroom, develops research-based curricula, and provides
access to flexible resources for worker training.

At the official OSTP roll-out of the president's fiscal year 2007 budget 5 days later in the small 4th floor auditorium of the Eisenhower Executive Office Building, Jack Marburger was all smiles, touting ACI. And as Norm Augustine and I had expected, he never mentioned the role Chuck Vest, Norm, or I had played or, for that matter, any of the key members of Congress.

Without question, ACI improved Jack's standing in the science community, and likely in the White House, as well. That was all to the good. Achieving science and technology policy objectives requires far more than a deep understanding of the science of science policy—paradoxically, a research area that Marburger had embraced.[22] It requires political connections, advocacy networks, and impeccable timing. And Bush's adopting ACI gave Jack a leg up he sorely needed.

By the time Bush gave his State of the Union Address and released his budget request, the 2006 election season had begun. Polls were showing that a blue wave was building. And Democrats were relishing the opportunity to retake control of Congress for the first time in a dozen years.

Smart money was on minority leader Nancy Pelosi to become the first woman speaker if Democrats won in November. A liberal from San Francisco, she was a superb tactician, but of more significance to the science community, she was a big booster of research. Before the calendar had begun to turn from 2005 to 2006, Pelosi already was traveling the countryside to promote the Democratic Party's vision for economic growth and global leadership. She summarized the vision, which focused on innovation and competitiveness, in a speech at Harvard's Kennedy School of Government on December 2, 2005.[23]

The Democrat's "Innovation Agenda: A Commitment to Competitiveness To Keep America #1" might have differed in detail from Bush's ACI, but in spirit, it was very much the same.[24] It called for enhancing STEM education and doubling funding for basic research in the physical sciences. It also called for improving broadband access, especially in rural and underserved communities; increasing investments in research and development to promote energy independence and clean energy; and providing a healthier environment for small business innovation and risk-taking entrepreneurship.

Competitiveness as a science policy imperative had clearly awakened from its 20-year slumber. After the new Democratic-controlled Congress was sworn in on January 4, 2007, it developed an even more vibrant life. Bart Gordon, who had been the ranking Democrat on the House Science Committee, spearheaded the effort to develop a legislative plan. As chairman of the renamed House Science and Technology Committee, he was a moderate accustomed to working across the political aisle. He had partnered with Republican Sherry Boehlert in calling for the Academies study, and with Boehlert now retired, he looked to Vern Ehlers to take the Republican lead on a competitiveness bill.

Ehlers had been elected in 1993 and had established his policy bona fides with two signature efforts beginning in 1995, following the Republican victory in the 1994 congressional election. Newt Gingrich, who had been elected House Speaker, was a science geek at heart, and in Ehlers he saw a thoughtful, knowledgeable, physicist who could command the respect of Republican House members on all manner of technological issues. His assessment would prove correct.

Gingrich immediately asked Ehlers to oversee the creation of the first online congressional database. For scientists, online communication (the Internet) and online data transfer (the World Wide Web) had already become essentials of professional life. For Congress, they represented a radical departure from prior practices.

Ehlers had scarcely any background in computer science, but his physics pedigree provided him with sufficient cover among his colleagues, almost all of whom were woefully ignorant of anything having to do with science and technology. The Library of Congress had direct responsibility for the project known as "Thomas," named for Thomas Jefferson, one of America's earliest science advocates. But Ehlers gained great credibility with his colleagues when Thomas was launched successfully, thereby enhancing his ability to help shepherd science bills through the House.

About 2 years after the Thomas project began, Gingrich gave Ehlers another assignment: develop an overarching plan to help guide American science and technology policy into the 21st century. Gingrich saw it culminating in the first comprehensive set of U.S. science and technology policy prescriptions since Vannevar Bush's 1945 report, *Science, The Endless Frontier.*[25] He also knew it would be the first such project the House had ever undertaken. It was a bold plan, and Ehlers was probably the right person to accomplish it. But in the end, it fell short of expectations, as some observers of Congress thought it inevitably would.

Using information gleaned from testimony, interviews, and documents submitted to the House Science Committee, Ehlers and his staff toiled for a year and a half before releasing the product of their work, *Unlocking Our Future: Toward a New National Science Policy.*[26] It contained a laundry list of recommendations, forty in all, without any priorities indicated. Many of them rehashed well-accepted principles; a few of them represented Republican must-haves and several contained new insights. It was a potpourri, as the following partial list demonstrates:

> *To maintain our Nation's economic strength and international competitiveness, Congress should make stable and substantial federal funding for fundamental scientific research a high priority.*
>
> *Because the federal government has an irreplaceable role in funding basic research, priority for federal funding should be placed on fundamental research.*

The federal government should continue to administer research grants that include funds for indirect costs and use a peer-reviewed selection process, to individual investigators.

Because innovation and creativity are essential to basic research, the federal government should consider allocating a certain fraction of these grant monies specifically for creative, groundbreaking research

In general, research and development in federal agencies, departments, and the national laboratories should be highly relevant to, and tightly focused on, agency or department missions.

In general, U.S. participation in international science projects should be in the national interest. The U.S. should enter into international projects when it reduces the cost of science projects we would likely pursue unilaterally or would not pursue otherwise.

University-industry partnerships should … be encouraged so long as the independence of the institutions and their different missions are respected.

The importance of stability of funding for large-scale, well-defined international science projects should be stressed in the budget resolution and appropriations processes.

University-industry partnerships should … be encouraged so long as the independence of the institutions and their different missions are respected

[T]he R&D tax credit should be extended permanently, and needlessly onerous regulations that inhibit corporate research should be eliminated.

As the principal beneficiaries, the states should be encouraged to play a greater role in promoting the development of high-tech industries, both through their support of colleges and research universities and through interactions between these institutions and the private sector.

At the earliest possible stages of the regulatory process, Congress and the Executive branch must work together to identify future issues that will require scientific analysis.

Scientists and engineers should be required to divulge their credentials, provide a resume, and indicate their funding sources and affiliations when formally offering expert advice to decision-makers.

To ensure that decision-makers are getting sound analysis, all federal government agencies pursuing scientific research, particularly regulatory agencies, should develop and use standardized peer review procedures.

Decision-makers must recognize that uncertainty is a fundamental aspect of the scientific process. Regulatory decisions made in the context of rapidly changing areas of inquiry should be re-evaluated at appropriate times.

Efforts designed to identify highly qualified, impartial experts to provide advice to the courts for scientific and technical decisions must be encouraged.

Curricula for all elementary and secondary years that are rigorous in content, emphasize the mastery of fundamental scientific and mathematical concepts as well as the modes of scientific inquiry, and encourage the natural curiosity of children must be developed.

Programs that encourage recruitment of qualified math and science teachers, such as flexible credential programs, must be encouraged.

To attract qualified science and math teachers, salaries that make the profession competitive may need to be offered. School districts should consider merit pay or other incentives as a way to reward and retain good K–12 science and math teachers.

More university science programs should institute specially-designed Masters of Science degree programs as an option for allowing graduate study that does not entail a commitment to the Ph.D.

Universities should consider offering scientists, as part of their graduate training, the opportunity to take at least one course in journalism or communication. Journalism schools should also encourage journalists to take at least one course in scientific writing.

Scientists and engineers should be encouraged to take time away from their research to educate the public about the nature and importance of their work. Those who do so, including tenure-track university researchers, should not be penalized by their employers or peers.

The recommendations largely followed a consensus view. But the document, as Ehlers, himself, admitted,[27] did not provide much in the way of a roadmap for achieving them. To get the report across the finish line, he noted, required satisfying a majority of his colleagues, each of whom had a slightly different set of priorities. The result, he said, was an unsatisfying all-of-the-above approach.

The report was far from a failure, but its shortcomings illustrate the difficulty of generating a science and technology policy document when there are tens or hundreds of players on the field, and when few of them have much knowledge of the subject. Nonetheless, Ehlers's effort gained him further respect among his House GOP colleagues and allowed him to garner Republican support for the America COMPETES Act of 2007.[28]

———

Two years had passed since Frank Wolf set the course for a competitiveness initiative. What had transpired probably exceeded his expectations. The innovation summit had been held on schedule; the Academies had issued a report in record time; the House Democrats had made it a central part of their legislative agenda; and, thanks to Norm Augustine, the Bush Administration had unexpectedly given the initiative a big financial boost. All that remained was policy legislation that would articulate a clear vision for the future.

The Association of American Universities and the National Association of State Universities and Land Grant Colleges (now known as the Association of Public and Land Grant Universities) lent their support on behalf of academia. The Council on Competitiveness, the Business Roundtable, and two new actors, the Task Force on American Innovation and the Information Technology and Information Foundation, provided industrial heft.

The Senate was first out of the chute. It had a 10-year history of authorizing increased support of the physical sciences, and on March 5, Majority Leader Harry Reid of Nevada introduced the "America Creating Opportunities to Meaningfully Promote Excellence in Technology, Education, and Science Act." Eighteen Democrats and seventeen Republicans joined him as original co-sponsors[29] of the COMPETES legislation. Less than 2 months later, the full Senate passed it by a vote of 88 to 6.

The House also moved swiftly, passing its own version of the bill on May 21, just 11 days after Bart Gordon and seven original co-sponsors[30]—four Republicans and three Democrats—had introduced the bill. Vern Ehlers immediately began working to corral GOP commitments, and on August 2, 143 Republicans joined 224 Democrats in voting for the language that had emerged from a House-Senate conference committee. That afternoon, the Senate approved the legislation by unanimous consent, and a week later President Bush signed it into law.

The final bill called on the president to convene a "Science and Technology Summit to examine the health and direction of the United States' science, technology, engineering, and mathematics [STEM] enterprises." It further ordered "the Director of the Office of Science and Technology Policy [OSTP] ... [to] enter into a contract with the National Academy of Sciences to conduct and complete a study to identify, and to review methods to mitigate new forms of risk for businesses beyond conventional operational and financial risk that affect the ability to innovate..." And to inspire more students to pursue STEM careers, it called on OSTP's director to encourage all elementary and middle schools to observe a STEM day twice a year.

Homing in on the essence of the issue, the act directed the president to establish a Council on Innovation and Competitiveness to develop "a comprehensive agenda for strengthening the innovation and competitiveness capabilities of the Federal Government, State governments, academia, and the private sector in the United States." It also directed the National Science and Technology Council (NSTC) "to identify and prioritize [annually] the deficiencies in research facilities and major instrumentation located at Federal laboratories and national user facilities at academic institutions that are widely accessible for use by researchers in the United States."

The bill contained a number of important policy prescriptions, but among the nation's scientists, the 3-year funding levels specified for the science agencies[31] attracted the greatest attention. Consistent with President Bush's American Competitiveness Initiative, the 2007 COMPETES Act set a course for doubling federal support of physical science basic research over 10 years. But those were just targets, and, using a familiar sleight of hand, the legislation only specified them for 3 years. More significantly, it would be up to appropriators to follow through, and often that's not a good bet.

Washington insiders know that policy prescriptions in authorization bills are far more enduring than proposed spending levels. But scientists generally aren't as savvy, and when future appropriations failed to live up to the 2007 COMPETES Act targets—which scholarly societies had touted to their members—consternation rippled through the research community. On the positive side, the legislation seared innovation and competitiveness into 21st century science and technology policy making.

Chapter 11

Recovery and reinvention 2009–2016

Following the 9/11 attack on the twin towers of New York's World Trade Center, George W. Bush's presidential popularity soared. Several polls pegged his approval rating at an astounding 92%.[1] But by the beginning of April 2008, with the international banking system practically frozen and the American economy on the verge of total collapse due to Wall Street excesses, his job approval had sunk to 30%.[2] It continued to drop during the following months, bottoming out at about 25% a few weeks before the November 2008 presidential election.

The majority of the public had not only turned against Bush, but also against Republicans in general. It was obvious that Barack Obama would be the next White House occupant and that Democrats would extend their majorities in the House and Senate. It was less obvious what impact the election would have on science.

What occurred during the first 2 months after the Democrats swept to victory highlights the relevance of the aphorism, *carpe diem*. Scientists generally shun politics. But there are exceptions. Donald Q. Lamb is one of them. At the time, he was the Robert A. Millikan Distinguished Service Professor of Astronomy and Astrophysics at the University of Chicago. More notably, for this story, he had worked on Obama's presidential campaign, and following the election, he joined the president-elect's transition team as a member of the Innovation and Science Subgroup of the Technology, Innovation, and Government Reform Policy Working Group.

As another physicist with an extracurricular passion for politics, I had gotten to know Don some years earlier. Still, I was surprised when he called me just days after the election with an enticing proposition. The context was Obama's desire to move quickly on stimulus spending to avoid a complete economic meltdown.

What did I think about making science part of the stimulus? That was Don's question. It was rhetorical, because he knew I would readily say it was a terrific idea. Despite the ACI and the COMPETES Act blueprint, American science was still in deep trouble. Major projects had been awaiting support for years,

Navigating the Maze. https://doi.org/10.1016/B978-0-12-814710-8.00011-5

and the opportunity to have them included in a significant economic initiative was too good to pass up.

In fact, Burton Richter, a Nobelist and former director of the Stanford Linear Accelerator Center, and I had tried once before. It was 8 years earlier, when the country was in a mild recession caused by the bursting of the dot-com bubble. But the Bush Administration had expressed little interest, and we shelved the concept. With Obama, perhaps the outcome would be different.

I called Burt and another colleague, Mike Telson, an electrical engineer who had spent two decades on Capitol Hill as a staff member of the House Budget Committee, and then 4 years as the chief financial officer of the Department of Energy during Clinton's second term. If we were to move ahead, there was one caveat: we had to put together a list of "shovel-ready" projects that could be started within 3 months and completed within 18 months. That, we were told, was when the money would run out. Tom Kalil, an economist and technology policy expert who had served 8 years in the Clinton White House and would become Obama's deputy director for technology and innovation at OSTP, would call the shots.

Generating a list of projects proved to be far more difficult than any of us imagined at the outset. Members of the Bush Administration who had already begun to pack up their files offered no assistance,[3] forcing us to play the role of private investigators. Jodi Lieberman, who helped staff the effort, spent hours on the telephone doggedly pursuing national laboratory directors, scientific project leaders, and anyone she could find who had useful information. It all came together at the end of November. We had assembled a list that met the transition team's timetable. It's $1.5 billion figure was significant, but not unreasonable, given what we had learned about the backlog of approved projects.

While we were coordinating with the Obama transition team, Nancy Pelosi, the House Speaker, was building support for a science initiative with her Democratic colleagues, among them, Rush Holt, Jr., from Princeton, New Jersey with a pedigree in physics, and Anna Eshoo, who represented a northern high-tech California district covering a major part of Silicon Valley. At the time we were unaware of Pelosi's discussions, and she was unaware of our detailed list. All that changed at the end of January. But before then, we were forced to go back and modify our list.

Transitions are rarely cake-walks, and while Obama's was smoother than some, major glitches occurred. Our work was caught up in one of them. It's important to remember that at the time many economists were warning that the country—and the world—risked entering a depression, perhaps rivaling the one in the 1930s, unless governments stepped in with massive spending. But the American political landscape was riven by ideological disagreements. The far right wasn't buying the doomsday scenario, and many conservatives were still fighting the war against Roosevelt's New Deal and Johnson's Great Society programs. Against this backdrop, Obama's advisers kept adjusting the size, scope, and timeline for an economic bailout.

What began with money out the door in 90 days and completion deadlines of 180 days became spending within 180 days and completion within 2 years, and finally money out the door in 1 year with a time horizon of 3 years. Our science list grew from $1.5 billion at Thanksgiving to $3.5 billion at Christmas, and finally to $5 billion by mid-January. That was when our work converged with Pelosi's efforts, but it happened somewhat fortuitously.

I was attending a Wednesday evening dinner at the Washington home of Rosa DeLauro, a member of the House Democratic leadership from Connecticut, whom I had known for many years. Anna Eshoo, a dinner regular and a good friend of Burt Richter, was also there. Having spent almost 10 weeks poring over projects, programs, and numbers, I was eager to share our information and suggestions with Anna. I had just finished describing the scope of the plan to her, when Nancy Pelosi walked up and joined the conversation.

"Nancy," Anna said, "you need to hear about this."

I was almost finished recapping what I had just told Eshoo, when Pelosi interrupted, "How much money do you want?'

"Five billion," I told her.

"Why don't you make it 10? Science has critical needs,[4] and this is a perfect time to address them. I'll get the money if you get me the list."

"When do you need it?" I asked.

"Today is Wednesday. By Friday evening." she said. "The House is going to take up a stimulus bill next week, and I want to make science part of it. Can you do it?"

Without thinking through what I was actually committing to, I replied, "Of course."

By the time I left the dinner, the magnitude of the task hit me. We had less than 48 hours to add $5 billion to our proposal. Could we manage it and maintain our credibility? In reality, we had no choice. The stakes were too high.

By Friday evening, our list was complete. In addition to shovel-ready projects, we expanded it to programs of finite (two-three year) duration in the major federal agencies supporting physical science, math, and engineering research and education. It deliberately did not include biomedicine, because we did not have enough knowledge to address it, and we were operating under a time gun. We assumed someone else would step in at the right moment, and as events would soon prove, we were right.

True to Pelosi's promise, Democrats introduced the America Recovery and Reinvestment Act (ARRA)[5]—the actual name of the economic stimulus bill— the following Wednesday. And 2 days later, on January 28, the House sent the bill to the Senate with 244 Democrats' votes and no Republican having voted in favor of it.[6] The science component of the 3-year emergency spending plan was not very large, $10 billion out of almost $475 billion, and the average House member probably never even noticed it. But it would become more visible in the Senate.

Unlike the House, which operates under a simple majority rule and limited time for discussion, the Senate imposes no time restriction on debate, unless three-fifths (60) of the members vote to terminate it. Known as cloture, the rule allows the minority party to influence and often modify legislation.

In 2009, at the start of the 111th Congress, there were only 99 members seated because the Minnesota election was still unresolved. The total included 56 Democrats and two independents who generally voted with them. For AARA to pass, both independents and at least two Republicans would have to vote for cloture. But there was an added uncertainty: Massachusetts Democratic Senator Edward Kennedy was suffering from brain cancer and might not be available to vote. That brought the Republican must-haves to three.

Maine's two senators, Susan Collins and Olympia Snow, both Republicans, were likely to do so. The only other Republican who might was Arlen Specter of Pennsylvania, another moderate known for parting ways with his party from time to time. But Specter had been one of the principal proponents of the appropriations bills that had doubled the budget of the National Institutes of Health (NIH) between 1998 and 2003. If the science stimulus didn't include NIH, he said he would not support cloture. The price tag he put on his vote was $10 billion, exactly matching what Pelosi had earmarked for the rest of science.

Specter's demand was not arbitrary. That year, NIH funding was on track to account for about half of all federal spending on research. What was on the table was a big boost for American science, and Specter's support was essential to closing the deal.

The new total for the science stimulus would be $20 billion. And even spread over 3 years, it represented a major surge for non-defense research and development (R&D),[7] which totaled $60.3 billion in 2009—exclusive of the ARRA add-on.

On February 10, with Kennedy still able to vote, the Senate passed the amended recovery bill 61 to 37. Three days later, on February 13, the House voted 246 to 183 in favor of the conference report, again without any Republican support. And late that afternoon, with Kennedy now incapacitated, the Senate cleared the legislation with no room for error, 60 to 38.

The surge in research funding was designed to be limited to 3 years. And for the most part, federal agencies used the money that way. Approved projects that had already been engineered got jump starts, among them Brookhaven National Laboratory's synchrotron X-ray light source upgrade, known as NSLS II, and SLAC National Laboratory's X-ray laser facility, known by its acronym, LCLS. The money that flowed to NIH as a result of Specter's intervention provided 2-year seed money for new university initiatives and specialized equipment. Only the National Science Foundation (NSF) broke the legislative trust by allocating the stimulus money to its general grant programs. The misstep would come back to bite the agency 3 years down the road, when the stimulus money ran out, and NSF had to ratchet back acceptances of new grant proposals.

In the end, the science stimulus was a windfall for American research programs. Its story illustrates the importance of seizing the day when unforeseen opportunities present themselves. And it demonstrates the importance of laying the groundwork for new policies well in advance of any opportunities. Without the prior focus on competitiveness and innovation, it is unlikely the science stimulus would have materialized.

The story is not without a significant portent. Although the overall recovery bill was guaranteed to split many lawmakers along ideological lines—dividing Keynesian pro-government interventionists from libertarian acolytes of Ayn Rand—the rift between Democrats and Republicans proved to be a yawning chasm. That not a single GOP House member voted for ARRA, and only three GOP senators supported the legislation, foreshadowed the rapidly growing hyper-partisanship that would plague Barack Obama's 8 years in office. Not even science—which, Jack Marburger had properly noted, should be bipartisan—would escape the rancor.

The path of the 2010 reauthorization of the America COMPETES Act[8] underscored the change in the political terrain. As he did in 2007, Vern Ehlers worked tirelessly to generate support for the COMPETES legislation from his Republican colleagues, but on May 28, only seventeen of them broke ranks and joined the full complement of 245 Democrats in sending the House bill on to the Senate.[9]

In an election year, the usually torpid upper chamber becomes even more so, and the COMPETES bill languished well past Election Day. For Democrats, the outcome at the polls was bad news. They lost control of the House by a wide margin, and their advantage in the Senate narrowed substantially. If the COMPETES reauthorization were to pass, it had to happen before the new Congress was seated.

Bipartisanship on science was still intact in the Senate, and after agreeing to one amendment, the upper chamber approved the bill by unanimous consent. The amended bill returned to the House floor for a final vote on December 17. This time only sixteen Republicans supported it. Clearly, Ehlers had not been able to find an antidote for the partisan venom that was poisoning science, and for the balance of Obama's two terms, no one else would be able to either.

There is little in Barack Obama's brief public record prior to his run for the presidency to suggest he would become one of the staunchest defenders of science of anyone to occupy the Oval Office. He didn't wait long to elevate its importance in his Administration, naming Harvard physicist John Holdren as his science advisor a month before taking office. He also moved with alacrity to fill high-level science and technology policy positions in the White House and across the federal agencies, selecting two physics Nobel Laureates for key posts: Steven Chu as Energy Secretary and Carl Wieman as OSTP Associate Director of Science.

It is too early to assess which of Obama's policies will endure, but his record is as long as it is broad. It is far too rich to capture in just a few pages, as the Administration's OSTP "Exit Memo"[10] illustrates. But there are two policy areas that deserve special mention: nuclear nonproliferation, bookended by the 2010 New Strategic Arms Reduction Treaty (New START)[11] and the 2015 Iran Joint Comprehensive Plan of Action (JCPOA)[12]—the Iran nuclear deal—and climate change, culminating in the Paris Agreement,[13] signed by 195 nations in December 2015. All three illustrate the outsized role science and technology play the in 21st century global arena. All three also illustrate the importance of compromise.

New START traces its lineage to START (I), a treaty that seemed completely implausible just a few years before Ronald Reagan and Soviet leader Mikhail Gorbachev struck a deal[14] to begin dismantling the massive nuclear weapons stockpiles each nation possessed. The date was December 8, 1987. Hardliners in both nations lined up against it, but Gorbachev held sway in the USSR, and Reagan—whom conservatives venerated as the Republican counterpart to the liberal icon, Franklin Delano Roosevelt—sold it to his GOP colleagues.

Work on the treaty language began in earnest shortly after the 1987 White House summit, but it took almost 4 years before it was ready for prime time. George H.W. Bush signed it on July 31, 1991, and sent it to the Senate for ratification that November, just as the Soviet Union was on the brink of final collapse. When the Russian tricolor flag replaced the hammer and sickle symbol of the USSR on December 26, 1991, Western democracies celebrated, but nonproliferation experts saw their tasks magnified substantially.

The Soviet nuclear arsenal was spread across a number of the socialist republics—among them, Ukraine, Uzbekistan, and Kazakhstan—which were being cut free of Russian control. The treaty Bush had just signed now needed buy-ins from those newly independent states. It also needed a plan to address the nuclear, biological, and chemical weapons stockpiles in countries that did not have the technical expertise to deal with them. Physicists, especially, who had previously been bomb builders during the Cold War, and the weapons labs (principally Los Alamos, Sandia and Livermore) where they worked, would have a new mission: dismantling weapons of mass destruction (WMDs) and disposing of the fissile material they used.

With the dissolution of the USSR clearly on the horizon, Congress passed the "Soviet Nuclear Threat Reduction Act of 1991" on November 25. It established the Cooperative Threat Reduction (CTR) Program, also known as Nunn-Lugar,[15,16] named for Democratic Senator Sam Nunn of Georgia and Republican Senator Richard Lugar of Indiana. It was Nunn's second attempt at passing such a bill. The first one, co-sponsored by Democratic Representative Les Aspin of Wisconsin, failed to garner broad support and never received a vote. But with Lugar solidly behind the legislation, the Senate reversed course and voted 86 to 4 for its passage.

The bipartisan legislation had four primary objectives: (1) Dismantling WMDs and all associated infrastructure in former Soviet states; (2) Securing WMD materials; (3) Increasing transparency and compliance with non-proliferation agreements; and (4) Enhancing military cooperation with the former Soviet states to accomplish the nonproliferation goals. Achieving those objectives represented a significant challenge, not only for diplomats in the State Department and military planners in the Defense Department, but also for scientists and technologists in the Energy Department, which had sole responsibility for nuclear weaponry.

Against this backdrop, ratification of START I was delayed for almost a year. When the treaty finally reached the floor of the Senate for a vote on October 1, 1992, the outcome was overwhelming: 93 in favor and only 6 opposed.[17] Work on a treaty supplement was already under way. Known as START II, it called for banning the use of multiple independently targetable reentry vehicles, called MIRVs, on intercontinental ballistic missiles, or ICBMs in defense speak. Again, the Senate ratified it overwhelmingly, 87 to 4, when it came up for a vote on January 26, 1996.[18]

But the bipartisan agreement on nuclear arms and disarmament would soon begin to fray. There was little disagreement about U.S. technical capabilities, but trust between the United States and Russia was starting to wane. In Moscow, the Duma delayed ratification of START II until April 14, 2000. But the effect of the vote was short lived. Spurred on by "neo-cons" who had his ear, George W. Bush gave Russia notice that the United States was going to withdrew from the 1972 Anti-Ballistic Missile Treaty.[19] In a tit-for-tat response, Russia announced its withdrawal from START II, although it remained committed to the terms of START I until the treaty was scheduled to expire in 2009.

By the time Barack Obama took office on January 20, 2009, the road to nonproliferation had become somewhat rocky. The overwhelming bipartisan support that START I and START II had enjoyed was no longer assured for a replacement treaty. With ratification requiring a two-thirds vote in the Senate, losing a handful of votes would not be not a problem. But if the projected losses approached 30, momentum away from the 67 votes needed could easily build.

That was the circumstance Obama faced when New START made its way to the Senate in 2010. Russia and the United States had signed the treaty in Prague on April 8, and after more than twenty hearings and extended debate, Democratic Senate Majority Leader Harry Reid put the treaty up for a vote just before the Christmas recess. Democrats had lost six Senate seats in the November midterm election, and delaying action until a new Congress was seated carried too much of a risk. Obama needed the vote sooner, rather than later, and he was prepared to cut a deal to assure ratification.

The months preceding consideration of the treaty revealed how much the non-proliferation winds had shifted. Even though Russia still retained a strong nuclear capability, America's scientific and technological superiority in the post-Soviet era was undeniable. Nonetheless, Jon Kyl, the Republican whip, and Mitch McConnell, the Republican minority leader, leveled withering attacks on New START, arguing that ratification would prohibit the United

States from maintaining the safety, security, and reliability of its nuclear arsenal. McConnell, who had pledged to keep Obama's presidency to one term, saw defeating the treaty as a way of denying the president a major foreign policy victory.

Kyl, an arch-conservative from Arizona was already on record of opposing non-proliferation agreements, having helped block ratification of the Comprehensive Test Ban Treaty[20] a decade before. Obama, who considered New START a signature achievement of his young presidency, was well aware of Kyl's ability to whip votes among his Senate Republican colleagues, and was on the lookout for something he could offer him. As summer was winding down, Kyl let it be known that he might be willing to negotiate. His price: $10 billion more for modernization of the U.S. nuclear arsenal—each year for the next 10 years. Kyl never promised to drop his opposition, but Obama blinked. He badly wanted the win, and he pledged to substantially boost his already massive modernization budget proposal of $80 billion over the next decade.

Having gone on record publicly advocating a larger figure, Obama couldn't back down, even after Kyl—and McConnell—voted against ratification. The final vote on December 22 was 71 in favor (four more than needed) and 26 opposed.[21] Had Obama rolled the dice, he probably would have won.

Obama's commitment to modernization of nuclear weapons was a boon to the three weapons laboratories. Fifteen years earlier, they were searching for missions that would justify their sizes and budgets. In 1995, a task force chaired by Robert Galvin, chairman of the Executive Committee of Motorola, issued a report on "Alternative Futures for the Department of Energy National Laboratories."[22] The "Galvin Report," as it was known inside the Washington Beltway, used an uncompromising lens to examine the national laboratory system in an era of dramatic geopolitical changes and significant technological advances. The following excerpt captures the tone of the report:

The National Security Role
Configuration of the Nuclear Weapons Laboratories

The current structure of the three nuclear weapons laboratories should be examined in light of the recently revised, official U.S. Nuclear Posture. The Department of Energy [DOE] should size its nuclear weapons laboratories support efforts over time to match DoD [Department of Defense] requirements. The restructuring must be accomplished in ways that preserve capabilities both for reduction to lower levels of support and for an expansion of support should the resumption of a threat to national security demand it. In addition, the restructuring must support the requirement to maintain confidence in the nuclear stockpile in a comprehensive test ban or under an extended moratorium. The restructuring will affect primarily weapons design capabilities, where the largest functional redundancy exists, and specifically Lawrence Livermore National Laboratory (LLNL); LLNL supports only four of eleven weapons designs currently in the U.S. stockpile.

The Task Force believes LLNL should retain enough nuclear weapons design competence and technology base to continue its activities in non-proliferation, counter-proliferation, intelligence support, and verification, to provide independent review for several years while alternative approaches to peer review are developed (see "Peer Review"), and to participate in weapons relevant experiments on the National Ignition Facility (NIF). LLNL would transfer, as cost-efficiency allows, over the next 5 years its activities in nuclear materials development and production to the other design laboratory. LLNL would transfer direct stockpile support to the other weapons laboratories as the requirements of science-based stockpile stewardship, support of the DoD nuclear posture, and the status of test bans allow. Under these conditions, the Task Force believes that the transfer can be made in 5 years. The Task Force notes that if the NIF is built at LLNL, this will reinforce the weapons design capability at that laboratory.

Details aside, the report's assessment of the DOE's weapons laboratories conveyed a simple message: There was unnecessary redundancy in weapons design, and one of the laboratories—LLNL—should focus its activities on nuclear non-proliferation and verification technologies. Galvin made that even clearer during his congressional testimony. The report was less critical of DOE's non-defense mission, summarizing its recommendations as follows:

The Science-Engineering Role
Summary of Recommendations

1. The Department of Energy should move to strengthen its efforts in fundamental science and engineering, both at the laboratories and in the universities.
2. The DOE should pay close attention to ensuring that a proper balance is maintained between the universities and the national laboratories in the performance of DOE-related basic research, both now and in the future.
3. Support for operating and maintaining large facilities in the DOE's Office of Energy Research should be budgeted separately from funds for specific programs.
4. The DOE should redouble its efforts to achieve better integration of basic research, technology development programs, and their applications, particularly in the area of environmental remediation.
5. Basic research at the laboratories should be more fully integrated into the national and international research community.
6. There should be additional stimulation of laboratory-university cooperation in basic research.

The Galvin Committee also examined the governance of the DOE's laboratories and summed up its findings with a single recommendation: "Over a period of one to 2 years, the Department and Congress should develop and implement a new modus operandi of Federal support for the national laboratories, based on a private sector style—"corporatized"—laboratory system."

Ultimately, the governance recommendation had the largest impact on the way the laboratories conducted their business. But the transformation didn't occur until Los Alamos National Laboratory suffered a security breach 5 years later. The violation never resulted in the loss of classified information, but it wrecked the life of one of its employees, diminished the role of universities in managing the laboratories, and dramatically increased the costs of administering them.

The bizarre case of nuclear physicist Wen Ho Lee, which led to the administrative unraveling, was the subject of a lengthy *Vanity Fair* exposé[23] in 2000. Edward Klein's story documented Lee's security violations and highlighted his strange and still unexplained behavior in 1999, when he downloaded and transferred classified files to personal tapes. It also focused attention on the FBI's highly questionable treatment of him, which included a 59-count indictment—including espionage—and more than 9 months in solitary confinement, before he was finally freed on a $1 million bond in September 2000. (In 2006, Lee pleaded guilty to improper handling of classified information and received a $1.6 million settlement from the federal government and several media organizations for his prison treatment and public disclosure of his name before any charges had been filed.) But it did not predict, nor could it have been able to predict, the ultimate fallout—the change in the way DOE laboratories were managed and where the weapons labs stood in the department's organization chart.

A legacy of World War II, the DOE laboratories were generally government owned, but contractor operated (GOCO). And many of the operators were universities. That was true of the weapons labs, as well as the multi-purpose labs. The University of California, for example, ran Los Alamos and Lawrence Livermore, and the University of Chicago managed Argonne National Laboratory and, in partnership with the Universities Research Association (URA), Fermi National Laboratory. The Galvin Report had implicitly questioned the role of the universities, but it had generated little more than an annoying buzz.

The Wen Ho Lee affair was different: it was a matter that grabbed the attention of lawmakers on Capitol Hill. When members of Congress catch a whiff of a federal scandal, especially one involving national security, their juices begin to flow. It was true during the Cold War. It remained true after the collapse of the Soviet Union, and it's still a guiding principle. Among elected officials, Wen Ho Lee's activities—quirky at best, dangerous at worst—stimulated far more than an olfactory response.

Lee might have been a culprit who willfully violated the law by downloading and removing classified material, or he might have been a victim of overzealous investigators. But as lawmakers saw it, Lee wasn't the real problem; the DOE was. The department's inability to read the tea leaves well before Lee downloaded classified materials and intervene before he committed his crime demonstrated administrative negligence. A legislative remedy was needed to prevent a repetition that could have more serious consequences.

Especially when a matter is urgent, Capitol Hill's policy tool is rarely a scalpel. It's a sledge hammer, and the Los Alamos security breach brought out a particularly large one. Momentum quickly developed to pull the weapons programs out of the DOE's portfolio, but more than a half century of precedent argued against giving the military direct control over the entire nuclear enterprise. The rationale for establishing the Atomic Energy Commission (AEC) under civilian control in 1946—to minimize the potential for using weapons of mass destruction—remained unchanged in 2000.

If the Department of Defense was off limits, was a 21st century carefully circumscribed analog to the AEC the answer? Isolating the weapons programs from the DOE's civilian responsibilities would have the virtue of closer control over classified nuclear research and development. But it would create several problems.

As Los Alamos and Livermore evolved, they expanded into the non-weapons research space, in part to make themselves more attractive to scientists of the highest caliber, and in part to make some of their expensive, one-of-a-kind facilities available for cutting-edge civilian research. Assigning the laboratories to a new, single-purpose agency would turn the clock back to the earliest days of the AEC. It would likely diminish the quality of the laboratory workforce and complicate civilian access to the lab's unique facilities. And, by constricting the lines of communication between lab scientists and the outside research community, it could hinder progress on such crucial issues as non-proliferation and counter-measures.

For months, the debate swirled across Washington: on Capitol Hill, inside the White House, and throughout the DOE. Finally, on July 22, 1999, the Senate acted,[24] adopting language 96 to 1 to establish a semi-autonomous agency within the DOE that would oversee all aspects of the department's nuclear programs. It would have a dedicated Undersecretary of Energy reporting directly to the Secretary of Energy. The language made it into the National Defense Authorization Act for Fiscal Year 2000,[25] which President Clinton signed into law on October 5, 1999. Acceding to the Senate's desires, the act created the National Nuclear Security Administration. No one was completely happy with the outcome, but in truth, no one had proposed a viable alternative.

In 1999, the weapons labs were the focus of the DOE's critics. But the academic culture of openness and freedom of inquiry, which critics believed contributed to the Los Alamos security breach, permeated the entire system. With some degree of regularity, detractors of the labs' modus operandi would raise their voices, but invariably the labs pushed back, arguing that their way of doing business was critical to the extraordinary scientific successes they achieved. And they had strong evidence to back up their responses: reams of patents and scores of Nobel prizes.

In 1995—the same year Galvin's Committee issued its findings—the Government Accountability Office (GAO), as it is now known, delivered a particularly harsh assessment, with its report carrying the unambiguous title,

"National Laboratories Need Clearer Missions and Better Management."[26,27] The summary of the 1995 report is pithy and cutting:

> *GAO found that: (1) the DOE laboratories do not have clearly defined missions and laboratory managers believe that the lack of DOE direction is compromising their ability to achieve national priorities; (2) DOE manages the laboratories on a program-by-program basis and has underutilized the laboratories' special multi-disciplinary abilities to solve complex, cross-cutting scientific and technology problems; (3) although DOE has developed a strategic plan to integrate its missions and programs in five main areas, it still may not be able to effectively manage the laboratories in the future; (4) the costly and inefficient day-to-day management of the laboratories inhibits a productive working relationship between the laboratories and DOE; (5) DOE does not balance laboratory research and administrative objectives; (6) the laboratories fear that rising research costs due to costly administrative requirements will limit their ability to compete for research projects, which in turn will hamper their commercial technology mission; (7) DOE has instituted contract reforms which it believes will lead to a more productive management approach; and (8) the laboratories can make vital contributions in many important areas such as weapons systems, energy conservation, environmental cleanup, and commercialized technologies with proper mission focus and management direction.*

As stinging as the report was, change came slowly. There is no doubt the labs could have been managed more effectively, but in addition to the extremely high-quality research they generated, the major facilities they built and operated were among the best in the world.

There was one problem plaguing the non-weapons labs, but it was one not easily solved. The bridge that allowed discoveries to be transformed into application was either weak or non-existent: the connections between the bench scientists and the industrial innovator were tenuous, at best. The impediments epitomized "The Valley of Death," a term Representative Vern Ehlers popularized in the 1998 House Science Committee policy report, *Unlocking Our Future*.[28]

Policymakers had long been aware of the failing, but solutions at the DOE had remained elusive. Hazel O'Leary, who was Secretary of Energy at the time the GAO leveled its 1995 broadside, concluded that encouraging industry to utilize lab facilities provided the most promising path forward. The Federal Technology Transfer Act of 1986,[29] she believed, contained the best available mechanism, the cooperative research and development agreement (CRADA). It addressed two significant industrial needs: commercializing research quickly and protecting intellectual property rights. O'Leary, whose background was in the legal profession and the utilities industry, moved quickly to expand its use, encouraging the multi-purpose laboratories, especially, to promote its virtues. Although its track record has been mixed, it has remained part of the DOE policy portfolio ever since.

Against the backdrop of the Galvin and GAO reports and the Wen Ho Lee affair at Los Alamos, the DOE moved to a more corporate laboratory governance model, with stricter controls over classified research and greater emphasis on productivity. Although universities retained some of their management responsibilities, they were now frequently paired with non-academic partners, such as Battelle and Northrup-Grumman. Even though almost two decades have passed since the DOE began "corporatizing" the national laboratories, it remains a matter of debate whether improvements in performance, to the extent they are discernible, justify the extraordinary increases in management costs that, in some cases, rose almost 10-fold.

In 2014, seeking an answer to that question, and desiring an evaluation of the overall structure and performance of the laboratory system, Congress directed the Secretary of Energy to establish a Commission to Review the Effectiveness of the National Energy Laboratories (CRENEL). Using an appropriations bill[30] as the vehicle to make its directive stick—as Frank Wolf had with the Innovation Summit a decade earlier—it charged the commission with examining all seventeen DOE laboratories "in terms of their alignment with the Department's strategic priorities, duplication, ability to meet current and future energy and national security challenges, size, and support of other Federal agencies...[and] to review the efficiency and effectiveness of the laboratories, including assessing overhead costs and the impact of DOE's oversight and management approach." Unlike DOE evaluations of the 1990s, the CRENEL report was reasonably positive.[31] In addressing why the DOE system was needed, for example, the report observed:

The National Laboratories are a unique scientific resource and national security asset, providing a vital experimental infrastructure to the Nation's research community and sustaining the nuclear weapons expertise critical to modern American security. In addition, the laboratories maintain a scientific and technical workforce, as well as a way of working, that fills a key need in the research and development process.

Whether through stewardship of open-access scientific user facilities, assessment of the nuclear arsenal, or fostering environments for cutting-edge research in energy, environmental management, and weapons science, the National Laboratories are an important component of the national S&T enterprise. Furthermore, the Nation often calls upon the scientific and technical expertise of the National Laboratories in times of emergent need, as has been done recently in response to the Fukushima Daiichi nuclear reactor accident and during the Iran nuclear negotiations, among others.

It was less generous in its observations about management issues. But instead of training its criticism on administrative laxity—as the Galvin and GAO reports had—it suggested the DOE had overreacted in responding to past criticisms:

The relationship between DOE and the laboratories has eroded, leading to ever-increasing levels of micromanagement and transactional oversight, which, in turn, have reduced the efficiency and effectiveness of laboratory operations. DOE and the laboratories must return to the spirit of the FFRDC [Federal Funded Research and Development Centers] model, focused on stewardship, accountability, competition, and partnership.

Instead, the National Laboratories are managed at multiple levels: day-to-day operations are overseen by the laboratory director and team in conversation with DOE through either DOE headquarters or site offices, which supply compliance guidance and strategic direction. Elements of departmental management can adversely impact the effectiveness and efficiency of the laboratories. For instance, mounting contract requirements, large numbers of assessments and data calls, and a lack of budgetary flexibility add undue administrative burdens on parts of the laboratory system. Addressing these concerns should be a priority for making the laboratories function better as a whole.

In contrast to the 1995 Galvin Report, which had suggested the laboratory system suffered from under-utilization and duplication of effort, the 2015 CRENEL Report reached the opposite conclusion:

The Commission does not believe there are too many laboratories, nor is there an undesirable degree of duplication. During its visits to all 17 laboratories, the Commission found each to be unique, conducting work of merit, and becoming of the title "National Laboratory." While work might appear duplicative at a high level, the Commission's closer look revealed that their capabilities and focus areas are diverse, complementary, and well-honed to meeting the missions of the Department. Every laboratory plays a key role: for instance, different synchrotrons address different types of scientific questions, while the existence of two NNSA [National Nuclear Security Administration] physics laboratories [Los Alamos and Lawrence Livermore] promotes both competition and a second opinion on high-stakes nuclear weapons work. Having grown out of historic mission decisions, the laboratories of today have evolved to serve not just the Nation but also their home regions and States through the fostering of a scientific community. Many also serve their regional economies.

In his State of the Union Address on January 29, 2002, George W. Bush labeled Iraq, Iran, and North Korea the "axis of evil." One common thread linked them—nuclear weapons. The Bush Administration believed Iraq had them, at least in some form, and Iran and North Korea were well on their way to developing them. Although they were never found,[32] weapons of mass destruction, WMDs in defense lingo, were the pretext for the American invasion of Iraq in 2003. By the time Barack Obama entered the White House in 2009, Iraq was still an extraordinary Middle East quagmire, but its WMD potential was no longer an issue. Iran, by contrast, posed a continuing a threat to the stability of the region, and its nuclear capabilities were progressing rapidly.

Economic sanctions imposed by the United Nations had taken a toll, but hardliners in Teheran refused to dial back their ambitions as they continued to expand their uranium enrichment program. Finally, in the spring of 2015, negotiators met in Lausanne, Switzerland. Seated around the table were representatives of the five permanent members of the U.N. Security Council—China, France, Russia, the United Kingdom, and the United States—plus Germany and the European Union. The stakes couldn't have been higher. At that point, according to intelligence estimates, Iran was within 3 months of having enough highly enriched uranium to produce at least one nuclear weapon.

Neither Israel nor Saudi Arabia was willing to stand idly by and allow their primary enemy cross the nuclear finish line. The potential for a preemptive strike on Iran's uranium enrichment and plutonium facilities was growing by the day. A failure in Lausanne could not have more profound consequences.

Seven issues dominated the negotiations, and they all involved technology: (1) An accurate accounting of Iran's uranium stockpile; (2) The enrichment levels of fissile material on hand; (3) The number and operating parameters of advanced centrifuges Iran was using for uranium enrichment; (4) The final status of the Arak heavy-water facility, which had the ability to produce and reprocess plutonium; (5) The disposal plans for existing fissile material; (6) The inspection regime needed to verify Iran's compliance with an agreement; and (7) How much time—called the "breakout time"—Iran would need to reconstitute its nuclear program and produce its first nuclear weapon if and when any agreement terminated.

Negotiating international agreements with adversaries is rarely easy, but when technical issues are paramount, the difficulties multiply. Secretary of State John Kerry was leading the American delegation, and if he had been following previous scripts, he would have had a team of nuclear experts assisting him. But in this instance, he really needed only one, his fellow Cabinet member, Secretary of Energy Ernest J. Moniz.

Ernie, known for his unusual coif, had cut his scientific teeth in theoretical nuclear physics at Stanford University and MIT. He had also accumulated significant policy experience, first as Bill Clinton's Associate Director for Science at OSTP, and then as his Under Secretary of Energy. Ernie's work on the Joint Comprehensive Plan of Action (JCPOA), as the Iran nuclear deal is more properly known, is one of the best examples of science diplomacy in the modern era.

In *The Guardian* article, "Ernest Moniz and the Physics of Diplomacy,"[33] published in the spring of 2015, Roger Pielke, Jr. captured the essence of his contribution:

He has been called President Obama's "secret weapon" and a "rock star", and his long hair has garnered comparisons to the distinctive coiffure of Javier Bardem, who played a psychotic assassin in the 2007 movie "No Country for Old Men." I write, of course, of Ernest Moniz, the US Secretary of Energy, who has played a pivotal role in the ongoing negotiations with Iran over its nuclear programme. Watching Moniz, we can learn a lot about successfully integrating science and politics.

Moniz, a PhD physicist from MIT, was brought into the negotiations ostensibly to provide a US counterpart to Ali Akbar Salehi, who leads Iran's Atomic Energy Organization. Salehi is another MIT-trained physicist who earned his PhD at the same time that Moniz began his teaching career at MIT. Moniz brings more than just scientific expertise to the negotiating table; he has considerable experience working in political settings, having put in earlier stints at the Office of Science and Technology Policy and at the Department of Energy as an undersecretary...

It would be a stretch to label Moniz's role in the Iran negotiations as that of a "science adviser," and it would diminish his contribution by suggesting he is offering "science advice." Instead, Moniz is fully involved in political questions with scientific and technical content. At the same time he is also fully engaged in the procedural aspects of the deal, involving issues such as surveillance, verification and enforcement. As some observers noted, "The men parsing the scientific details did not then have to summarise them in layman's language for the politicians who were negotiating the deal: they were themselves the politicians negotiating the deal."

When science becomes successfully integrated in a political process, the focus shifts away from questions about evidence and towards questions of action. Moniz is playing a supporting role in helping to advance the interests of his government. The negotiations had of course been long underway between the US State Department and the Iranian Foreign Ministry before Moniz was brought in to help finalise a proposed deal. Moniz was part of the supporting cast behind President Obama and Secretary of State John Kerry.

At no time in the process was the phrase "the science says that we must ..." ever uttered to justify one course of action over another. When President Obama was first elected in 2008, he promised to restore science to "its rightful place." In the Iran negotiations, he appears to have done so. For those wanting to see science playing a greater role in politics, the lesson is clear: success lies in integrating science with politics, and not in advocating some special role for "science advice."

Historians will debate the efficacy of JCPOA[34] years from now, weighing whether more restrictions on Iran's belligerent Middle East behavior could have been achieved; but however they judge it, they will almost certainly conclude that Moniz lent extraordinary credibility to the terms of the agreement. The disposition of fissile material; the dismantling and moth-balling of advanced centrifuges; the conversion of the heavy-water Arak reactor to a light-water facility, dedicated to medical isotope rather than plutonium production; and the extremely intrusive inspection regime all bore Ernie's stamp of approval.

On July 14, 2015, all eight parties to the agreement signed off on the final terms. Iran agreed to suspend its nuclear weapons activities, dismantle three quarters of its centrifuges, ship its partially enriched uranium abroad, terminate its plutonium production, and commit to monthly inspections of its facilities. In return, United Nations sanctions were to be lifted, and Iranian assets held by foreign banks were to be released. "Adoption Day,"[35] as the State Department labeled it, occurred 3 months later on October 18.

But Barack Obama, facing stiff opposition from congressional Republicans, never asked the Senate to ratify the agreement. Two years and seven months later, the absence of treaty approval allowed President Donald J. Trump to announce his intention to withdraw the United States from JCPOA.[36] Unlike Obama, Trump had neither a science advisor nor a science savvy Cabinet member to advise him. Unlike Obama, who insisted on detailed technical briefing papers, Trump seemed content to rely on his gut. If you're lucky, sometimes that can work. But, more often than not, successfully navigating the science and technology policy maze in the 21st century requires trustworthy information, fact-based analysis, sound political judgment, impeccable timing, and considerable wisdom.

By the time Barack Obama entered the White House in January 2009, global temperatures had resumed their upward climb,[37] polar icecaps were receding,[38] ocean levels were rising[39], and extreme weather events—severe storms with devastating floods and extended droughts with damaging wildfires—were occurring with increasing frequency.[40] Atmospheric carbon dioxide had reached highs never seen before in the past 800,000 years, and the trend was ever upward.[41]

These were facts, not speculations, and they led the overwhelming majority—more than 97%—of scientists to conclude that (1) global warming was a reality; (2) climate change was a consequence; (3) green-house gases (among them, most problematically, carbon dioxide) were the cause; and (4) human activity was a major, if not the dominant, contributor. For policymakers, it should have been a slam dunk: contain greenhouse gas production and save the planet. The story of what actually unfolded illustrates what happens when science gets too far ahead of the lay public, and when a science and technology issue becomes etched into partisan belief systems.

Climate change is a very complicated business. But the essentials of global warming are not difficult to understand, even for someone with little technical background. Unfortunately, scientists emphasized the complexity of the matter and dismissed the ability of non-scientists to grasp the essentials. Their attitude reflected an arrogance that all too often is a rite of passage to membership in the scientific priesthood. And the fallout had extraordinary consequences for sound policy.

Climate change became synonymous with party affiliation. If you were a Republican, you probably viewed it with skepticism. If you were a Democrat, you probably saw it as an ominous threat. It wasn't always so. In 1989, as the release of once-confidential memos revealed, Republican President George

H.W. Bush received and took seriously the advice he was receiving on the issue. On February 9, 1989, shortly after Bush assumed office, Frederick M. Bernthal, the Assistant Secretary of State for Oceans and International Environmental and Scientific Affairs, wrote the following to Under-Secretary-Designate for Economic Affairs Richard T. McCormack:[42]

> *Here is the background material you requested at this morning's staff meeting…*
>
> *Climate Change: The Administration will need to develop a strategy on how to address the climate change issue. Secretary [of State James] Baker has already signaled strong U.S. support for the recently organized IPCC [United Nations Intergovernmental Panel on Climate Change] process, including development of policy options for limiting emissions or adapting to climate change. The elements of such a strategy would include research to reduce scientific uncertainties, short-term measures such as additional research on energy sources and efficiency and an approach to calls for an international climate convention.*

Scroll forward 20 years, which is an eternity in politics, and you would be hard pressed to find more than several dozen Republican office holders who completely bought into the proposition that human activity was the major source of Earth's changing climate.[43] Some of the skeptics expressed nuanced views, but a number unabashedly called it a scientific hoax. Where they hailed from had a lot to do with the intensity of their doubts.[44] The most skeptical represented heavily Republican states and districts, where more than 70% of GOP voters rejected the scientific consensus.

By contrast, Barack Obama saw climate change as an existential threat to the United States and the world. He made that clear when he selected his science adviser, John Holdren. A Harvard physicist, Holdren had cut his policy teeth on environment, energy, and nuclear non-proliferation issues. But climate change had become almost a singular focus for him.

Profiting from large Democratic majorities in the House and Senate, Obama had an opportunity that few other modern occupants of the White House enjoyed. He could shape the legislative agenda to his liking—at least at the out-set—and climate change was high on his to-do list. However, shoring up the banking industry following the 2008 financial collapse and stanching the consequent economic bleeding had to come first. Afterward, he could turn his attention to other matters. He knew well that he had, at best, 18 months to achieve any legislative success before the mid-term elections intruded on the congressional calendar. And considering how slowly the Senate moved, that was about the time it would take to get one big thing done.

With the Copenhagen IPCC conference scheduled for December, climate change was a prime contender. But in January 2007, several weeks before he announced his candidacy, Obama had publicly committed to passing universal health care by the end of his first term. As *Politico* reported some years later, that promise was little more than an impromptu "check-the-box, news-cycle expedient."[45] But the setting where he delivered the pledge was not a smoke-filled room. It was the Families USA health care conference, one of the largest in the country. He simply couldn't walk back his commitment, even if he had wanted to.

Two years later, when it came time to choose his one big thing, Obama believed he had no choice. He had to deliver on health care. Climate change would have to wait. But participating in the Copenhagen conference without having anything tangible to point to on greenhouse gas emissions was a non-starter. Getting legislation through the Senate with health care hanging in the balance, he knew would be almost impossible. But if he pushed hard enough, he was optimistic the House, with its rules prohibiting extended debate, might just be able to do both. That became the Administration's strategy, and it led to a number of unforeseen consequences.

Ultimately, Obama delivered on his health care promise, signing the Patient Protection and Affordable Care Act (ACA) on March 23, 2010, 14 months into his first term. The legislation had followed a tortuous course through Congress, and when the House and Senate finally passed it, they did so without a single Republican vote.[46]

While the ACA was caught up in policy debates and political wrangling, the American Clean Energy and Security Act of 2009,[47] as the climate change legislation was officially known, was moving swiftly through the House of Representatives. Championed by Henry Waxman, the new chairman of the Energy and Commerce Committee, and Ed Markey, the chairman of the Select Committee on Energy Independence and Global Warming, the bill made it to the House floor just 6 weeks after it had been introduced. The final vote, which came on June 26, 2009, reflected the ever-widening partisan divide on climate change. Democrats voted overwhelmingly (211-44) in favor of the Waxman-Markey bill, while Republicans voted even more overwhelmingly (8-168) against it. Over the course of two decades, climate change had moved from a serious policy issue with broad concern expressed on both sides of the aisle, to a brash party-defining label, having more to do with tribalism than sound science.

Nonetheless, Obama pocketed the House victory and trumpeted it in Copenhagen as evidence of how seriously his administration considered the issue. He had hoped Congress would return to it following the 2010 midterm elections, but that was not to be.

It's worth pausing for a brief look at the science behind climate change.[48] Setting aside the complexities that make specific predictions difficult, the outlines of the science are actually quite easy to understand.

Sunlight comes in an array of visible colors, as light passing through a glass prism reveals. Violet lies at the shorter "wavelength" end of the spectrum, and red, at the longer end. The other colors, following the sequence in a rainbow, appear between. But sunlight also contains "radiation" that is not visible: ultraviolet, which has wavelengths shorter than violet, and infrared, which has wavelengths longer than red.

To understand the origin of the light, it is necessary to recognize that every hot object emits radiation, and the hotter the object is, the more the shorter wavelengths contribute. A warm cooktop, with barely any light visible, for example, emits radiation mostly in the infrared. We don't see the radiation, but we feel it—as heat. As the cooktop gets hotter, it begins to become visible, first emitting red light, and then as its temperature steadily increases, shifting toward orange. The temperature of an object actually determines the exact wavelength at which the intensity reaches its maximum value. The relationship is known as Wien's Law.[49]

With a blistering surface temperature of about 5800 K—corresponding to almost 9950° F—the Sun produces light with a maximum intensity in the green part of the spectrum. After traveling 93 million miles, the sunlight eventually reaches the Earth's atmosphere, where it interacts with atoms and molecules in the air. Those constituents prevent most of the ultraviolet, and a fraction of the infrared wavelengths from reaching the Earth's surface. Ozone absorbs the most harmful ultraviolet rays, and other molecules, known as greenhouse gases, absorb a portion of the infrared rays.

The visible light fares better, but it doesn't get a free pass to the Earth's surface. Clouds reflect some of it, and atmospheric molecules scatter much of the rest. Shorter wavelengths scatter more than longer ones, making the sky appear blue, and sunsets, red. The light that does make it through the atmosphere heats the Earth's surface, helping to make our planet habitable.

The sunlight's tale doesn't end there. The Earth's surface reemits radiation, but mostly in the infrared, because its temperature is about 288 K (59° F), 20 times lower than the Sun's. Some of the emitted infrared radiation escapes into space, but greenhouse gases absorb the rest, trapping heat and raising the overall temperature of our planet.

How hot the Earth becomes depends on which greenhouse molecules are present, and how many there are. Venus, for example, which has a very dense atmosphere of mostly carbon dioxide—a greenhouse gas—is the hottest planet in the Solar System, even though it is not the closest to the Sun. And if the Earth's atmosphere didn't contain any greenhouse gases, the average temperature would be 255 K (0° F).

Not all greenhouse gases are equally potent, and not all of them are equally problematic. For example, water vapor and methane absorb infrared radiation extremely well. But neither stays in the atmosphere very long. By contrast, carbon dioxide doesn't absorb infrared radiation as readily, but once it enters the atmosphere, it stays there for centuries.

A few more facts complete the climate science chronicle. (1) We know that burning fossil fuels—coal, oil, and natural gas—is the principal contributor to atmospheric carbon dioxide, which reached more than 410 parts per million in 2018, the highest level in almost a million years, the span of accessible data. (2) We know that the Earth was a much warmer planet 65 million years ago, because dinosaurs roaming the continents were cold-blooded. (3) We know that plants absorb carbon dioxide from the air, and through photosynthesis, convert it into carbon. (4) We know that coal, oil, and natural gas are the result of carbon sequestered when plants died and became buried. (5) Finally, we know that if we continue to burn the sequestered carbon, we will ultimately return the Earth to the conditions that existed 65 million years ago. How fast we get there will depend on the strategies we pursue.

In short, we can choose our policies but not our science. We can limit fossil fuel emissions by relying more on wind, solar, and nuclear energy, and becoming more energy efficient. Or we can opt to continue emitting carbon dioxide at a high rate and learn to live with the consequences. Altering our behavior is a choice; altering climate science is not.

The Waxman-Markey bill was based on a policy choice: Restricting carbon dioxide emissions. To achieve the goal, it incorporated a policy prescription: A Cap and Trade mechanism, which environmentalists favored, rather than a straight carbon tax or a carbon fee combined with a rebate—a "fee-bate"—which most economists advocated. The carbon Cap and Trade policy, on which we'll elaborate in this section, followed the successful approach used to reduce emission of sulfur dioxide (SO_2) from coal-fired utility plants, especially in the Ohio Valley, which was the principal cause of harmful acid rain in the Northeast.

The SO_2 plan began with an inventory of current emissions to establish a baseline for the total. It then awarded each power plant an allowance for the first year, reflecting the plant's fraction of the existing total. Finally, it established annual percentage decreases for the total, reducing each plant's allowance by the same percentage. If a plant exceeded expectations—by switching to natural gas, for example, or capturing SO_2 in the smokestack before it escaped—it could sell (trade) its unneeded allowance to a plant that was unable to meet the required reduction.[50]

Environmentalists favored the Cap and Trade approach because the cap controls total emissions in a completely predictable fashion. The trade mechanism allows the free market to set a price for pollutants and treat them like any other commodity, in principal optimizing the cost of achieving the required reductions. For SO_2, Cap and Trade worked well. The limited sources of the pollutant made the inventory simple, and monitoring plant compliance relatively easy. Emissions declined, and acid rain abated.

The Waxman-Markey bill hoped to replicate the approach for carbon dioxide (CO_2). But regulating CO_2 through a Cap and Trade policy would be considerably more difficult. The sources of the emissions were numerous and disparate. Power plants, for example, which can be tracked easily, accounted for only 40% of CO_2 emissions in 2009.[51] Transportation, which is far more distributed and difficult to monitor, accounted for 34%, while residential housing, commerce and industry, also difficult to inventory, were responsible for 26%. In addition, many American companies either produced products overseas, or used components fabricated in other countries. Folding CO_2 emissions into such a global system would add another layer of complexity.

Many economists saw another major problem with Cap and Trade. It was difficult to predict the implementation costs at the outset, and until the allowance trading market stabilized, the cost uncertainties could have a deleterious impact on the overall economy. A tax or fee-bate system,[52] they believed, was simpler to implement, with much greater economic predictability. It had two flaws, however. While the costs to the economy were far more predictable, the reduction in emissions was far less certain. In addition, slapping a tax or a fee on energy, even with a rebate to mitigate its impact, might prove politically unrealizable. With the European Union having already adopted Cap and Trade to regulate CO_2 in 2005, Obama lent his support to the Waxman-Markey legislation.

The 2010 mid-term election was a disaster for Obama's presidency. The American Recovery and Reinvestment Act (ARRA) had bailed out Wall Street, but in the minds of many voters, it had left Main Street behind. That perception, and the Administration's inability to explain the benefits of the Affordable Care Act, proved to be albatrosses around the necks of Democrats. When the votes were tallied that November, Republicans had racked up major victories. Having wondered only 2 years earlier whether they could ever regain the majority in either chamber of Congress, they rode an electoral wave led by Tea Party populists and picked up sixty-three House seats, the largest gain by a party in 70 years. Any hope Obama had for achieving climate change legislation had effectively vanished, especially because Republicans had taken control of many state legislatures and governorships. Those victories would allow them to gerrymander districts and solidify their control of the House for the next decade.

Obama still had some options to exercise on climate change and energy policy. The 2007 Energy Independence and Security Act[53] had set the course for significant improvements in vehicle efficiency. It increased the Corporate Average Fuel Economy (CAFE) Standards from 27.5 miles per gallon, then in existence, to a minimum of 35 miles per gallon for passenger vehicles in 2020. It also required CAFE standards for passenger and nonpassenger vehicles manufactured between 2021 and 2030 to be the maximum feasible. To that end, it directed the Secretary of Transportation to ask the National Academy of Sciences for an evaluation of the standards for passenger vehicles and trucks and report its findings to Congress. In other words, it gave Obama wide latitude to require major improvements in vehicle efficiency.[54] The only drawback was that any new standard above 35 miles per gallon for cars was not legislatively binding. A subsequent president could roll it back—and the next one did, or at least tried to.

Obama took advantage of the wide berth he was given, and on July 29, 2011, his White House announced a new standard:[55] 54.4 miles per gallon for cars and light trucks by 2025. The report that accompanied the announcement, "Driving Efficiency: Cutting Costs for Families at the Pump and Slashing Dependence on Oil,"[56] explained the rationale and provided a historical context. It's worth noting that nowhere in the title do the words "climate change" or "global warming" appear. By that time, it was clear to anyone paying attention to the political land-scape that such words would have been toxic to the voters who had installed the new House majority. Saving families money and patriotically achieving energy independence would play much better in the American heartland. The Obama communications shop didn't always get it right, but in this instance, it properly recognized that framing a message is as important as the message itself.

Obama won re-election handily in 2012, but Democrats were only able to add two seats to their Senate voting majority (55-45),[57] far from the two-thirds needed to ratify a treaty, or even the three-fifths needed to terminate debate. With Republicans still maintaining a hefty (234-201) majority in the House, Obama knew he would have little chance to advance his agenda legislatively. The 2014 midterm election made matters worse for him. Democrats lost even more ground in the House (247-188), and Republicans gained control of the Senate (54-46), largely as a result of the unpopularity and botched rollout of the Affordable Care Act. The deck was stacked against Obama, but he still had a few cards to play. For 3 years, as hyper-partisanship froze Washington, he used executive orders and regulatory authority to achieve his policy objectives.

As Obama entered his last year in the White House, he was acutely aware that as diminished as his second-term legislative influence had been, it was about to shrink even more. He was on the threshold of becoming a lame duck president. But an international convocation in late 2015 gave him one last shot at making a significant mark on climate change.

The venue was in Le Bourget, outside Paris at the 21st Conference of the Parties of the United Nations Framework Convention on Climate Change (UNFCC). Representatives of the participating nations began hammering out an agreement in November, and by the middle of the next month, they had reached a consensus. On December 12, all 196 nations signed onto the accord.[58]

The participants had settled on a single objective: To keep the average global temperature from reaching 2° C (3.6° F) above preindustrial levels, and to strive to keep the increase to 1.5° C (2.7° F). Well aware of the failure of previous negotiations in Kyoto (1997), Bali (2007), and Copenhagen (2009) to secure binding action on CO_2 reductions, the Paris negotiators abandoned the idea of forcing signatories to set binding targets by certain dates. Instead, they settled on a protocol that was more palatable, especially to industrialized nations. Article 4 spelled out the new approach. The following excerpt highlights the major elements:

1. In order to achieve the long-term temperature goal…, Parties aim to reach global peaking of greenhouse gas emissions as soon as possible, recognizing that peaking will take longer for developing country Parties, and to undertake rapid reductions thereafter in accordance with best available science, so as to achieve a balance between anthropogenic emissions by sources and removals by sinks of greenhouse gases

in the second half of this century, on the basis of equity, and in the context of sustainable development and efforts to eradicate poverty.

2. Each Party shall prepare, communicate and maintain successive nationally determined contributions that it intends to achieve. Parties shall pursue domestic mitigation measures, with the aim of achieving the objectives of such contributions.

3. Each Party's successive nationally determined contribution will represent a progression beyond the Party's then current nationally determined contribution and reflect its highest possible ambition, reflecting its common but differentiated responsibilities and respective capabilities, in the light of different national circumstances.

4. Developed country Parties should continue taking the lead by undertaking economy-wide absolute emission reduction targets. Developing country Parties should continue enhancing their mitigation efforts, and are encouraged to move over time towards economy-wide emission reduction or limitation targets in the light of different national circumstances.

5. Support shall be provided to developing country Parties..., recognizing that enhanced support for developing country Parties will allow for higher ambition in their actions.

6. The least developed countries and small island developing States may prepare and communicate strategies, plans and actions for low greenhouse gas emissions development reflecting their special circumstances.

7. Mitigation co-benefits resulting from Parties' adaptation actions and/or economic diversification plans can contribute to mitigation outcomes under this Article.

8. In communicating their nationally determined contributions, all Parties shall provide the information necessary for clarity, transparency and understanding...

9. Each Party shall communicate a nationally determined contribution every 5 years...

10. The Conference of the Parties serving as the meeting of the Parties to this Agreement shall consider common time frames for nationally determined contributions...

11. A Party may at any time adjust its existing nationally determined contribution with a view to enhancing its level of ambition...

12. Nationally determined contributions communicated by Parties shall be recorded in a public registry maintained by the secretariat.

13. Parties shall account for their nationally determined contributions...

14. In the context of their nationally determined contributions, when recognizing and implementing mitigation actions with respect to anthropogenic emissions and removals, Parties should take into account, as appropriate, existing methods and guidance under the Convention...

15. Parties shall take into consideration in the implementation of this Agreement the concerns of Parties with economies most affected by

the impacts of response measures, particularly developing country Parties.

16. Parties…shall notify the secretariat of the terms of that agreement, including the emission level allocated to each Party within the relevant time period, when they communicate their nationally determined contributions. The secretariat shall in turn inform the Parties and signatories to the Convention of the terms of that agreement.

17. Each party to such an agreement shall be responsible for its emission level as set out in the agreement referred to in paragraph 16 of this Article in accordance with paragraphs 13 and 14 of this Article…

18. If Parties acting jointly do so in the framework of, and together with, a regional economic integration organization which is itself a Party to this Agreement, each member State of that regional economic integration organization individually, and together with the regional economic integration organization, shall be responsible for its emission level as set out in the agreement communicated under paragraph 16 of this Article in accordance with paragraphs 13 and 14 of this Article…

19. All Parties should strive to formulate and communicate long-term low greenhouse gas emission development strategies,…taking into account their common but differentiated responsibilities and respective capabilities, in the light of different national circumstances.

By eliminating binding targets, signatories hoped that critics of previous accords would be mollified. Secretary of Energy Ernie Moniz, who joined Secretary of State John Kerry in leading the United States delegation, wasn't as sanguine. He knew the Administration had to frame the climate agreement with an eye toward the American political climate. Otherwise, congressional Republicans were bound to find fault with it. Instead of emphasizing the threat of climate change, he stressed the importance and benefits of innovation. He made that clear when he addressed the Paris participants:[59] "At the risk of—using a Brazilian analogy—being a one-note samba, innovation is one of the foundations for increased ambition as one re-examines targets in the years ahead."

After he returned home, he convened a meeting at the Department of Energy's (DOE) Independence Avenue headquarters—known as much for its Brutalist style of architecture as its extensive bureaucracy—and repeated the message to a group of forty policy wonks, lobbyists, and high-level bureaucrats. The Paris Agreement was all about energy innovation, he said, in which the United States had the talent and scientific infrastructure needed to lead the world. "Mission Innovation" was going to be the DOE's signature message when he testified in the coming weeks on Capitol Hill.

Moniz combined all the attributes of a successful Beltway science and technology policymaker. He was smart but not arrogant; persuasive but not antagonistic; well-versed in history but not intimidated by precedent; forward-looking but not starry-eyed. He was also extremely knowledgeable and analytical; outgoing and engaging; and above all, very politically aware. If anyone could market science and technology policy, it was Ernie.

A few weeks after the meeting at the DOE Forrestal building, he testified on the Administration's budget request before several House and Senate committees. He sold Mission Innovation[60] as "essential for economic growth enabled by affordable and reliable energy, for energy security, for U.S. competitiveness, and for a transition to a low carbon energy future."[61] He mentioned climate change and the president's Climate Action Plan on several occasions, but embedded the references in a much broader agenda. He also championed the use of nuclear power, a priority of many conservatives.[62]

For the most part, members of Congress on both sides of the aisle responded cordially, if not favorably. But in the end, the appropriations outcome reflected the predictably reflexive position of budgetary hawks. As the American Institute of Physics Federal Science Budget Tracker reported at the time,[63] "The Department of Energy and its science programs stand out as winners in the budget request, as the department would play a central role in President Obama's proposal to double clean-energy R&D over 5 years as part of the Mission Innovation initiative. However, key appropriators have largely dismissed this proposal, instead choosing to cut or flat fund renewable energy, reverse cuts to fossil energy, and increase spending on nuclear energy, the Office of Science, and ARPA-E."

The United States signed the Paris Agreement. But not even Ernie Moniz's astute marketing strategy could overcome the political obstacles on Capitol Hill. Opposing climate change was a badge of courage for most Republicans, and spending money—however it was disguised—on something they didn't believe in, was simply a non-starter. Once again, politics trumped science, science policy, and the science of science policy. It was ever thus, and it will be ever thus. On that note, so ended the Obama years.

Chapter 12

Loose change

During the course of two centuries or more, science and technology have transformed American life in ways almost unimaginable. The pace of the transformation has accelerated remarkably in the past few decades, so much so, that as a species we are facing difficulties adapting. In his book, *Thank You for Being Late: An Optimist's Guide to Thriving in the Age of Accelerations*,[1] Pulitzer Prize winning author and *New York Times* columnist Thomas Friedman described the significant challenges we face today, and prescribed a number of bromides that might ease our growing discomfort. He identified the major disruptive technologies—the iPhone, integrated circuits, search engines and the World Wide Web, the Internet, the cloud, DNA sequencing, and cognitive computation (also known as artificial intelligence)—that have fueled the accelerations. In the end, he argued that without a commitment to "lifelong learning" many of us will find it increasingly difficult to cope.

Shining a spotlight on the technologies that have driven the accelerations, as Friedman has done compellingly in his book, is invaluable. But spotting the policies that enabled the technological revolution is essential for successful planning. While most of the advances Friedman identified are creations of the last decade of the 20th century—large-scale integrated circuits being a notable exception—in truth, the era of rapid change began well before. And much of it had to do with policies that dramatically altered industrial research and development and, in the process, ushered in an era of extraordinary scientific creativity, entrepreneurship, and science as a global enterprise.

Prior to World War II, industry played the leading role in the American research theater. Government, to the extent it had any billing, was very much a supporting actor. The end of the war brought with it dramatic change. Industry did not cede the spotlight, but the federal government was no longer in the wings. Within several decades it would command an important part of the stage, becoming a major sponsor of long-term, fundamental research.

The federal government's dominance in that arena resulted from the confluence of two policy streams. One was targeted: establishing federal agencies devoted to research and funding their programs generously, as Chapter 4 detailed. The other was ancillary, but just as consequential: enforcing antitrust laws more rigorously and changing the tax code. No description of the science

Navigating the Maze. https://doi.org/10.1016/B978-0-12-814710-8.00012-7

and technology policy maze would be complete without an accounting of the impact of those two policy decisions.

The war effort allowed American manufacturers to recover from the devastations of the Great Depression. It also caused them to appreciate the importance of technology, which had propelled the Allies to victory. By the time the Axis Powers surrendered, there were few corporate leaders who did not recognize that innovation held the keys to future profitability of their companies. Corporate giants of the post-war era, stalwarts, such as AT&T, General Electric (GE), and General Motors (GM); and newbies, such as IBM, Hewlett-Packard (H/P), and Xerox, saw "vertical integration" as the operating model that would lead to enduring corporate empires. Building on Vannevar Bush's proposition that basic research was critically important to technological progress, they established central laboratories that amalgamated basic research, applied research, development, testing, and evaluation—everything needed prior to production and marketing. Many of them achieved great prominence, but none more so than AT&T's Bell Laboratories. When it came to discovery and innovation, it was *la crème de la crème*.

Bell Labs attracted the best minds, and it gave them extraordinary latitude to pursue their most creative ideas. And that faith paid off. Its scientists invented the transistor—the foundation of semiconductor electronics—and the charge-coupled device (CCD)—the backbone of high-quality digital imaging. They also developed radio astronomy, the laser,[2] and information theory. And they created Unix—the platform used by Apple's Mac OS (operating system)—as well as the ubiquitous C programming language. Seven Nobel Prizes in physics and one in chemistry went to scientists—fourteen in all—working at Bell Labs. There was no other place in the world quite like it, and it is unlikely there will ever be another, at least in the private sector.

Explaining the extraordinary success of Bell Labs is like fitting pieces of a jigsaw puzzle together to reveal a great work of art. In his book, *The Idea Factory: Bell Labs and the Great Age of American Innovation*,[3] Jon Gertner does just that. He captures the culture that personified Bell Labs, which at its zenith in the late 1960s, employed 15,000 people at its sprawling New Jersey campus. Its staff included more than 1200 Ph.D.s, mostly in physics, chemistry, and materials science. Walter Isaacson's commentary, "Inventing the Future,"[4] which appeared in *The New York Times Sunday Book Review* in 2012, provides a succinct summary of Gertner's work.

Although Bell's management valued long-term research highly, it maintained an unwavering focus on AT&T's business needs: developing, manufacturing, and capitalizing on technologies that connected people across the country and across the world reliably and at a reasonable cost. Throughout the years, directors of Bell Labs recognized that outstanding scientists and engineers were an invaluable commodity, and they gave them the freedom to discover and invent, trusting that their work would improve AT&T's bottom line. They also understood that research breakthroughs could be transformational; that developing new technologies required patience; and that complex problem-solving profited from collaborative efforts that cut across scientific and engineering disciplines.

The culture of Bell Labs reflected a critical characteristic of modern science and technology that was not well appreciated at the time: how complex the relationships are among basic research, applied research, innovation, and development. Donald Stokes is generally credited with clarifying the nature of the connections in his 1997 treatise, *Pasteur's Quadrant: Basic Science and Technological Innovation.*[5] Until then, many science and technology policy professionals viewed the connections as a linear progression: basic research leading to applied research; applied research leading to innovation and development; and development leading to production.

Michael Armacost, president of the Brookings Institution, which published the book, framed the issue with these words, when he wrote in the Preface:

> *More than fifty years ago, Vannevar Bush released his enormously influential report,* Science the Endless Frontier, *which asserted a dichotomy between basic and applied science. This view was at the core of the compact between government and science that led to the golden age of scientific research after World War II—a compact that is currently under severe stress. In this book, Donald E. Stokes challenges Bush's view and maintains that we can only rebuild the relationship between government and the scientific community when we understand what is wrong with this view.*
>
> *Stokes begins with an analysis of the goals of understanding and use in scientific research. He recasts the widely accepted view of the tension between understanding and use, citing as a model case the fundamental yet use-inspired studies by which Louis Pasteur laid the foundations of microbiology a century ago. Pasteur worked in the era of the "second industrial revolution," when the relationship between basic science and technological change assumed its modern form. During subsequent decades, technology has been increasingly science-based—with the choice of problems and the conduct of research often inspired by societal needs.*
>
> *On this revised, interactive view of science and technology, Stokes builds a convincing case that by recognizing the importance of use-inspired basic research we can frame a new compact between science and government.*

Replace the word "societal" with "business" in the last line of the second paragraph, and you have the Bell Labs paradigm.

Stokes presented a compelling proposition for reducing the complex research and development relationships to a two-dimensional space, contending that it was more relevant and accurate in today's world than the "linear model," which he attributed to Vannevar Bush. A few clarifying words about Stokes' paradigm in a moment, but first a comment about his assertion that Bush got the relationships wrong. Stokes wrote:

The belief that understanding and use are conflicting goals—and that basic and applied research are separate categories—is captured by the graphic that is often used to represent the "static" form of the prevailing paradigm, the idea of a spectrum of research extending from basic to applied:

This imagery in Euclidean one-space retains the idea of an inherent tension between the goals of understanding and use, in keeping with Bush's first great aphorism ["Basic research is performed without the thought of practical ends."], since scientific activity cannot be closer to one of these poles without being farther away from the other.

The distinction of basic from applied research is also incorporated in the dynamic form of the postwar paradigm. Indeed the static basic-applied spectrum associated with the first of Bush's canons is the initial segment of a dynamic figure associated with Bush's second canon, the endlessly popular "linear model," a sequence extending from basic research to new technology:

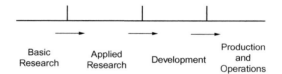

The belief that scientific advances are converted to practical use by a dynamic flow from science to technology has been a staple of research and development (R&D) managers everywhere. Bush endorsed this belief in a strong form—that basic advances are the principal *sources of technological innovation , and this was absorbed into the prevailing vision of the relationship of science to technology. Thus an early report of the National Science Foundation commented in these terms in this "technological sequence" from basic science to technology, which later came to be known as "technology transfer" ....*

Stokes was correct in asserting that the linear model does not capture the essence of research and development. But he was wrong in claiming that the model was "the staple of R&D managers." It certainly was not true at Bell Labs, the most successful industrial R&D enterprise of the 20th century. Stokes was also wrong in asserting that Vannevar Bush propounded the "linear model."

First, Bush never made the case for such a paradigm explicitly. And second, the extent to which he implied such a relationship exists must be viewed in the context of the times. Prior to World War II, the federal government's support of academic research was vanishingly small. Bush's objective was to alter the equation substantially. Emphasizing the importance of basic research was, above all, essential to a successful political strategy. And in the end, it worked.

Harley Kilgore might have fought Bush over the structure of the National Science Foundation—ultimately prevailing—but he, too, understood the political necessity of stressing basic research in selling Truman on the importance of science in the federal government's portfolio.

Despite his overreach, Stokes made a compelling case for a new way of thinking about research, innovation, and development. He captured the essence of his proposed paradigm with two simple diagrams. The first, the "Quadrant Model of Scientific Research," provided a new static picture:

Research is inspired by:

Consideration of use?

	No	Yes
Yes	Pure basic research (Bohr)	Use-inspired basic research (Pasteur)
No	Data and taxonomy	Pure applied research (Edison)

Quest for fundamental understanding?

In his original work, Stokes left the lower left quadrant empty. Subsequent science and technology policy scholars recognized that data and taxonomy are essential for research of any kind, and added those categories.

Stokes replaced the linear dynamic model with one showing far greater complexity:

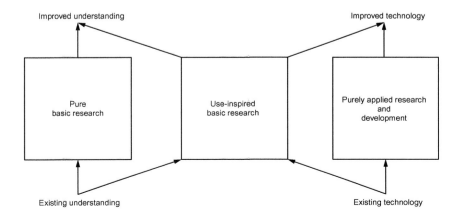

Arguably, even his new model was too simplistic, because it failed to incorporate the critical role existing technology plays in "pure basic research." The work of Arno Penzias and Robert Wilson, two Bell Labs physicists who won the Nobel Prize in 1978, illustrates that point perfectly. Using a state-of-the-art Bell Labs radio-antenna, originally developed to detect signals reflected from balloon satellites, they made the somewhat serendipitous discovery of "cosmic microwave background radiation." Their observation was momentous. It provided the smoking gun cosmologists needed to nail down the "big bang" theory of the origin of the universe.

Bell Laboratories was blessed with another benefit: extraordinary financial backing by its parent company, AT&T, which seemingly had limitless supplies of cash to spend on it. It was not a fortuitous circumstance. In 1921, Congress passed the Willis-Graham Act,[6] which, in the interests of promoting universal communications access, exempted telephone companies from federal anti-trust laws. During coming decades, AT&T, which was the largest telephone company at the outset, either bought most of its competitors outright, or brought them under the Bell System operating umbrella. By the end of World War II, "Ma Bell," as the collective enterprise was known, consisted of 22 local Bells, AT&T "Long Lines," Western Electric, and Bell Labs. The overwhelming majority of Americans were customers of Ma Bell.

It was a cozy relationship; too cozy, critics said. The local Bells and AT&T Long Lines provided the telephone service. Western Electric manufactured all the equipment local Bell and Long Lines customers were allowed to use—unless they paid additional usage fees. And Bell Labs was the powerful R&D arm of the vertically integrated corporation. The federal government set the phone rates, based on the company's operating costs, and guaranteed AT&T a predictable profit. Every member of Ma Bell's family was required to pay a fixed fraction of its revenues to Bell Labs. In some sense, every dime Bell Labs spent was a cost AT&T claimed in its negotiations with federal rate regulators. On such a playing field, the company's premier R&D facility was a freebie.

Overreach is often the downfall of the king of the mountain, and in 1949, as telecommunications began to embrace a raft of related technologies, AT&T's monopolistic behavior drew the attention of the Justice Department. To stave off more draconian measures, the company agreed to limit its ownership to 85% of the American service networks. The "Consent Decree" allowed AT&T to prosper, and by not altering the Bell Labs financing arrangement, it ensured that the company's R&D arm would thrive, as well.

The 1949 agreement notwithstanding, critics continued to hammer away at Ma Bell's undue influence over rapidly developing telecommunications technologies. They argued that AT&T, by virtue of its size and control over the market, was stifling innovation. A vertically integrated monopoly might be good for the company, but it was not good for the country, so Judge Harold Greene ruled decades later. It came at the end of one of the most significant trials involving American science and technology policy.

With fiber optics and cellular mobile communication technologies already visible on the horizon, the Justice Department filed an antitrust action against AT&T in 1974 for restraint of trade, focusing the complaint on Western Electric's monopoly over the equipment AT&T used throughout its business. William O. "Bill" Baker, a renowned chemist, had recently become president of Bell Labs when the Justice Department hit AT&T with the lawsuit. It came as Bell Labs was just about to turn fifty, and Baker could see that it probably would not be around to celebrate its centennial if the Justice Department prevailed in breaking up the company.[7]

It's easy to see how he reached his pessimistic prediction. If the local Bells—or "Baby Bells" as they would be called in the aggregate following the divestiture— were free to purchase equipment on the open market, and if neither they nor Western Electric was obligated to support Bell Labs, the rationale for AT&T's signature R&D facility would disappear, and its business model would fail.

It took them 8 years, but in early January 1982, AT&T and the Justice Department came to an accommodation. The local Bells would be cut free, Western Electric would have to compete in the open market, and AT&T would be allowed to enter other telecommunications arenas. When the divestiture took place on January 1, 1984, Bell Labs was split in two. The entity that retained the hallowed name became a wholly owned unit of AT&T Technologies, as Western Electric was now called. The other, redesigned to serve the needs of the newly independent Baby Bells, got the moniker Bellcore.

In the immediate aftermath of the settlement, Bell Labs continued to generate extremely high-quality research. For their work in the 1980s, two Lab scientists received physics Nobel Prizes, Steven Chu in 1997 for "laser cooling" of atoms and Horst Störmer in 1998 for discovering the fractional quantum Hall effect. But the divestiture eventually took its toll. In 1996, AT&T jettisoned its manufacturing and research arms, establishing Lucent Technologies as an independent company. A decade later, Lucent merged with Alcatel, a French electronics company, and by 2008, the once vaunted Bell Laboratories claimed only four physicists on its staff. So it was of little consequence when Alcatel-Lucent declared an end to all research in materials science, semiconductors, or any kind of "basic research."

Bell Labs' run as a powerhouse of scientific discovery and innovation lasted a little more than 80 years. We usually don't think of the judiciary as a major player in science and technology policy. But Judge Greene's decision to accept the Justice Department's antitrust briefs was just that. The ensuing divestiture began Bell Labs' long slide into oblivion, as Bill Baker had predicted at the time; but it opened up the field of telecommunications to thousands of new actors. Innovators flourished, venture capitalists made large bets on new technologies, entrepreneurs grew into billionaires, and companies such as Apple, Google, and Facebook became the new faces of the 21st century. All of that might have happened naturally, but it's more than speculation that Judge Greene accelerated the start of the "Age of Accelerations."

If you worked or visited Bell Labs before the long slide downward began, you understood that bringing extremely smart people together from different science and technology disciplines all under one roof, giving them significant resources, and allowing them to make decisions with minimal interference from distant management, was the best way to solve big, complex problems.

Steven Chu, the physics Nobel Laureate, took that conviction with him after he left Bell in 1987. He seized the opportunity to put the philosophy to work almost immediately after he arrived at Stanford University that same year. Even though he was a physicist, he initiated an interdisciplinary research effort to tackle difficult questions in biomedicine. Seven years later, after becoming director of Lawrence Berkeley National Laboratory, he challenged the LBNL staff to solve an extremely complex problem: finding efficient ways to turn sunlight into liquid fuels. Success would be a game changer for climate change.

In 2009, he had an opportunity to make an indelible mark in a much larger arena. That was the year he became Secretary of Energy. And in his first budget request, he proposed establishing a series of Energy Innovation Hubs, each having the feel of Bell Labs experience, but on a smaller scale. Each Hub would focus on one of eight grand energy challenges; each would receive 5 years of guaranteed funding; and each would be untethered as much as possible from the Washington bureaucracy.

Chu knew what he wanted, but he was a political novice, and his attempt to change the energy equation turned out to be far more difficult than he had imagined. He sold the Hubs as a set of mini-Bell Labs, but his sales pitch fell flat on Capitol Hill. Members of Congress respected his pedigree and admired his Nobel Prize, but in private many of them confided that he came across as condescending and arrogant. They also criticized him for not adequately prepping his high-level staff, whom they said failed miserably in explaining what the Hubs could achieve that existing Department of Energy programs couldn't. Even with Democrats fully in charge of the federal government, Chu found himself sailing into the wind with little room to tack.

Chu was brilliant, innovative, ethical, and a devoted public servant; but he was poorly schooled in politics. He had wonderful policy ideas, but he never fully understood how to minimize opposition to them in a town that relishes skewering every new kid on the block, no matter how smart and how honored. The highpoint of Chu's tenure in Washington was probably his Senate confirmation hearing on January 13, 2009. Accompanied by his family, Chu elicited nothing but praise—even awe—from members of the Energy and Natural Resources Committee in the art-deco Dirksen Hearing Room.[8] On that day, Chu had star quality.

If his Washington confirmation hearing engendered awe, an announcement a continent away on August 30, 2011 provoked utter disdain. That was the day Solyndra closed its doors for the last time.[9,10] Two years earlier, the Freemont, California solar panel manufacturer had been the recipient of a $535 million loan guarantee as part of the America Recovery and Reinvestment Act (ARRA).[11] It happened when a set of Obama Administration policy imperatives converged.

The White House wanted to get money out the door quickly to stimulate the badly failing American economy. It wanted to wean the nation off fossil fuels to combat climate change. It wanted to bolster the emerging American solar panel industry. And to satisfy proponents of renewable energy, it wanted to bring federal government's treatment of solar and wind power into harmony with its treatment of nuclear power. It very hastily placed a big, but risky, bet on Solyndra.

Chu, who embraced all the imperatives, decided to give the company his personal seal of approval, appearing in person with California Governor Arnold Schwarzenegger at the Freemont groundbreaking ceremony. His decision to do so, at the time, provoked consternation among Beltway cognoscenti. If the Department of Energy's (DOE) gamble on Solyndra didn't pay off, Chu would own the fallout of the hurried decision.

Experienced policymakers and bureaucrats had long recognized that visibly endorsing a poorly vetted a project could cost them their jobs if the project turned sour and was more than a blip on the ledger. But Chu was not experienced, and Solyndra was far from a mom-and-pop venture. It had raised more than $800 million in venture capital, and the federal government was on the hook for more than half a billion dollars if the company failed to deliver.

Had Chu designated one of his low-ranking subordinates to represent the DOE at the groundbreaking, he might have survived beyond Obama's first term. But he was so closely tied to the Solyndra fiasco, that not even his scientific accolades could save him. He resigned shortly after Obama took the oath of office in 2013.

Steve Chu's rocky tenure at the Department of Energy showed that science and technology policy is not very different from many other innovative endeavors in life. Having a first-rate idea or product is just the beginning. To succeed, you need to know how to market it with a compelling and enticing story. You need to understand your customer—in Chu's case, members of Congress—and you need to tailor your idea—Energy Innovation Hubs, for example—to meet their needs. You also have to understand the political landscape and know how to protect yourself from the barbs that will inevitably be hurled your way. Chu, whom I know personally, is a truly remarkable scientist and an outstanding public servant, but unfortunately, he never quite accommodated himself to the ways of Washington.

Steve Chu shared the 1997 Nobel prize for developing a process known as "laser cooling." The seemingly arcane subject involves using laser radiation to bring a beam of atoms traveling thousands of mile per hour to a virtual standstill. In that final state, the atoms behave as if they are almost frozen in place at a temperature close to "absolute zero," a condition that allows atomic clocks to measure time with extraordinary precision.

Achieving low temperatures—although not nearly as low as Chu and others reached—had been part of the physics research toolkit for decades. But prior to the development of laser cooling, getting there required the use of helium refrigerators. Which brings us to a strange policy saga that began in 1960 and lasted nearly half a century. First, a few preliminaries.

Helium is a noble gas. It is light and chemically inert, and it liquifies at the lowest temperature of any element, never becoming solid, even at a temperature of absolute zero under normal atmospheric conditions. Its properties make it almost essential today for a variety of critical applications, among them, most significantly, semiconductor manufacturing, magnetic resonance imaging (MRI), advanced nuclear reactors, radiological weapons detectors, space applications, and many areas of fundamental physics research.

A century ago, as we noted in Chapter 4, other than party balloons, helium's key application involved dirigibles. And to that end, only two things about it mattered. It was lighter than air, and it didn't burn. Recognizing its potential military importance, Congress established a federal program[12] in 1925 to capture helium as it emerged as a byproduct of natural gas production and to store it in an underground reserve. And there it sat for several decades, attracting little attention from policymakers.

By 1960, however, other helium applications had become increasingly apparent, leading Congress to authorize a significant upgrade to the infrastructure for helium recovery, purification, and storage. But instead of appropriating the money needed for the improvements, legislators directed the Interior Department to borrow funds from the federal Treasury. The "Helium Act Amendments of 1960,"[13] which mandated the new program, further required the Interior Department to repay the borrowed money, including compound interest, within 35 years. To raise the funds, the department would simply sell gas from the Reserve on the open market.

The tab finally came due in 1995, by which time the Helium Reserve owed the Treasury more than $1.4 billion, even though the original cost was less than $20 million. Compound interest, as any investor or lender knows, is a potent financial instrument. It was a truly bizarre situation, because rarely, if ever, does the federal government mandate an authorized program to pay interest on the money the government used to establish the program in the first place.

Nonetheless, the 1960 law required the Interior Department to do just that. But meeting the $1.4 billion obligation in 1 year would require dumping a significant fraction of the gas in the Reserve all at once. That, in turn, would distort the helium market, which, by 1995, had attracted a significant number of private producers. The economic impact on those companies was uppermost in the minds of lawmakers as they debated what eleventh-hour actions to take. With the Office of Science and Technology Policy sitting on the sideline, they never focused on future scientific and technological needs.

The result was the "Helium Privatization Act of 1996,"[14] which provided a market accommodation period, but still required the Interior Department to sell off gas in the Reserve, starting in 2005. Moreover, once the loan, including accrued interest, had been paid off, the Reserve would have to close, even if gas remained in the repository. By 2013, as the payout neared completion, the Reserve still contained a large quantity of helium, amounting to about 40% of the quantity produced annually at all of the natural gas wells.

The 1996 law was clear: The Reserve had to be closed, and the remaining helium effectively forfeited. But scientists, who were dismayed by the waste of a precious resource and concerned about what future helium price hikes might mean for research, decided to make their voices heard. They argued that helium supplies were finite, that demand would continue to increase, and that squandering the Reserve was technologically and economically shortsighted. In brief, they said, closing the Reserve was bad science and technology policy.[15] They made their case forcefully and compellingly. The result was the "Helium Stewardship Act of 2013,"[16–18] which maintained the Reserve, finally taking note of the needs of science and technology, and not simply the needs of legislators to remedy a misguided appropriations workaround dating back 43 years.

Even so, Mark Elsesser, Manager of Science Policy at the American Physical Society, who has followed the issue closely, notes that research is not out of the woods. Increasing industrial and medical demand and potential constriction of future helium supplies will likely drive prices up, placing many university research programs at risk.[19] To mitigate such an outcome, Elsesser helped initiate a program with the Defense Logistics Agency that enables academic helium users to partner with DLA and benefit from the agency's lower negotiated prices.[20]

During the decades that spanned the helium saga, the world of science and technology changed dramatically. The demise of Bell Labs received immense publicity, but it was only one of the many dramatic transformations that swept over America's industrial research enterprise in the final decades of the 20th century. Apart from the pharmaceutical sector, most major companies dramatically reduced their spending on long-term research—basic, use-inspired, or applied. "Vertical integration" and "central laboratories," two catchphrases of the post-World-War II era, were consigned to the dustbins of history.[21]

Ford, GE, GM, H/P, IBM, Sylvania, Xerox, and a host of other iconic corporations abandoned their full-service R&D facilities. Instead, they scoured the globe for innovations developed by scientists and engineers wherever they were located. They bought up smaller companies for their patent rights, and they struck agreements with universities to license innovations stemming from federally funded research, just as the 1980 Bayh-Dole Act[22] envisioned. The corporate game plan no longer involved supporting research for the long haul. In the new environment, speed to market was paramount, and long-term R&D had to be sacrificed.

Significant modifications to the tax code, and profound changes in the behavior of Wall Street traders played big parts in the transformation. The tax code part of the story involves the differential treatment of earned and investment income. The first is subject to a marginal rate that increases as income rises. The second, known as the capital gains rate, is fixed, and applies to income on investments held for a specified period of time, historically between 6 months and 2 years.

Between the early 1950s and the late 1980s, the highest marginal tax rate on earned income fell dramatically.[23] At the same time, the tax rate on capital gains remained relatively flat, fluctuating between 20% and 40% over the course of five decades.[24] Even though high-income earners rarely, if ever, paid the maximum marginal rate—92% in 1952, for example—the large disparity between earned income and long-term capital gains rates during the 1950s, 1960s, and 1970s was an incentive for shareholders to exercise patience with their investment portfolios.

The 1980s brought about dramatic changes in their behavior. The gap between the two rates was narrowing quickly, and by 1988 it had actually shrunk to zero. There was no longer any tax incentive to hold onto stock longer the company's near-term forecast.

Technology also began to exert a major influence, as electronic trading grew in importance exponentially. Computer terminals replaced the trading pits on Wall Street, where men—and they were invariably men—who had "seats" on the Exchange used to shout out bids and then swap buy and sell order "slips."

At almost warp speed, investment firms built their own trading floors and populated them with thousands of their own traders. Mathematicians and physicists, by the hundreds, abandoned science careers in academia for high-paying jobs at "hedge funds." There, they developed complex trading instruments and algorithms that had little to do with the long-term projections of companies whose stocks and bonds might be on their radar screens. The "quants," as they were called, wrote the codes; the computers did the rest, and they did it at ever-increasing speeds.

As computers became faster, real-time human decision-making became an impediment to profitmaking. High-frequency trading (HFT) completed financial transactions in fractions of a second, far faster than any individual could manage. HFT needed really smart minds at the beginning and really high-technology at the end. It's dominance in today's trading place is staggering. In 2014, more than three quarters of all stock trades took place automatically at lightning speed.

In such an environment, it's easy to see why corporate projections of future earnings—known as guidance, in financial parlance—extending much beyond a few quarters were far less relevant than they had been a decade or two earlier. Fast, faster, and fastest is what mattered—on Wall Street and in corporate board rooms. To the extent that companies still had functioning R&D engines, almost all of the workings were labeled with the D for Development.

There were exceptions: pharmaceuticals, most prominently, and those imbued with the culture of California's Silicon Valley. Apple, Google, Facebook, Intel, and AMD, for example, were young enough to retain their entrepreneurial character and commitment to innovation. But in the rest of industrial America, companies increasingly looked for a path to new products that began far outside the factory gate or the corporate campus. More often than not it originated in universities or national laboratories, where long-term scientific research was the currency of the realm. The challenge was how to convert those breakthroughs into innovations and profitable industrial products. Optimizing "technology transfer," as the process is known in the policy world, and avoiding Ehlers' Valley of Death, has turned out to be far more difficult than policymakers originally imagined. It remains a work-in-progress.

James Simons, a stellar mathematician, was one of the original quants. In 1982, he founded Renaissance Technologies and, after earning tens of billions of dollars during the next few decades, he turned his attention to philanthropy, using a sizable portion of his profits to fund scientific research in universities, and at his own Flatiron Institute in lower Manhattan. Simons might be unique among quants in focusing on philanthropic support of science later in life. But the Simons Foundation, which he established in 1994, is by no means a lone ranger in the science philanthropy world. Niche players before the 2010 congressional election, which ushered in the age of American populism, science philanthropies began to grow in prominence as dysfunction became the Washington norm. They also began to grow in number.

The America COMPETES Act of 2007[25] had made a compelling case for federal support of basic science, arguing that innovation, economic growth, and global competitiveness depended on it. The 2007 legislation had set down markers for federal science agencies and their budgets, but 5 years later, it was clear the commitments would not be fulfilled anytime soon.

Assessing the gloomy Washington outlook and seeing the risks federal inaction posed to American scientific and technological leadership, six foundations launched the Science Philanthropy Alliance (SPA)[26] in 2012. They viewed SPA, not as a substitute for the dominant role the federal government played in supporting basic science, but rather as a means of filling critical research gaps. Within half a dozen years, SPA's membership had grown to twenty-four and its philanthropic reach had expanded significantly. In 2017, according to SPA's survey,[27] private support of basic research at major universities topped $2.3 billion, most of it for work in the life sciences. In that same year, by contrast, federal spending on basic research totaled about $34 billion.[28] Foundations and individual philanthropists were far from achieving major billing, but they were no longer bit players.

It's worth considering how the rise of private science giving might affect basic research. There are four obvious entries on the positive side of the ledger. Private giving can be opportunistic and effectively target deficiencies in the federal portfolio; it can take risks that federal bureaucrats shun; it can help smooth out federal budgetary swings; and it can partner with the federal government on major initiatives.

But the negative side of the ledger also needs to be examined. As private giving increases, legislators might well see it as a substitute for federal appropriations, allowing science budgets to be trimmed, especially in times of ballooning deficits. Philanthropies and foundations can be more opportunistic than federal agencies, but they can also be more capricious: they're beholden to their small number of donors and board members, rather than millions of voters and taxpayers. They do not have to use a peer-review process, which, although far from perfect, usually provides protection from frivolous scientific ventures.

To skew the ledger more toward the positive side, private givers might consider a carrot to encourage good legislative behavior and a stick to discourage bad behavior. First the carrot—to promote higher federal support of basic research, philanthropies could offer to match appropriations increases up to a specified dollar amount. Now the stick—to deter appropriators from trimming science spending after they have assessed philanthropic commitments, private givers could refrain from developing their annual budgets until the appropriations process has ended.

There is little argument that federal support of basic research measured against the national economy has been stagnant for decades, hovering around 0.4% of the gross domestic product (GDP).[29] That trend does not portend well for the future of a nation whose prosperity is increasingly tied to technological innovation.

The warning signs have been visible for some time. For example, in the "Global Innovation Index 2018 Report,"[30] compiled by INSEAD, Cornell University and the World Intellectual Property Organization (WIPO), the United States still ranks only sixth—almost unchanged over the last decade—behind Switzerland, the Netherlands, Sweden, the United Kingdom, and Singapore, and barely ahead of Finland, Denmark, and Germany. And according to a number of economic analyses, wages of the average American in the 21st century have been suppressed, in part, by lagging productivity[31] traceable to flagging innovation.

But innovation is double-edged sword. It can improve the lot of the average worker by increasing productivity and take-home pay, as it did for three decades following the end of World War II. By so doing, it can ameliorate income and wealth disparity, which Thomas Piketty explored at length in his treatise, *Capital in the Twenty-First Century*.[32] But it can also produce technological disruptions that lead to permanent workforce displacement, as PricewaterhouseCoopers's John Hawksworth and Richard Berriman detail in a 2018 PwC report[33] on the potential impact of automation in the 21st century. Managing these countervailing influences will be one of the biggest challenges for science and technology policymakers in the coming years.

We will look at the challenge in the context of Donald Trump's 2016 successful election in the Epilogue. But before that, we need to look at two more issues that are vital to policymaking in the current era: the tension between globalism and nationalism and the disparities in STEM (science, technology, engineering, and math) education outcomes.

The last decade of the 20th century and the first two decades of the 21st century have seen extraordinary socio-economic transformations sweep across the continents, and most of them have been driven by advances in technology. Globalization, once a term confined to the realm of economics or foreign affairs, entered the popular idiom following Thomas Friedman's 2005 international best seller, *The World Is Flat.*[34] Telecommunications and, more generally, information technology made national boundaries fuzzier. Manufacturing became a global enterprise: cars assembled in Detroit might use parts fabricated in Mexico, China, Japan, or Germany. Service centers could provide customers with assistance 24 hours a day because they were located across twenty-four time zones.

Everyone, everywhere could connect with anyone, anywhere using smart phones, tablets, and laptops. If email was too cumbersome, you could send text messages or use WhatsApp. If you craved communities, you could find them on Facebook, Instagram, or LinkedIn. If you wanted to spread the gospel, you could tweet or post. If you needed to hear somebody's voice or see somebody's face, you could Skype across the world or use FaceTime.

Science, which was the initiator of the transformation, became a global enterprise, itself. Researchers worked in international teams at facilities located on almost every continent. When they could, they collaborated remotely. They shared their work on electronic bulletin boards, often before they published it in peer-reviewed journals that used bits instead of ink. Students from one country studied at universities in another country: sometimes they emigrated and sometimes they returned home. The face of the world was the face of science.

The changes happened in less time than it takes a baby to reach physical maturity. But, as any parent knows, reaching emotional maturity takes longer. That comparison fairly well describes the mismatch between the 21st century technological revolution and the human capacity to adapt to it. Globalism surrounds us, but tribalism is still alive and well.

The tension between nationalism and internationalism is profound, more so in today's technologically interconnected world than at any time in the last half century. Policymakers must confront it in every area they deal with, from foreign affairs, defense, and trade to science, intellectual property, and taxation. And they must do so while navigating a maze filled with lobbyists, politicians, corporate leaders, financial titans, social activists, teachers, academicians, research scientists, and laboratory directors, all while contending with a twenty-four-hour news cycle and the drumbeat of social media. Finding a viable route through the Los Angeles freeway system at peak commuter time is easy compared with solving a problem that has the potential to disrupt the world order.

Of no lesser importance, but perhaps somewhat more tractable, is STEM education. As the PricewaterhouseCoopers (PwC) analysis suggests, automation could displace up to two in five American workers within 15 years.[35] It doesn't mean that the American workforce will have to shrink by 40%. There will be new jobs, but they will require different skills than the old ones, and most of those skills will fall under the STEM umbrella. It's also worth noting that the job losses will affect white collar as well as blue collar workers, as the impact of artificial intelligence (AI) becomes more pervasive.

Creative destruction, the term economists use to describe what is in store for us, is not a new phenomenon. For more than two centuries it has driven America's economic growth. But there are two differences this time around. First, the new skills will require more STEM proficiency, something a vast number of workers lack. And second, the average worker is likely to confront the impact of creative destruction more than once in a lifetime. As Tom Friedman emphasizes, coping with the persistent impact of technology will require a commitment to lifelong learning.[36]

If the PwC forecast is even remotely correct, politicians who are pushing universities to focus their curriculum on training students for existing job opportunities will be on the wrong side of history. The jobs of today will almost certainly not be the jobs of tomorrow. Providing students with broad-based STEM skills is far better education policy than focusing narrowly on the proficiencies job recruiters might be seeking today.

Lifelong learning and improving STEM education are both achievable goals. But there is another problem that requires more study, and probably a more comprehensive policy solution. It's the impact of child poverty on STEM proficiency.

Every 3 years, the Organization for Economic Cooperation and Development (OECD) conducts an international survey of educational proficiencies among 15-year-old students. In math and science, American students invariably perform abysmally—ranking 40th in math and 25th in science in the 2015 survey among the 72 participating nations.[37] The poor overall performance might seem shocking, but it tells only one part of the story. The other part lends some clarity to the scores and indicates what needs to be done.

The Program for International Student Assessment (PISA), as the survey is called, also samples the economic, cultural, and social status (ESCS) of the students taking the exam. And the ESCS results are incredibly revealing when it comes to child poverty. Using the eligibility for a free or reduced-price school lunch as a measure of privation, it's possible to sort the PISA scores accordingly. And the disparities are dramatic.

In science,[38] for example, students in schools where 75% or more attendees were eligible for the lunch program scored 446 points on the PISA exam, well below the OECD average of 496. By contrast, students in schools where 10% or fewer were eligible scored 553, a disparity of more than one hundred points. The results for math[39] show a similar ESCS gap.

The problem in the United States is more acute than in many other OECD countries, because American child poverty rates are far higher than they are in in similarly advanced nations. In weak economic times, almost one in three American children grow up in families living below the poverty line. In better times, it's one in five.[40] By contrast, in Denmark its one in 34; in Finland, one in 27; and throughout most of Western Europe, on average about one in 9 or 10.[41]

As STEM skills become more and more important, consigning 20% of the population to a grim future would not only be a drag on the American economy, it could pose a threat to the American democracy, as income disparity continues to widen beyond what it is today. Finding solutions to the problem should be on the critical path of policymakers and elected officials. *Our Kids*,[42,43] by the Harvard sociologist Robert Putnam should be required reading for all of them.

As the second decade of the 21st century draws to a close, several other science and technology issues remain unresolved. The dramatic advances in information technology and artificial intelligence, which continue to create the disruptive societal accelerations Thomas Freidman wrote about in *Thank You for Being Late*,[43] pose extraordinary challenges for policymakers well beyond the workforce dislocations PwC has predicted. They go to the very heart of science as a historical province of the elite.

In a populist era, being part of an exclusive social class is not a good place to be. But changing the culture of the scientific community is a tall order, especially if the community doesn't see it as a necessity. And so far, scientists haven't recognized it as a significant problem.

Technology has also created a major challenge to the way research findings are shared. Bits in the cloud, rather than ink on a page, in principle, make it possible for anyone to ferret out the latest scientific discoveries with just a few clicks of a computer mouse—but only if the results are freely available. If taxpayer money has been used to support scientific research, the resulting discoveries should be available at no additional cost to any member of the public who wants to see them, or so proponents of the movement known as Open Access argue. Scientific publishers, who regard themselves as guardians of the "truth," respond that "peer review" of scientific manuscripts by experts is essential to keep poorly-tested theories or bogus results from seeing the light of day.

Open science advocates parry the riposte by noting first that peer review is far from perfect, and second, that complete openness allows far more scrutiny by a wider range of experts. Balancing the pros and cons of open science—which also includes making vast amounts of scientific data freely available—requires policymakers to execute a high-wire act they have rarely encountered before.

Making scientific findings more comprehensively and more readily available also places a greater premium on scientific reproducibility, because greater openness carries with it greater scrutiny. With a populace that has become highly skeptical of institutions of all kinds—government, universities, industries, and the science and technology enterprise as a whole—just a few highly visible missteps can do irreparable damage to public trust in science. If that trust disappears, so, too, will support for using taxpayer money to pay for research.

Grand challenges exist, but so do grand opportunities. Two of them are related to breakthroughs in medicine. CRISPR-Cas9,[44] or simply CRISPR, is a genome editing tool that made headlines in 2014. It is faster, less expensive, and more accurate than other existing editing methods, and it holds the promise of generating preventative cures for myriad diseases at reasonable costs.

But life in the science policy world is never simple, especially where health is concerned. Using CRISPR in the research laboratory is one thing. Using the technique on human subjects is quite another. Even if it is shown to be abundantly safe, medical ethicists and policymakers will have to grapple with the question of whether it is proper to use CRISPR to enhance desirable traits, such as intelligence or physical appearance, or restrict its use to modifying genes associated with illnesses such as cancer, heart disease, and mental disorders.

Precision medicine,[45] is less fraught. It relies on assembling information on gene variability, environmental influences, and the effect of personal lifestyles in order to tailor the treatment of each patient individually, rather than apply a one-size-fits-all protocol. A holistic approach to medicine is certainly not new, but combining it with "big data," as precision medicine does, is new, and it promises to be a big deal.

One more observation: American society is far more diverse today than it was half a century ago. And the developing world is far more developed now than it was just two decades ago. How we deal with the changes will determine the future of America and the world. If we adopt policies that promote inclusion, we can make science a unifying proposition. By so doing, we can improve the human condition far beyond the benefits technology, alone, can provide.

This brings us to the end of our navigation through the maze that has characterized American science and technology policy for more than 225 years. I have tried to highlight the essentials, as I perceive them; but some critics undoubtedly will take issue with my choices. By using historical narratives, I have attempted to weave a fabric that is rich in texture, colorful in appearance, and enduring in utility.

I will summarize the tour with a few final thoughts. Science and technology policies have shaped the nation and the world as we know it today. They have had a remarkable run. In pursuit of successful outcomes they have, their practitioners have marshalled facts, data, analyses, and forecasts, which, taken together, constitute what George W. Bush's science adviser, Jack Marburger, called the science of science policy. But they have achieved their greatest successes when they have applied political savvy, exploited personal relationships, and timed their efforts perfectly. And in some cases, they have benefitted from serendipity, or in the vernacular, dumb luck.

Today, the maze is far more complex than it was when our nation was founded. There are far more vehicles trying to navigate it, and there are far more drivers of those vehicles. Whether science and technology will continue to maximize societal benefits and minimize societal harms will depends on how effectively the drivers navigate the maze. I hope this book will help them achieve that goal.

Chapter 13

Epilogue

The Trump era

On January 20, 2017, Donald J. Trump, a New York real estate developer and reality TV star, was sworn in as the 45th president of the United States. He had won the election the previous November, losing the popular vote by 3 million to Hillary Rodham Clinton, but beating her handily in the Electoral College. Many pundits called it a stunning upset. In truth, it wasn't. It's true that Clinton had run a poor campaign, that shortly before Election Day FBI director James Comey had revealed the reopening of its investigation into Clinton's emails, and that Russia had run an effective meddling effort to help elect Trump. If any one of them hadn't happened, the outcome might have been different.

But, in truth, the election hinged on the votes of a large segment of the American population that had been left behind, even as the nation as a whole had prospered. Trump's twin messages, "Make America Great Again" and "America First," resonated strongly with a disaffected, disillusioned, and despairing sector of the American population. A few astute political observers predicted that the ensuing populist wave would carry Trump into the White House, but they constituted a small minority.

It's not my purpose to expound on the election in general, but rather to highlight how science and technology might have affected the outcome and how the Trump White House began to dismantle accepted science and technology policies in the first five hundred days of his presidency. The connection between technology and the disaffected voters who elected Trump in states such as Ohio, Michigan, Pennsylvania, and Wisconsin is quite strong. It's true that the North American Free Trade Agreement (NAFTA) and other trade pacts led to American manufacturing job losses. It's also true that environmental policies accelerated coal's transition to the back burner, and that immigrants—some of them undocumented—crossed the southern border to fill a lot of low-paying jobs.

But as the 2008–2010 auto bailout amply demonstrated, robots can perform many of the traditional assembly-line tasks faster, cheaper, and more reliably. Workers discovered they were expendable not only in Detroit, but in manufacturing plants across the nation. And Barack Obama's Advanced Manufacturing Initiative might have helped American industry compete more effectively in global markets, but it did not deliver tens of thousands of new low- or medium-skilled jobs. To be fair, that was never its primary

Navigating the Maze. https://doi.org/10.1016/B978-0-12-814710-8.09993-9

283

objective. The initiative, endorsed by the President's Council of Advisors on Science and Technology, focused much more on innovation and global competitiveness. Manufacturers cheered. Workers didn't, and 8 years after the financial meltdown and the resulting economic crisis, they expressed their dissatisfaction and anger at the ballot box.

The collapse of the Eastern coal mining industry also owes its demise to science and technology. Trump railed against environmental regulations; he said they stole jobs from patriotic Americans. Miners and their families swarmed to his side, when he asserted that policies established to keep waters pure, air clean, and carbon emissions down were too stringent and needed to be rolled back. In truth, though, coal mining in Appalachia began to disappear when high-tech Western surface mining became more economically advantageous. Eastern (Pennsylvania, Ohio, West Virginia, and Kentucky) deep mining simply couldn't compete.

Cheaper natural gas also began to displace coal as the major source of energy for electricity generation. That, too, was the result of technology. Hydraulic fracturing—also known simply as "fracking"—which opened up vast reserves of formerly inaccessible gas, is an extraordinary example of how disparate scientific advances can give birth to a new industry. Geology, chemistry, physics, computer science, and materials science converged to allow drilling companies to tap shale gas reserves, using horizontal drilling and sophisticated mapping.

Policymakers and elected officials were asleep at the switch. They either were ignorant of the technological transformations that were sweeping across middle America, or they chose to ignore them. The result was an electoral outcome in the nation's heartland that few expected. Give Donald Trump high marks for sensing the anguish many voters were feeling. But give him low marks for implementing policies to deal with it.

Trump entered the White House as a disrupter, and science and technology did not draw a pass from his intentions to shake things up. While he kept France Córdova as director of the National Science Foundation and Francis Collins as director of NIH, he made major changes at the Environmental Protection Agency (EPA), the Department of the Interior, the Department of Energy, and the office of Management and Budget. He chose Scott Pruitt, a well-known climate-change denier, as EPA administrator; Ryan Zinke, a land developer, as Secretary of the Interior; Rick Perry, who admitted he didn't know what DOE did, as Secretary of Energy; and Mick Mulvaney, a slash and burn Tea Party conservative, as Director of the Office of Management and Budget and later White House acting chief of staff.

Finally, he allowed the position of Director of the Office of Science and Technology Policy (OSTP) to remain vacant for 18 months—twice as long as any of his predecessors—before nominating Kelvin Droegemeier, a well-respected meteorologist, to the post. Droegemeier would be the first presidential adviser without credentials applicable to nuclear weapons and nonproliferation policy. But as a former OSTP senior staffer remarked cynically, "It probably won't matter. The president isn't going to consult him anyway."

Trump tore up the Paris Climate Agreement and the Iran Nuclear Deal (Joint Comprehensive Plan of Action—JCPOA). He attempted to place restrictions on legal immigration and imposed a ban on travel from seven mostly Moslem nations. He threatened to curtail granting student visas to Chinese graduate students, and he pressed his case for moving ahead with tariffs on high-tech goods. In short, he was delivering on his America First agenda. He was not pursuing a traditional isolationist policy, but rather a highly nationalistic policy—one that appealed strongly to his voting base, but threatened the global nature of science.

On tax and spending matters, he helped ram through Congress a $1.5 trillion tax cut, reducing the corporate rate to 21%, the lowest it has been in 80 years, and ballooning the federal deficit. As part of the 2017 tax package, he initially supported eliminating the industrial Research and Experimentation (R and E) credit, which Congress had made permanent only a year before. And to tackle the mushrooming federal deficit, he twice proposed cutting federal research spending by up to 30%. Congress rejected the proposals after striking a 2-year budget agreement in early 2018 that established spending limits for fiscal years 2018 and 2019.

As I am writing the Epilogue at the close of 2018, Trump's science and technology legacy is far from complete. But to inform my observations, I spoke with three prominent Washington observers: Rush Holt, Jr., president and CEO of the American Association for the Advancement of Science (AAAS); Glenn Ruskin, director of external affairs and communications for the American Chemical Society; and Mary Woolley, president of Research!America (R!A), a preeminent medical and health research advocacy organization. I asked all three to reflect not only on Trump, himself, but also on the mood of the country and how it relates to science. Here is what they said in brief.

Rush Holt (AAAS): Trump's approach to science and science policy unfortunately mirrors much of the general public's. It dismisses evidence if doesn't fit a chosen argument. It rejects a hallmark of American science and technology policymaking stretching back more than two centuries—the importance of empirical thinking and fact-based decision making. It reflects a lack of scientific curiosity, probably characterizing the same lack among most of his supporters.

We probably haven't reached a tipping point, but we need to find better ways to give people the ability to evaluate evidence and to appreciate how much science has contributed to their lives.

Glenn Ruskin (ACS): The American Chemical Society has attempted to be constructively vocal, assuming good intentions on the part of the president, even if he does not articulate them. That said, the White House and OMB have repeatedly declined to acknowledge any correspondence from ACS, including arguments that link science to economic growth and improvements in the nation's infrastructure.

The ACS membership is divided on how much the Chemical Society should push back on Trump. Half believe ACS's response has been too tepid, and half believe it has been too aggressive. Allowing science to become embroiled in politics has always been a third rail for ACS members. But many believe that Trump is no friend of science and wonder whether it's necessary to draw a line in the sand before a tipping point is reached. They are concerned that "fake news"—Trump's characterization of the mainstream media—will mutate into "fake science." And that would be the tipping point.

Mary Woolley (R!A): Trump has the opportunity to take on a major medical issue and make a mark the way Nixon did with cancer and Reagan eventually did with HIV-AIDS. Trump, if he wanted to, could make opioids part of his legacy. Unfortunately, he responds less to scientific information than the echo chamber provided by social media and electronic media. By so doing, he his subverting the role of science in effective policymaking. The public probably recognizes the benefits of medical research more than he does.

It is too soon to evaluate the impact of Trump and his dedicated followers might have on the science and technology policies that will shape America and the world in the years to come. But without question, the Trump era will be remembered for its disruptions and doubts.

References

Preamble: Opening doors and expanding horizons

1. U.S. DOT, Bureau of Transportation Statistics, Table 2-9: U.S. Air Carrier (a) Safety Data (www.bts.gov/sites/bts.dot.gov/files/docs/browse-statistical-products-and-data/national-transportation-statistics/220806/ntsentire2018q1.pdf).
2. See for example, "Flint Water Crisis," *The New York Times*, October 8, 2016 (www.nytimes.com/news-event/flint-water-crisis).
3. "How Your Refrigerator Has Kept Its Cool Over 40 Years of Efficiency Improvements," Marianne DiMascio, American Council for an Energy Efficiency Economy Blog, September 11, 2014 (http://aceee.org/blog/2014/09/how-your-refrigerator-has-kept-its-co).
4. "LED Basics," Office of Energy Efficiency and Renewable Energy, U.S. Department of Energy (www.energy.gov/eere/ssl/led-basics#how_efficient).
5. T. Baer and F. Schlachter, "Lasers in Science and Industry: A Report to OSTP on the Contribution of Lasers to American Jobs and the American Economy" (www.laserfest.org/lasers/baer-schlachter.pdf).
6. "Science the Endless Frontier," A Report to the President by Vannevar Bush, Director of the Office of Scientific Research and Development, July, 1945, U.S. Government Printing Office, Washington, D.C. (www.nsf.gov/od/lpa/nsf50/vbush1945.htm).
7. "Banking Act of 1933 (Glass-Steagall)," Julia Maues, Federal Reserve History (www.federalreservehistory.org/essays/glass_steagall_act).
8. "Financial Services Modernization Act of 1999, Commonly Called Gramm-Leach-Bliley," Joe Mahon, Federal Reserve History
   (www.federalreservehistory.org/essays/gramm_leach_bliley_act).

Introduction: What is science and technology policy

1. The principal engineering fields are biomedical, chemical, civil, electrical, environmental, mechanical, software and systems, including industrial.
2. Mathematics is a very broad discipline that is usually divided among its specialties as follows: algebra, analysis (including calculus), combinatorics, computation (including numerical analysis), geometry and topology, operations, and probability and statistics.
3. The energy and telecommunications sectors each accounted for 5.9 percent of the U.S. GDP in 2015.
4. Defense accounted for 3.3 percent of the U.S. GDP in 2015; transportation accounted for 3.0 percent.
5. Agriculture accounted for 1.3 percent of U.S GDP in 2014; agriculturally related industries, such as food services, beverages and apparel, contributed a total of 5.7 percent.

Chapter 1: The early years 1787–1860

1. J. Wooley and G. Peters, *The American Presidency Project*, "William J. Clinton, XLII President of the United States: 1993-2001, Remarks on Presenting the National Medals of Science and Technology, March 14, 2000—www.presidency.ucsb.edu/ws/?pid=58246.

2. Jeffrey M. Smith, who worked at the White House Office of Science and Technology Policy and prepared President Clinton's remarks, had learned of the existence of the inscription on the first American coin and persuaded the Smithsonian to lend it to him for the day so the president could use it. As Jeff Smith tells the story, after the medal ceremony ended, the president absently put the coin in his jacket pocket and left the room without handing it back. Smith ran after the president, catching up to him as he got into an elevator. "Mr. President, Mr. President, the coin, the coin!" Jeff remembers calling out. Suddenly realizing he still had the $300,000 museum piece, Clinton reached into his pocket, pulled it out and tossed it nonchalantly to Jeff, just as the elevator doors were closing.

3. It's interesting to note that the words, "In God We Trust," did not appear on U.S. coins until 1864.

4. Harvard, founded in 1636, legitimately lays claim to being the oldest institution of higher learning in the United States. The University of Georgia, which received its charter in 1785 but did not begin admitting students until 1801, and the University of North Carolina at Chapel Hill, which received its charter in 1789 and took in its first class of students in 1795, are generally acknowledged as the earliest public institutions of higher learning.

5. See for example, "The Federalist Papers," The Avalon Project *Documents in Law, History and Diplomacy*, Lillian Goldman Law Library, Yale Law School, Yale University (http://avalon.law.yale.edu/subject_menus/fed.asp).

6. The American Philosophical Society's Mission Statement can be found at https://amphilsoc.org/about/missionstatement.

7. Taken from the original bylaws of the American Academy of Arts and Sciences (www.amacad.org/content.aspx?d=1424).

8. *Science and the Founding Fathers* (W.W. Norton, 1995), by I. Bernard Cohen, a noted historian of science, provides valuable, although occasionally controversial, insights into the influence of science on the writers of the Declaration of Independence and the American Constitution.

9. "From George Washington to the United States Senate and House of Representatives, 8 January 1790," *Founders Online,* National Archives, last modified December 28, 2016, http://founders.archives.gov/documents/Washington/05-04-02-0361. [Original source: *The Papers of George Washington*, Presidential Series, vol. 4, *8 September 1789–15 January 1790*, ed. Dorothy Twohig. Charlottesville: University Press of Virginia, 1993, pp. 543–549.]

10. Election law at the time awarded the vice presidency to the runner-up in the presidential election.

11. Amendment XII, which the states ratified in 1804, remedied the constitutional defect by requiring electors to cast a single vote for a combined slate of president and vice president.

12. See, for example, https://www.monticello.org/site/jefferson/louisiana-purchase.

13. See, for example, https://www.monticello.org/site/jefferson/lewis-and-clark-expedition.

14. The Archives of the Smithsonian provide a detailed history of the institution (https://siarchives.si.edu/history/general-history).

15. S. Newcomb, "Memoir of Joseph Henry" (Read before the National Academy of Sciences, April 21, 1880—www.princeton.edu/ssp/joseph-henry-project/joseph-henry/henry-joseph-newcomb.pdf).

16. The Smithsonian Institution Archives contains James Smithson's handwritten will and testament: siarchives.si.edu/history/exhibits/stories/last-will-and-testament-october-23-1826.

17. See Smithsonian Institution, Board of Regents, Minutes of December 3, 1846 (https://siarchives. si.edu/collections/siris_sic_11479).

18. The American Philosophical Society, of which Joseph Henry was a distinguished member, houses a collection of Henry's letters from 1836 to 1878. A concise biography can be found at www.amphilsoc.org/collections/view?docId=ead/Mss.B.H39p-ead.xml.

19. The Smithsonian Libraries contains an account of the Exploring Expedition, available at http:// www.sil.si.edu/DigitalCollections/usexex/navigation/ScientificText/USExEx19_14select.cfm.

20. See for example, N. Kollerstrom, *J. Hist. Astr.*, **23**, 185–192 (1992) (dioi.org/kn/halleyhollow. htm).

21. See Peter W. Sinnema, "Branch: Britain, Representation and Nineteenth-Century History," June 2012 (www.branchcollective.org/?ps_articles=peter-w-sinnema-10-april-1818-john-cleves-symmess-no-1-circular).

22. The Smithsonian "Castle" was designed by the architect James Renwick, and at the time of its construction, it occupied a piece of land separated from downtown Washington by the Washington City Canal, which connected today's Anacostia River with the Potomac River. The canal was filled in, and a new street, now known as Constitution Avenue, replaced it. In 1865, a fire destroyed a portion of the Castle requiring major reconstruction. The building, which was completely restored to its Victorian grandeur in 1968 and 1969, is on the registry of Historic Landmarks. See siarchives.si.edu/history/smithsonian-institution-building-castle.

23. See "150 Years of Advancing Science: A History of AAAS Origins 1848–1899," AAAS Archives and Records Center (archives.aaas.org/exhibit/origins2.php).

24. A.H. Dupree, *Science in the Federal Government*, The Johns Hopkins University Press (Baltimore, Md., 1986), pp. 115–148.

25. C.M. Green, *Washington: A History of the Capital, 1800-1950* (Princeton University Press, Princeton, New Jersey, 1976), p. 167.

26. D.S. Greenberg, *The Politics of Science—New Edition* (The University of Chicago Press, Chicago, 1999), p. 55.

Chapter 2: The civil war era and its legacy years 1860–1869

1. See, for example, *Scientists and National Policy Making*, R. Gilpin, ed., (Columbia University Press, New York, 1964).

2. See F. Seitz, *A Selection of Highlights from the History of the National Academy of Sciences— 1863–2005*, University Press of America (Lanham, Md., 2006) and www.nasonline.org/about-nas/history/highlights/.

3. A.H. Dupree, *Science in the Federal Government*, The Johns Hopkins University Press (Baltimore, Md., 1986), pp. 115–148.

4. A.H. Dupree, Proc. Am. Phil. Soc. **101** (5), October 31, 1957, pp. 434–440.

5. They still don't.

6. Two novels are particularly evocative: D. Brown, *Inferno* (Doubleday—Random House, LLC, New York, 2013) and *Angels and Demons* (Washington Square Press—Simon and Schuster, New York, 2000).

7. The scientists chose the name Lazzaroni tongue in cheek. The word was 18th century Neopolitan slang for the homeless idlers and beggars who used the Hospital of St. Lazarus in Naples as a refuge.

8. See, for example, H.A. Neal, T.L. Smith and J.B. McCormick, *Beyond Sputnik: U.S. Science Policy in the 21st Century* (The University of Michigan Press, 2008) and references therein.

9. www.happyvermont.com/2015/06/25/visiting-strafford-vermont/.

10. 7 U.S. Code § 301—Land Grant Aid of Colleges, Thirty-Seventh Congress. Sess. II. Ch. 130.1862, pp. 503, 504 (http://uscode.house.gov/statviewer.htm?volume=12&page=503#) or alternatively (www.ourdocuments.gov/doc.php?flash=true&doc=33&page=transcript).
11. The most recognizable Land Grant institutions in science and technology today include the University of California—Berkeley, Cornell University, the University of Illinois at Champaign-Urbana, the University of Maryland at College Park, the Massachusetts Institute of Technology, Ohio State University, Michigan State University, Pennsylvania State University and Texas A&M University.
12. A.H. Dupree, *Science in the Federal Government*, The Johns Hopkins University Press (Baltimore, Md., 1986), pp. 144, 145.
13. See *Report of the National Academy of Sciences for the Year 1887-Appendix D* Govt. Printing Office, Washington, DC, 1888).
14. "Penn Biographies—Alexander Dallas Bache (1806-1867)," Penn University Archives and Records Center (www.archives.upenn.edu/people/1800s/bache_alexdr_dallas.html).
15. A.H. Dupree, *op. cit.*, p. 147.
16. See "The Homestead Act of 1862," Educator Resources, National Archives (www.ourdocuments.gov/doc.php?flash=true&doc=33&page=transcript).
17. The novel is currently available as a Penguin Classic paper back: Mark Twain and Charles Dudley Warner, *The Gilded Age: A Tale of Today* (Penguin Publishing Group, New York, 2001).
18. "Pacific Railway Act of 1862," Thirty-Seventh Congress Sess. II. Ch. 120, 1862, pp.489–498 (memory.loc.gov/cgi-bin/ampage?collId=llsl&fileName=012/llsl012.db&recNum=520).
19. 7 U.S. Code § 2201—"Establishment of Department," Thirty-Seventh Congress. Sess. II. Ch. 72, 1862, pp. 387, 388 (uscode.house.gov/statviewer.htm?volume=12&page=387).
20. 7 U.S. Code § 361a—"Agricultural Experiment Stations," Forty-Ninth Congress. Sess. II. Ch. 314, 1887, pp. 441, 442 (uscode.house.gov/statviewer.htm?volume=24&page=440).
21. 7 U.S. Code § 343—"Cooperative Extension Activities," Sixty-Third Congress, Sess. II, Ch. 79, 1914, pp. 372–374 (uscode.house.gov/statviewer.htm?volume=24&page=440).
22. 7 U.S. Code § 2202—"Executive Department; Secretary," Fiftieth Congress, Sess. II Ch. 122, p. 659 (uscode.house.gov/statviewer.htm?volume=25&page=659).

Chapter 3: The gilded age 1869–1900

1. See, for example, "John Quincy Adams, VI President of the United States, 1825–1829, First Annual Message, December 6, 1825" (The American Presidency Project, The University of California Document Archive—www.presidency.ucsb.edu/ws/?pid=29467).
2. A.H. Frazier, *United States Standards of Weights and Measures: Their Creation and Creators* (Smithsonian Institution Press, City of Washington, 1978—repository.si.edu/bitstream/handle/10088/2439/SSHT-0040_Hi_res.pdf).
3. "A Brief History of the Naval Observatory," (Naval Oceanography Portal—www.usno.navy.mil/USNO/about-us/brief-history).
4. In 1830, John Quincy Adams successfully ran for the House of Representatives, the only ex-president to do so. He served there until his death in 1848, following a massive cerebral hemorrhage on the floor of the House.
5. A.E. Theberge, *The Coast Survey 1807–1867—History of the Commissioned Corps of the National Oceanic and Atmospheric Administration*, Vol. I, (NOAA Central Library, National Centers for Environmental Information, U.S. Dept. of Commerce, 1998—www.lib.noaa.gov/noaainfo/heritage/coastsurveyvol1/CONTENTS.html).
6. *Ferdinand Hassler (1770–1843): A Twenty-Year Retrospective, 1987–2007*, NIST Spec. Pub. 1068, ed. by H. Hassler and C.A. Burroughs (U.S. Dept. of Commerce, Tech. Admin, NIST, March 2007—www.nist.gov/sites/default/files/documents/nvl/HasslerSP1068.pdf).

7. A.H. Dupree, *op. cit.*, p. 52.

8. On May 29, 1832, shortly before the House of Representatives approved the bill that named Hassler head of the Survey for the second time, Rep. Aaron Ward of New York accused him of squandering federal resources during his first stint as superintendent. But his years in scientific purgatory—except for his short stint from 1830 to 1931 as head of the Office of Weights and Measures—left Hassler far savvier politically. He provided a careful justification for the Survey's expected expenditures, building his case on the commercial benefits of precision instrumentation and meticulous measurements. His friendship with the occupant of the White House, President Andrew Jackson, certainly did not hurt his cause. On February 12 of the previous year, Hassler had gained Jackson's attention, when he had positioned himself under the south colonnade of the White House to observe a solar eclipse. [F.R. Hassler, "Results of the Observation of the Solar Eclipse of 12th February 1831, etc.," Trans. Amer. Phil. Soc., New Series, **4** (1834) p. 131].

9. See, for example, G. Gugliotta, "New Estimate Raises Civil War Death Toll," *The New York Times*, April 12, 2012 (www.nytimes.com/2012/04/03/science/civil-war-toll-up-by-20-percent-in-new-estimate.html).

10. Although Andrew Johnson was a Southerner, he had been scathing in his opposition to secession by the Confederate states, perhaps even more so than Lincoln. Upon acceding to the presidency, Johnson had been fully expected to take a hard line on Reconstruction and punish members of the Confederate army and former slave owners. But he unexpectedly reversed course and in fairly short order offered general amnesty to most of the former rebels. His continued flouting of policies favored by congressional hardliners eventually led to his impeachment by the House of Representatives early in 1868. After a relatively short trial, the Senate voted to exonerate him but only by a margin of a single vote. ["The Impeachment of Andrew Johnson (1868) President of the United States," (The United States Senate: Art and History, Senate Historical Office—www.senate.gov/artandhistory/history/common/briefing/Impeachment_Johnson.htm)].

11. United States Military Academy West Point History (USMA Library—www.usma.edu/library/SitePages/history.aspx).

12. G. Thomas, "The Founders and the Idea of a National University: Constituting the American Mind" (Cambridge University Press, New York, 2015), p. 87.

13. Military Peace Establishment Act, Seventh Congress, Sess. I, Ch. 11, 1802. p. 137 (www.loc.gov/law/help/statutes-at-large/7th-congress/c7.pdf).

14. A.H. Dupree, *op. cit.*, pp. 195-208.

15. R.C. Cochrane, *The National Academy of Sciences: The First Hundred Years, 1863–1963* (National Academy of Sciences, Washington, 1978), pp. 127–129.

16. See J.W. Powell, *The Exploration of the Colorado River and Its Canyons* (Dover Publications, Inc., Mineola, New York, 1961—Originally published by Flood and Vincent under the title *Canyons of the Colorado*), p. 131.

17. R.C Cochrane, *op. cit.*, pp. 100–127.

18. *Ibid.*, p. 130.

19. *Ibid.*, p. 131.

20. A.H. Dupree, *op. cit.*, pp. 209, 210.

21. *Ibid.*, p. 215–231.

22. R.C. Cochrane, *op. cit.*, pp. 146, 147.

23. The Coast and Interior Survey along with the Navy's Hydrographic Office would constitute one bureau; the existing Geological Survey, a second; meteorological activities including the Signal Service's weather mapping function, a third; and a central physical observatory and laboratory devoted to solar and terrestrial radiation and other "investigations in exact science," including

new electrical standards and the existing activities of the Coast Survey's Office of Weights and Standards, the fourth.

24. The nine-member commission would comprise the president of the National Academy of Sciences, the Secretary of the Smithsonian, "two civilians of high scientific reputation" appointed by the president of the United States, an officer of the Army Corps of Engineers, a Navy professor of mathematics also appointed by the president, the superintendent of the Coast and Geodetic Survey, the director of the Geological Survey and the head of the meteorological bureau.

25. In 2014, the last year for which detailed data are available, American universities spent $63.7 billion on science and engineering research and development, of which federal funding accounted for 58 percent. [National Science Board, 2016, *Science and Engineering Indicators 2016*, Arlington, VA: National Science Foundation (NSB-2016-1)—www.nsf.gov/statistics/2016/nsb20161/#/report/chapter-5/expenditures-and-funding-for-academic-r-d].

26. "Testimony Before the Joint Commission to Consider the Present Organization of the Signal Service, Geological Survey, Coast and Geodetic Survey, and the Hydrographic Office of the Navy, With a View to Secure Greater Efficiency and Economy of Administration of the Public Service in Said Bureaus, Authorized by the Sundry Civil Act Approved July 7, 1884, and continued by the Sundry Civil Act Approved March 3, 1885" (Government Printing Office, Washington, 1886), p. 1078 (https://books.google.com/books?id=Hocj26LFyPwC&pg=PA1019&lpg=PA1019&dq=powell+testimony+allison+commission+1886&source=bl&ots=2bWtBLax94&sig=nW-R4Q3zxLH2IZlo9xv638wsc6M&hl=en&sa=X&ved=0ahUKEwi1oKKN-OPVAhUK7IMKHVnQDyo4ChDoAQgpMAE#v=onepage&q=powell%20testimony%20allison%20commission%201886&f=false)

27. *Ibid.*, p. 1082.

28. See K.J. Holmes, *Nature* **501**, 310 (2013) and references therein.

29. A.H. Dupree, *op. cit.*, pp. 233–235.

30. A map showing how Western state lines might have been drawn based on drainage and irrigation considerations can be found at bigthink.com/strange-maps/489-how-the-west-wasnt-won-powells-water-based-states.

31. In 1890, Powell's budget for the Geological Survey's topographical study west of the 100th meridian (the West) was slashed from a proposed $700,000 to a mere $162,500. (Fifty-First Congress. Sess. I. Ch. 837, 1890, p. 391.)

32. R.C. Cochrane, *op. cit.*, p. 157.

33. *Ibid.*, pp. 159-161.

34. J.M. Michael, "The National Board of Health: 1879-1883," *Public Health Rep.* **126** (1), January–February 2011, pp. 123–129 (www.ncbi.nlm.nih.gov/pmc/articles/PMC3001811/).

35. James Carroll and Walter Reed, for whom the National Military Center in Bethesda, Md. is named, are credited with the discovery.

36. "An Act to Prevent the Introduction of Infectious or Contagious Diseases into the United States and to Establish a National Board of Health," Forty-Fifth Congress. Sess. III. Ch. 202. March 3, 1879.

37. "Annual report of the National Board of Health, 1883" (U.S. Government Printing Office, Washington, DC, 1884.)

38. A.H. Dupree, *op. cit.*, pp. 263–267.

Chapter 4: A new century—A new America 1900–1925

1. G. Gamm and S.S. Smith, "Steering the Senate: The Consolidation of Senate Party Leadership, 1879–1913," Congress and History Conf., Pol. Sci. Dept. and Sch. of Hum., Arts, and Soc. Sci., Massachusetts Institute of Technology, Cambridge, Mass., May 30-31, 2003.

2. Article I Sec. 2 of the Constitution stipulates that: "The Number of Representatives shall not exceed one for every thirty Thousand, but each State shall have at Least one Representative..." Until 1911, Congress routinely increased the size of its membership to account for population growth. The Apportionment Act of 1911 (Pub. Law 62–5, 37 Stat. 13 set the size of the House of Representatives at 435, effective in 1913 with the election of the 63rd Congress.

3. The House office complex comprises the Cannon (1909), Longworth (1933) and Rayburn (1965) buildings.

4. The Senate complex comprises the Russell (1908), Dirksen (1958) and Hart (1982) buildings.

5. A.H. Frazier, *United States Standards of Weights and Measures: Their Creation and Creators* (Smithsonian Institution Press, City of Washington, 1978—repository.si.edu/bitstream/handle/ 10088/2439/SSHT-0040_Hi_res.pdf).

6. "A Solar Eclipse for the Ages," Home, News, National Centers for Environmental Information, National Oceanographic and Atmospheric Administration (www.ncei.noaa.gov/news/1900- total-solar-eclipse).

7. See, for example, "Tesla—Master of Lighting: Life and Legacy—War of the Currents," PBS. org (www.pbs.org/tesla/ll/ll_warcur.html) and "The War of the Currents: AC vs. DC Power," U.S. Department of Energy, November 18, 2014 (energy.gov/articles/war-currents-ac-vs-dc- power).

8. A.H. Dupree, *Science in the Federal Government*, The Johns Hopkins University Press (Balti- more, Md., 1986), pp. 273.

9. For a comprehensive history of the National Bureau of Standards, see R.C. Cochrane, *Measures for Progress: A History of the National Bureau of Standards—America in Two Centuries, an Inventory* (Arno Press, New York, 1976).

10. Public Law 177—"An Act to Establish the National Bureau of Standards," Fifty-Sixth Con- gress, Sess. II. Ch. 872. March 3, 1901, pp. 1449, 1450 (http://uscode.house.gov/statviewer. htm?volume=31&page=1449).

11. Theodore Roosevelt became president following William McKinley's death from an assassin's bullet on September 14, 1901.

12. Transformers trace their American origin to Joseph Henry's work in the 1830s, although prac- tical devices did not enter the marketplace until the 1880s.

13. 30 U.S.C. § 1—"An Act to Establish in the Department of the Interior a Bureau of Mines," Sixty- First Congress, Sess. II. Ch. 240, May 16, 1910, pp. 369, 370 (uscode.house.gov/statviewer. htm?volume=36&page=369#).

14. F.W. Powell, *The Bureau of Mines: Its History, Activities and Organization* (The Institute for Government Research, D. Appleton and Company, New York, 1922), p. 5.

15. 50 U.S.C. § 161—"An Act Authorizing the Conservation, Production, and Exploitation of Helium Gas, a Mineral Resource Pertaining to the National Defense, and to the Development of Commer- cial Aeronautics, and for Other Purposes," Sixty-Eighth Congress, Sess. II, Ch. 426. March 3, 1925, pp. 1110, 1111 (uscode.house.gov/statviewer.htm?volume=43&page=1110#).

16. "World War One: How the German Zeppelin Wrought Terror," BBC News, England, August 4, 2014 (www.bbc.com/news/uk-england-27517166).

17. "Building the Panama Canal, 1903-1914—Milestones 1899-1913," Office of the Historian, U.S. Department of State (history.state.gov/milestones/1899-1913/panama-canal).

18. See, for example, "Our History," U.S. Forest Service, U.S. Department of Agriculture (www.fs. fed.us/learn/our-history).

19. "John Muir's Yosemite," T. Perrottet, *Smithsonian Magazine*, July 2008 (www. smithsonianmag.com/history/john-muirs-yosemite-10737).

20. "Hetch Hetchy Environmental Debates," The Center for Legislative Archives, National Archives, Washington, D.C. (www.archives.gov/legislative/features/hetch-hetchy).

21. Sixty-Third Congress, Sess. I, H.R. 7207 [Report 41], August 3, 1913 (archive.org/stream/hetchhetchygrant00unit/hetchhetchygrant00unit_djvu.txt).

22. A.H. Dupree, *op. cit.* p. 252.

23. Public Law 57-161, 43 U.S.C. § 391—"An Act Appropriating the Receipts from the Sale and Disposal of Public Lands in Certain States and Territories to the Construction of Irrigation Works for the Reclamation of Arid Lands," Fifty-Seventh Congress. Sess. I. Ch. 1093, June 17, 1902, pp. 388–390 (uscode.house.gov/statviewer.htm?volume=32&page=388).

24. See, for example, "Water in the West," Bureau of Reclamation Historic Dams and Water Projects, National Park Service, U.S. Department of the Interior (www.nps.gov/nr/testing/ReclamationDamsAndWaterProjects/Water_In_The_West.html).

25. "Agency History" (www.census.gov/history/www/census_then_now), United States Census Bureau.

26. When the Department of Commerce and Labor split into separate departments, the Bureau of the Census moved to the Department of Commerce, where it has remained since.

27. See Smithsonian National Air and Space Museum at https://airandspace.si.edu/.

28. "Samuel Pierpont Langley, 1834-1906," Smithsonian History, Smithsonian Institution Archives (siarchives.si.edu/history/samual-pierpont-langley).

29. In 1955, the Smithsonian Astrophysical Observatory merged with the Harvard College Observatory and moved to Cambridge, Massachusetts.

30. "Charles Doolittle Walcott, 1850–1927," Smithsonian History, Smithsonian Institution Archives (siarchives.si.edu/history/charles-doolittle-walcott).

31. A. Roland, "The National Advisory Committee for Aeronautics 1915–1958," *Model Research* (NASA SP-4103, Washington, D.C., 1985) pp. 1–26 (articles.adsabs.harvard.edu//full/1985NASSP4103.....R/0000001.000.html).

32. A.R. Buchalter and P.M Miller, "The National Advisory Committee for Aeronautics, An Annotated Bibliography," Monographs in Aerospace History #55 (NASA SP-2014-4555, Washington, D.C., 2014), pp. 20–22 (history.nasa.gov/monograph55.pdf).

33. A.H. Dupree, *op. cit.*, p. 286.

34. "National Advisory Committee for Aeronautics: Letter from the Board of Regents of the Smithsonian Institution Transmitting a Memorial on the Need of a National Advisory Committee for Aeronautics in the United States," February 1, 1915, Referred to the Committee on Naval Affairs, 63rd Congress, 3rd Session, House of Representatives, Document No. 1549 (books.google.com/books?id=oJ03AQAAIAAJ&pg=RA1-PA35&lpg=RA1-PA35&dq=smithsonian+institution+board+of+regents+aeronautics+february+1+1915&source=bl&ots=P8pp9tTcNL&sig=i5bmLaxem7SprZWQhMzvM6fRYa8&hl=en&sa=X&ved=0ahUKEwj14M3J3J3WAhUFw4MKHQ_1AFYQ6AEIOjAF#v=onepage&q=smithsonian%20institution%20board%20of%20regents%20aeronautics%20february%201%201915&f=false).

35. Public Law 271, 63rd Cong., 3rd Sess., March 3, 1915 (38 Stat. 930).

36. "National Aeronautics and Space Act of 1958," Public Law 85-568, 72 Stat. 426; signed into law, July 29, 1958 (Record Group 255, National Archives and Records Administration, Washington, D.C.; NASA Historical Reference Collection, History Office, NASA Headquarters, Washington, D.C.—history.nasa.gov/spaceact.html).

37. www.nih.gov/about-nih.

38. www.wrnmmc.capmed.mil/About%20Us/SitePages/Home.aspx.

39. "Kinyoun's Early Years—Dr. Joseph Kinyoun The Indispensable Forgotten Man," History, National Institute of Allergy and Infectious Diseases, NIH (www.niaid.nih.gov/about/joseph-kinyoun-indispensable-man-early-years).

40. "NYU Langone History," NYU Medical School History, Lillian and Clarence de la Chapelle Medical Archives, NYU Health Sciences Library (archives.med.nyu.edu/content/nyu-langone-history).
41. "Bellevue Hospital Medical College: A Guide to the Records," Bellevue Hospital Medical College, Lillian and Clarence de la Chapelle Medical Archives, NYU Health Sciences Library (archives.med.nyu.edu/collections/bellevue-hospital-medical-college-guide-records).
42. "Andrew Carnegie's Story," Carnegie Corporation of New York (www.carnegie.org/interactives/foundersstory/#!).
43. "The Hygienic Laboratory—Dr. Joseph Kinyoun The Indispensable Forgotten Man," History, National Institute of Allergy and Infectious Diseases, NIH (www.niaid.nih.gov/about/joseph-kinyoun-indispensable-man-hygienic-laboratory).
44. "The Marine Hospital Service—Dr. Joseph Kinyoun The Indispensable Forgotten Man," History, National Institute of Allergy and Infectious Diseases, NIH (www.niaid.nih.gov/about/joseph-kinyoun-indispensable-man-marine-hospital-service).
45. "Army Medical Museum and Library," Historic Medical Sites in the Washington, DC Area—Celebrating the Bicentennial of the Nation's Capital, History of Medicine, U.S. National Library of Medicine, NIH (www.nlm.nih.gov/hmd/medtour/armymuslib.html).
46. See, for example, P. McSherry, www.spanamwar.com/casualties.htm. The total does not include 260 American sailors who lost their lives when the USS Maine sank in Havana harbor, the event that triggered the war.
47. *Ibid.*
48. "An Act Making Appropriations for Sundry Civilian Expenses of the Government for the Fiscal Year Ending June Thirtieth, Nineteen Hundred and Two—Under the Treasury Department, Public Buildings, Marine Hospitals," Fifty-Sixth Congress, Sess. II, Ch. 853, March 3, 1901, p. 1137.
49. See for example, "Origins of the National Institutes of Health—The 1900's Bring Change," History of Medicine, U.S. National Library of Medicine, NIH (www.nlm.nih.gov/exhibition/nih_origins/change.html).
50. A.H. Dupree, *op. cit.*, p. 269.
51. "An Act To Change the Name of the Public Health and Marine Hospital Service to the Public Health Service, To Increase the Pay of Officers of Said Service, and for Other Purposes," Sixty-Second Congress, Sess. II, Ch. 288, August 14, 1912, p. 309.
52. "Hamlet, Act II," William Shakespeare.
53. The Geneva protocols prohibited both, beginning in 1925.
54. J.K. Taubenberger and D.M. Morensi, "1918 Influenza: the Mother of All Pandemics," Emerg Infect Dis, Centers for Disease Control and Prevention **12** (1), 15-22 (2006)—www.webcitation.org/5kCUlGdKu?url=http://www.cdc.gov/ncidod/EID/vol12no01/05-0979.htm.
55. S.D. Collins, "The Influenza Epidemic of 1928–1929 with Comparative Data for 1918-1919," *Am. J. Pub. Health*, **XX** (2), 119–129 (1930).
56. Seventy-First Congress, Sess. II. Ch. 320, May 26, 1930, pp. 379, 380.
57. The name was changed to National Institutes of Health (plural) in 1948.
58. A.H. Dupree, *op. cit.*, pp. 306-307.
59. For a comprehensive history of the Superconducting Super Collider project, see M. Riordan, L. Hoddeson and A.W. Kolb, *Tunnel Visions: The Rise and Fall of the Superconducting Super Collider*, The University of Chicago Press (Chicago, 2015).
60. History, U.S. Naval Research Laboratory. www.nrl.navy.mil/about-nrl/history.
61. S. Olson, "The National Academy of Sciences at 150," PNAS **111**, Suppl. 2, 9327–9364 (2014). www.pnas.org/content/111/Supplement_2/9327.full.

62. A.H. Dupree, *op. cit.*, pp. 308–312.

63. "Report of the National Academy of Sciences for the Year 1916," p. 22 (United States Government Printing Office, GPO, Washington, D.C., 1917). babel.hathitrust.org/cgi/pt?id=osu. 32435061127254;view=1up;seq=24.

64. For a detailed description of the creation of the National Research Council, see R.C. Cochrane, *The National Academy of Sciences: The First Hundred Years, 1963–1963* (National Academy of Sciences, Washington, 1978), pp. 200–241.

65. S. Olson, *op. cit.*, p. 9329.

66. A.H. Dupree, *op. cit.*, pp. 311–315.

67. The Manhattan Project was originally given the code name, "Manhattan District," by the Army Corps of Engineers, or simply, "Manhattan," since many of the original nuclear scientists and engineers worked in laboratories and universities in New York City. For a compelling read, see N.P Davis, *Lawrence and Oppenheimer* (Simon and Schuster, New York, 1968). For the official history of the Manhattan Project, see H. D. Smyth, *Atomic Energy for Military Purposes, the Official Report on the Development of the Atomic Bomb under the Auspices of the United States Government"* (Princeton University Press, Princeton, 1945). archive.org/details/ atomicenergyform00smytrich. The official history is often called simply, "The Smyth Report."

68. R.C. Cochrane, *op. cit.*, pp. 235–236.

69. Executive Order 2859—National Research Council of the National Academy of Sciences, Executive Orders, Federal Register, National Archives. www.archives.gov/federal-register/ codification/executive-order/02859.html.

70. V. Bush, *Science the Endless Frontier—A Report to the President on a Program for Postwar Scientific Research, July 1945* (Reprinted as part of the Tenth Anniversary Observance by the National Science Foundation, Washington, D.C., 1960). archive.org/stream/ scienceendlessfr00unit/scienceendlessfr00unit_djvu.txt.

71. H. Ford with S. Crowther, *My Life and Work* (Doubleday, Garden City, 1923), p. 73. archive. org/stream/mylifeandwork00crowgoog#page/n86/mode/2up/search/multitude.

72. For a concise history of the period, see R.F. Weingroff, "Federal Aid Road Act of 1916: Building the Foundation," *Public Roads*, **60** (1), Summer 1996 (Federal Highway Administration Research and Technology, U.S. Department of Transportation. www.fhwa.dot.gov/ publications/publicroads/96summer/p96su2.cfm).

73. "Statistical Abstract of the United States: 1926," U.S. Census Bureau (www.census.gov/library/ publications/1927/compendia/statab/49ed.html).

74. The Tenth Amendment reads, "The powers not delegated to the United States by the Constitution, nor prohibited by it to the States, are reserved to the States respectively, or to the people."

75. See caselaw.findlaw.com/us-supreme-court/204/24.html.

76. Article I, Section 8, Clause 3.

77. "History of the Interstate Highway System," Federal Highway Administration, U.S. Department of Transportation. www.fhwa.dot.gov/interstate/history.cfm.

Chapter 5: From depression to global engagement 1925–1945

1. H. Hoover, "The Nation and Science," *Science*, **LXV** (1672), 26-29 (1927).

2. A.H. Dupree, *op. cit.*, pp. 342.

3. V. Bush, *op. cit.*

4. See, for example, R.V. Bruce, *Bell: Alexander Graham Bell and the Conquest of Solitude* (Little Brown and Company, Boston, 1976), p. 181.

5. Seth Shulman, in his investigative book, *The Telephone Gambit* (W.W. Norton and Company, New York, 2008) makes a compelling argument that Bell stole his design from Elisha Gray, a well-known and well-regarded electrical engineer who co-founded the Western Electric Company. Gray, who had filed a "caveat" with the Patent Office three weeks prior to Bell's filing, challenged Bell's patent in court, but lost his case, even though, as Shulman points out, the evidence suggested he had an extremely strong case.

6. A photograph of the Alexander Graham Bell's original patent filing can be viewed at unwritten-record.blogs.archives.gov/patent-number-174-465alexander-graham-bellrecord-group.

7. R.V. Bruce, *op. cit.*, p. 229.

8. *Ibid.*, p. 231.

9. *Ibid.*, p. 340.

10. "Volta Laboratory and Bureau," Washington, DC: A National Register of Historic Places Travel Itinerary, U.S. National Park Service. www.nps.gov/nr/travel/wash/DC14.HTM.

11. R.V. Bruce, *op. cit.*, p. 354.

12. For a summary of science in the Depression and New Deal Era, see A.H. Dupree, *op. cit.*, pp. 345-368.

13. J.K. Wright and G.F. Carter, *Isaiah Bowman 1878—1850: A Biographical Memoir* (National Academy of Sciences, Washington, D.C., 1959).

14. books.google.com/books?id=10MrAAAAYAAJ&pg=PP8&lpg=PP8&dq=Research+–+A+National+Resource+Charles+H.+Judd&source=bl&ots=qvQO475pYR&sig=ZOjUQhg ccJP8xA-xitCp5jV8kS4&hl=en&sa=X&ved=0ahUKEwj1nb2qqorXAhUM8IMKHWr3Aq 0Q6AEIKDAA#v=onepage&q=Research%20–%20A%20National%20Resource%20 Charles%20H.%20Judd&f=false.

15. "An Act to Provide for, Foster, and Aid in Coordinating Research Relating to Cancer; to Establish the National Cancer Institute; and for Other Purposes," Seventy-Fifth Congress, Sess. I. Ch. 565, Public Law No. 244 ("National Cancer Act of 1937," About NCI, Legislative History, National Cancer Institute, NIH. www.cancer.gov/about-nci/legislative/history/national-cancer-act-1937).

16. J.L. Pennick, Jr., C.W. Pursell, Jr., M.B. Sherwood and D.C. Swain, eds., *The Politics of American Science: 1939 to the Present, Revised Edition* (The MIT Press, Cambridge, Mass., 1972), p. 50.

17. V. Bush, "Report of the President of the Carnegie Institution of Washington for the Year Ending October 31, 1939," *Yearbook No. 38, July 1,1938–June 30, 1939*, pp. 5-6 (Carnegie Institution of Washington, Washington, D.C. 1939). archive.org/details/yearbookcarne38193839carn.

18. V. Bush, Reprinted as part of the Tenth Anniversary Observance by the National Science Foundation, Washington, D.C., 1960). archive.org/stream/scienceendlessfr00unit/scienceendless fr00unit_djvu.txt.

19. J.L. Pennick, Jr., *et al.*, *op. cit.*, p. 10.

20. *Ibid.*, pp. 10-14.

21. N.P Davis, *Lawrence and Oppenheimer* (Simon and Schuster, New York, 1968) and H. D. Smyth, *Atomic Energy for Military Purposes, the Official Report on the Development of the Atomic Bomb under the Auspices of the United States Government"* (Princeton University Press, Princeton, 1945) provide excellent historical renderings of the Manhattan Project.

22. For Belgium, the issue would become moot after it's surrender to German forces on May 28, 1940.

23. N.P Davis, *op. cit.*, p. 97.

24. S.L. Schwartz, ed., *Atomic Audit: The Costs and Consequences of U.S. Nuclear Weapons Since 1940* (Brookings Institution Press, Washington, DC, 1998)—The Costs of the Manhattan Project: www.brookings.edu/the-costs-of-the-manhattan-project.
25. H.D. Smyth, *op. cit.*, pp. 247–254.

Chapter 6: Donning the mantle of world leadership 1945–1952

1. *Life*, August 27, 1945, p. 27 (100photos.time.com/photos/kiss-v-j-day-times-square-alfred-eisenstaedt).
2. N.P Davis, *Lawrence and Oppenheimer* (Simon and Schuster, New York, 1968), p. 250.
3. V. Bush, *Science the Endless Frontier—A Report to the President on a Program for Postwar Scientific Research, July 1945* (Reprinted as part of the Tenth Anniversary Observance by the National Science Foundation, Washington, D.C., 1960). archive.org/stream/science endlessfr00unit/scienceendlessfr00unit_djvu.txt.
4. D. McCullough, *Truman* (Simon and Schuster, New York, 1992), p. 355.
5. A concise description of the America First Committee can be found in Krishnadev Calamur's article in *The Atlantic*, which provides a context for President Donald Trump's use of the slogan in his 2016 campaign and the messaging of the Trump White House (K. Calamur, "A Short History of "America First'," *The Atlantic*, January 21, 2017. www.theatlantic.com/politics/archive/2017/01/trump-america-first/514037).
6. J.L. Penick, Jr., C.W. Pursell, Jr., M.B. Sherwood and D.C. Swain, eds., *The Politics of American Science 1939 to the Present*, revised edition, The MIT Press, Cambridge, Mass., 1974), pp. 82–95.
7. Science Policy Research Division, Congressional Research Service, Library of Congress, "The National Science Board: Science and Policy Management for the National Science Foundation 1968-1980"—Report for the Subcommittee on Science Research and Technology, Transmitted to the Committee on Science and Technology, U.S. House of Representatives, Ninety-Eighth Congress, First Session, Serial E (U.S. Government Printing Office, Washington, 1983), p. 43.
8. "The Technology Mobilization Act," S. 2721, Seventy-Seventh Congress, Sess. II, 1942; submitted and referred to the Senate Subcommittee on War Mobilization of the Military Affairs Committee but never subjected to a vote.
9. R.C. Cochrane, "The Post War Organization of Science," *The National Academy of Sciences: The First Hundred Years, 1863-1963* (The National Academy Press, Washington, DC, 1978), Ch. 14 (www.nap.edu/catalog/579/the-national-academy-of-sciences-the-first-hundred-years-1863).
10. J.L. Penick, Jr. *et al.*, *op. cit.*, p. 83.
11. *Ibid.*, p. 84.
12. R.C. Cochrane, *loc. cit.*
13. "The National Science Board: Science and Policy Management for the National Science Foundation 1968-1980," *op. cit.*, p. 44.
14. R.C. Cochrane, *loc. cit.*
15. In an ironic historical twist, seven decades after the American Association for the Advancement of Science took its stance opposing Kilgore's bill, it chose as its CEO Rush Holt, Jr., a physicist, former member of the House of Representatives from New Jersey and the son of the man Kilgore had defeated in the West Virginia Democratic primary in 1940.
16. "The Government's Wartime Research and Development, 1940-44, Part II, Findings and Recommendations," Senate Subcommittee on War Mobilization, Report, Seventy-Ninth Congress, Sess. I, 1945, pp. 26-29, as quoted by J.L. Penick, Jr. *et al.*, *op. cit.*, pp. 103-105.

17. H.S. Truman, "Special Message to the Congress Presenting a 21-Point Program for the Reconversion Period," Public Papers 1945-1953 **128**, (Harry S. Truman Presidential Library). www.trumanlibrary.org/publicpapers/index.php?pid=136&st=&st1.

18. See M. Lomask, "The Birth of NSF," *Mosaic*, November/December 1975, 20–27 (1975). www.mosaicsciencemagazine.org/pdf_track.php?mode=A&pk=338.

19. "Record of the 79th Congress (Second Session) 1946, *Editorial Research Reports 1946* **II** (CQ Researcher, CQ Press, Washington, DC. http://library.cqpress.com/cqresearcher/document. php?id=cqresrre1946080300-H2_1.

20. *Congressional Record: Proceedings and Debates of the 79th Congress, Second Session*, Senate, Tuesday July 2, 1946 (Government Printing Office, Washington, DC, 1946) pp. 8104, 8105. www.gpo.gov/fdsys/pkg/GPO-CRECB-1946-pt7/content-detail.html.

21. M. Lomask, *op. cit.*, p. 26.

22. R.C. Cochrane, *op. cit.,* pp. 462, 463.

23. *Ibid.*, pp. 454-456.

24. "A Short History of the National Institutes of Health—New Institutes," Office of NIH History and Museum, National Institutes of Health. history.nih.gov/exhibits/history/docs/page_07.html.

25. For an insightful analysis of the 1945-1950 NSF travails, see W.A. Blanpied, "Inventing US Science Policy," (About NSF, History, National/Science Foundation). www.nsf.gov/about/history/nsf50/science_policy.jsp.

26. D. McCullough, *op. cit.*, p. 493.

27. "News and Notes," *Science*, **105**, 171 (1947). science.sciencemag.org/content/105/2720/171.

28. "Comparison of the Provisions of S. 525 and S. 526," *Science*, **105**, 253–254 (1947). science.sciencemag.org/content/105/2723/253.

29. M. Lomask, *op. cit.*, p. 26.

30. R.C. Cochrane, *op. cit.*, pp. 433–454.

31. Thomas's bill in its final form had a 24-member board.

32. The Constitution provides that the president has a 10-day time window either sign or refuse to sign a bill. If the president fails to act within that time-window, the bill becomes law automatically provided Congress is in session. But if Congress is not in session, presidential inaction kills the legislation. It is called a "pocket veto," and the president is not required to provide any explanation for his lack of action, although Truman did so.

33. J.R Steelman, *Science and Public Policy, Vol. I, A Program for the Nation: A Report to the President* (Government Printing Office, Washington, DC, 1947).

34. "Legislative History of the National Science Foundation Act of 1950, Appendix II" (Report No. 796 Accompanying H.R. 4846, 81st Congress, June 14, 1949). www.nsf.gov/pubs/1952/b_1952_8.pdf.

35. D. McCullough, *op. cit.*, pp. 652, 661.

36. *Ibid.*, pp. 618-620, 628, 642.

37. M. Lomask, *op. cit.*, p. 27.

38. "National Science Foundation Act of 1950," Public Law 507; 42 U.S.C. § 16. See legcounsel.house.gov/Comps/81-507.pdf.

39. "Special Message to Congress on Atomic Energy," Public Papers, Harry S. Truman, 1945-1953, No. 156 (Harry S. Truman Presidential Library and Museum). www.trumanlibrary.org/publicpapers/index.php?pid=165&st=&st1.

40. See, for example, V.C. Jones, *United States Army in World War II, Special Studies—Manhattan: The Army and the Atomic Bomb*, (Center of Military History, United States Army, Washington, DC, 1985), pp. 574–578. history.army.mil/html/books/011/11-10/CMH_Pub_11-10.pdf.

41. R.C. Cochrane, *op. cit.*, pp. 454–456.

42. Public Law 585, 79th Cong., 2nd Sess., Ch. 724, 60 Stat., pp. 755–775; 42 U.S.C. § 1801 *et seq.* (1946) (www.loc.gov/law/help/statutes-at-large/79th-congress.php).

43. B.S. Old, "The Evolution of the Office of Naval Research," *Physics Today*, **14** (8), 30-35 (1961).

44. Public Law 588, 79th Cong., 2nd Sess., Ch. 727, 60 Stat., pp. 779, 780; 10 U.S.C. § 5150 *et seq.* (1946) (www.loc.gov/law/help/statutes-at-large/79th-congress.php).

45. V. Bush, *op. cit.*

46. H. Varmus, "Squeeze on Science," *Washington Post*, A33, October 4, 2000 (see www.aip.org/fyi/2000/harold-varmus-support-nsf-and-doe-office-science).

47. J.L. Penick, Jr. *et al.*, *op. cit.*, pp. 169–171.

48. Public Law 692, 81st Cong., 2nd Sess., Ch. 714, 64 Stat., pp. 443–447; 42 U.S.C. § 203 *et seq.* (1950) (www.loc.gov/law/help/statutes-at-large/81st-congress.php).

49. "About NIH—Who We Are," National Institutes of Health, U.S. Department of Health and Human Services (www.nih.gov/about-nih/who-we-are).

50. See R. Rhodes, *The Making of the Atomic Bomb* (Simon and Schuster, New York, 1986).

51. Oppenheimer actually split his time between the University of California–Berkeley and the Californian Institute of Technology in Pasadena. But he was more at ease at Berkeley because James Millikan, Caltech's top physicist despised him, possibly because of his Jewish heritage. (N.P. Davis, *op. cit.*, p. 52).

52. N.P. Davis, *op. cit.*, pp. 12-14.

53. For a fascinating description of the German nuclear effort during World War II, see D. Irving, *The German Atomic Bomb: The History of Nuclear Research in Nazi Germany* (Simon and Schuster, New York, 1968).

54. R. Rhodes, *Dark Sun: The Making of the Hydrogen Bomb* (Simon and Schuster, New York, 1995).

55. M. Telson, former chief financial officer of the Department of Energy, private communication.

56. The following list comprises the 17 national laboratories on the federal roster as of January 2018 with dates reflecting when they started their research activities rather than when they received their official national laboratory status: National Energy Technology Laboratory [Pittsburgh, Pennsylvania (1910); Morgantown, West Virginia (1946); Sugar Land, Texas (2000); Fairbanks, Alaska (2001); Albany, Oregon (2005)]; Lawrence Berkeley National Laboratory (1931); Los Alamos National Laboratory (1943); Oak Ridge National Laboratory (1943); Argonne National Laboratory (1946); Ames Laboratory (1947); Brookhaven National Laboratory (1947); Sandia National Laboratory (1948); Idaho National Laboratory (1949); Princeton Plasma Physics Laboratory (1951); Lawrence Livermore National Laboratory (1952); Savannah River National Laboratory (1952); SLAC National Accelerator Laboratory (1962); Pacific Northwest National Laboratory (1965); Fermi National Accelerator Laboratory (1967); National Renewable Energy Laboratory (1977); Thomas Jefferson National Accelerator Facility (1984).

57. "The Galvin Report," Secretary of Energy Advisory Board Task Force on Alternative Futures for the Department of Energy National Laboratories, R. Galvin, chair, February 1995. scipp.ucsc.edu/~haber/UC_CORP/galvin.htm.

58. "The CRENEL Report," Commission to Review the Effectiveness of the National Laboratories, Final Report, T.J. Glauthier and J.L. Cohon, co-chairs, October 28, 2015. energy.gov/labcommission/downloads/final-report-commission-review-effectiveness-national-energy-laboratories.

59. "The Cox Report," Report of the Select Committee on U.S. National Security and Military/Commercial Concerns with the People's Republic of China (U.S. House of Representatives, 105th Cong., 2nd Sess., Report 105–851, January 3, 1999).

60. See, for example, E. Klein, "The Hunting of Wen Ho Lee," *Vanity Fair*, December 2000, pp. 142 ff.

Chapter 7: Growing pains 1952–1974

1. J. Wooley and G. Peters, *The American Presidency Project*, "Presidential Job Approval, F. Roosevelt (1941)—Trump," (The University of California, Santa Barbara). www.presidency.ucsb.edu/data/popularity.php?pres=33.
2. D. McCullough, *op. cit.*, pp. 892,893.
3. *Ibid.*, pp. 712, 844.
4. *Ibid.*, p. 887.
5. David Eisenhower, private communication.
6. D. McCullough, *op. cit.*, pp. 839–855.
7. Brownell had visited Eisenhower at NATO Headquarters in Paris earlier in the year, but was unable to secure his commitment to run [H. Brownell, *Advising Ike: The Memoirs of Attorney Herbert Brownell* (University Press of Kansas, Lawrence, Kan., 1993), pp. 94–97]. The follow-up phone call sealed the deal [J.E. Smith, "Ike Reconsidered: Conference Public Program," Roosevelt House Public Policy Institute at Hunter College, March 12, 2013, Video Transcript: www.youtube.com/watch?v=NJDu35FfcrQ (1:21:45–1:22:50)].
8. To observe the fervor with which the 1952 Republican Convention greeted MacArthur, watch www.c-span.org/video/?3985-1/gen-douglas-macarthur-keynote-address.
9. D. McCullough, *op. cit.*, pp. 903, 904.
10. J. Wooley and G. Peters, *op. cit.*, "Dwight D. Eisenhower, XXXIV President of the United States: 1953-1961, Address Accepting the Presidential Nomination at the Republican National Convention in Chicago, July 11, 1952". www.presidency.ucsb.edu/ws/index.php?pid=75626.
11. "NSF Appropriations and Requests By Account: FY 1951–FY 2017" (www.google.com/url?sa=t&rct=j&q=&esrc=s&source=web&cd=1&ved=0ahUKEwjvw9T__vPYAhXB2VMKHeYbCmwQFggpMAA&url=https%3A%2F%2Fdellweb.bfa.nsf.gov%2FNSFRqstAppropHist%2FNSFRequestsandAppropriationsHistory.pdf&usg=AOvVaw3nwQklKa_JG7ryxihwQttm).
12. *The New York Times*, October 4, 1957, p. 1 (www.nytimes.com/learning/general/onthisday/big/1004.html).
13. 72 Stat. 426-2, Public Law 85-568, July 29, 1958, H.R. 12575, pp. 426-438; 42 U.S.C. § 2451 *et seq.* (1958) (www.gpo.gov/fdsys/pkg/STATUTE-72/pdf/STATUTE-72-Pg426-2.pdf).
14. See www.darpa.mil for an overview of DARPA's mission, history and activities.
15. 72 Stat. 1580, Public Law 85-684, September 2, 1958, H.R. 13247, pp. 1580-1605; 20 U.S.C. § 401 *et seq.* (1958) (www.gpo.gov/fdsys/pkg/STATUTE-72/pdf/STATUTE-72-Pg1580.pdf).
16. See, for example, "The National Science Board: A History in Highlights 1950-2000, The 1970s, Supporting the Director as Science Advisor," (U.S. National Science Foundation, 2000). www.nsf.gov/nsb/documents/2000/nsb00215/nsb50/1970/supp_dir.html – and J.S. Rigden, *Rabi, Scientist and Citizen* (Harvard University Press, Cambridge, Mass., 2000), p. 251.
17. D.D. Eisenhower, *At Ease: Stories I Tell to Friends* (Doubleday, New York, 1967), pp. 157-168.
18. 70 Stat. 374, Public Law 84-627, June 29, 1956, H.R. 10660, pp. 374-387; 23 U.S.C. § 48 *et seq.* (1956) (www.gpo.gov/fdsys/granule/STATUTE-70/STATUTE-70-Pg374/content-detail.html).
19. For a brief history of the Highway Trust Fund, see "Funding Federal-aid Highways," U.S. Department of Transportation, Federal Highway Administration, Office of Policy and Governmental Affairs, Publication No. FHWA-PL-17-011, January 2017. www.fhwa.dot.gov/policy/olsp/fundingfederalaid/07.cfm.

20. See, for example, J. Fox "The Great Paving: How the Interstate Highway System Helped Create the Modern Economy—and Reshaped the FORTUNE 500," *Fortune Magazine*, January 26, 2004 (archive.fortune.com/magazines/fortune/fortune_archive/2004/01/26/358835/index.htm).

21. J. Wooley and G. Peters, *op. cit.*, "1-Inaugural Address, January 20, 1953". www.presidency.ucsb.edu/ws/index.php?pid=9600.

22. J. Wooley and G. Peters, *op. cit.*, "421-Farewell Radio and Television Address to the American People, January 17, 1961". www.presidency.ucsb.edu/ws/index.php?pid=12086.

23. N.P. Davis, *op. cit.*

24. A. Pais with supplemental material by R. Crease, *J. Robert Oppenheimer: A Life* (Oxford University Press, New York, 2006).

25. U.S. Atomic Energy Commission, *In the Matter of J. Robert Oppenheimer: Transcript of Hearing Before Personnel Security Board and Texts of Principal Documents and Letters* (MIT Press, Cambridge, Mass., 1954), pp. 837, 838.

26. J. Wooley and G. Peters, *op. cit.*, "Lyndon B. Johnson, XXXVI President of the United States: 1963-1969, 19-Remarks Upon Presenting the Fermi Award to Dr. J. Robert Oppenheimer, December 2, 1963". www.presidency.ucsb.edu/ws/index.php?pid=26076.

27. J. Wooley and G. Peters, *op. cit.*, "John. F. Kennedy, XXXV President of the United States: 1960-1963, 1-Innaugural Address, January 20, 1960". www.presidency.ucsb.edu/ws/index.php?pid=8032.

28. J.M. Logsdon, "John F. Kennedy and NASA," *NASA 50th Magazine* (NASA, May 25, 2015). www.nasa.gov/feature/john-f-kennedy-and-nasa.

29. W.S. Bainbridge, ed., *Leadership in Science and Technology: A Reference Handbook, Vol. 1* (Sage Publications, Los Angeles, 2012), p. 451.

30. J. Wooley and G. Peters, *op. cit.* "430-Address, at the Anniversary Convocation of the National Academy of Sciences, October 22, 1963". www.presidency.ucsb.edu/ws/?pid=9488.

31. Universal Coordinated Time, or UTC, is based on Greenwich Mean Time referenced to extraordinarily accurate atomic clocks.

32. The author, who was opposed to the Vietnam War, was nonetheless the object of one such experience in New Haven, Conn. in 1970, simply because he was a physicist on the Yale University faculty.

33. J. Markon revisited the 1970 bombing of the University of Wisconsin mathematics building in "After 40 Years, Search for University of Washington Suspect Heats Up Again," *Washington Post*, September 21, 2010 (www.washingtonpost.com/wp-dyn/content/article/2010/09/21/AR2010092106588.html).

34. Z. Wang, *In Sputnik's Shadow: The President's Science Advisory Committee and Cold War America* (Rutgers University Press, New Brunswick, New Jersey, 2008), p. 291.

35. 83 Stat. 204, Public Law 91-121, November 19, 1969, S. 2546, Sec. 203, p. 206. www.gpo.gov/fdsys/search/pagedetails.action?collectionCode=STATUTE&browsePath=1969%2FPUBLIC LAW&granuleId=STATUTE-83-Pg204&packageId=STATUTE-83&fromBrowse=true.

36. As a testament to Fulbright's enduring commitment to science, the U.S. State Department regards its International Fulbright Science and Technology Award as one of its most prestigious honors.

37. 84 Stat. 1676, Public Law 91-604, December 31, 1970, H.R. 17255, pp. 1676-171; 42 U.S.C. § 7401 *et seq.* (1970). www.gpo.gov/fdsys/pkg/STATUTE-84/pdf/STATUTE-84-Pg1676.pdf.

38. T. Friedman, *The World Is Flat: A Brief History of the Twenty-first Century* (Farrar, Strauss and Giroux, New York, 2005).

39. M. Greenstone, "The Impacts of Environmental Regulations on Industrial Activity: Evidence from the 1970 and 1977 Clean Air Act Amendments and the Census of Manufacturers,"

Working Paper 8484 (National Bureau of Economic Research, Cambridge, Mass, 2001). www.nber.org/papers/w8484.

40. 86 Stat. 816, Public Law 92-500, October 18, 1972, S. 2770, pp. 816–903; 33 U.S.C. § 1251 *et seq.* (1972) (www.gpo.gov/fdsys/pkg/STATUTE-86/pdf/STATUTE-86-Pg816.pdf).

41. *Time*, Friday, August 1, 1969.

42. For a retrospective of the Clean Water Act, see A. Snider, "Clean Water Act: Vetoes by Eisenhower Presaged Today's partisan Divide," *E&E News—Greenwire*, Thursday, October 18, 2012. www.eenews.net/stories/1059971457.

43. Nixon might have been in favor of controlling water pollution, but he didn't see why it had to be so expensive. His opposition to the Clean Water Act's $24-billion price tag trumped any concern he harbored about the problem.

44. Air resistance, or "drag," increases rapidly with speed, causing vehicles to burn more fuel per mile traveled.

45. A. Buck, "A History of the Energy Research and Development Administration," (U.S. Department of Energy: Office of Management, Office of the Executive Secretariat, Office of History and Heritage Resources, March 1982).

46. 88 Stat. 1233, Public Law 93-438, October 11, 1974, H.R. 11510, pp. 1233-1254; 42 U.S.C. § 5801 *et seq.* (1974) (www.gpo.gov/fdsys/pkg/STATUTE-88/pdf/STATUTE-88-Pg1233.pdf).

47. 89 Stat. 871, Public Law 94-163, December 22, 1975, S. 622, pp. 871-969; 42 U.S.C. § 6201 *et seq.* (1975) (www.gpo.gov/fdsys/pkg/STATUTE-89/pdf/STATUTE-89-Pg871.pdf).

48. B. Richter *et al.*, "How America Can Look Within to Achieve Energy Security and Reduce Global Warming," *Rev. Mod. Phys.* **80**, S1 (2008), pp. 28-51. doi.org/10.1103/RevModPhys.80.S1.

49. 85 Stat. 778, Public Law 92-218, December 23, 1971, S. 1828, pp. 778–786; 15 U.S.C. § 1022 *et seq.* (1971) (www.gpo.gov/fdsys/pkg/STATUTE-85/pdf/STATUTE-85-Pg778.pdf).

50. See "National Cancer Act of 1971," About NCI, National Cancer Institute, National Institutes of Health. www.cancer.gov/about-nci/legislative/history/national-cancer-act-1971#bill.

51. F. Luntz, *Words That Work: It's Not What You Say, It's What People Hear* (Hyperion, New York, 2007).

52. For a concise summary, see W. Burr, ed., "Missile Defense Thirty Years Ago: Déjà Vu All Over Again?" *National Archive Electronic Briefing Book No. 36* (National Security Archive, George Washington University, December 18, 2000). nsarchive2.gwu.edu/NSAEBB/NSAEBB36/index.html.

53. Z. Wang, *op. cit.*, pp. 297–305.

54. W.S. Bainbridge, *op. cit.*, p. 460.

55. For details about the Watergate scandal, see B. Woodward and C. Bernstein, *All the President's Men* (Simon and Schuster, New York, 1974) and its follow-up, *The Final Days* (Simon and Schuster, New York, 1976).

56. General Electric moved its headquarters from Fairfield, Conn. to Boston, Mass. in 2017, and Fujifilm purchased a controlling share of Xerox in 2018.

57. J. Sadowski, "Office of Technology Assessment: History, Implementation, and Participatory Critique," *Technology in Society* **42**, 9–20 (2015).

58. 86 Stat. 797, Public Law 92-484, October 13, 1972, H.R. 10243, pp. 797-803; 41 U.S.C. § 5 *et. seq.* (1972) (www.gpo.gov/fdsys/pkg/STATUTE-86/pdf/STATUTE-86-Pg797.pdf).

59. For historical notes and a compendium of OTA reports, see U.S. Congress, Office of Technology Assessment, The OTA Legacy: 1972–1995 (Washington, DC: April 1996). www.princeton.edu/~ota/.

60. J. Sadowski, *op. cit.*, p. 16.

61. A.B. Carter, "Directed Energy Missile Defense in Space—A Background Paper" (Washington, D. C.: U.S. Congress, Office of Technology Assessment, OTA-BP-ISC-26, April 1984), p. 81.

Chapter 8: A fresh start 1974–1992

1. G.R. Ford, private communication.
2. N.A. Rockefeller, private communication.
3. 90 Stat. 459, Public Law 94-282, May 11, 1976, H.R. 10230, pp. 459-473; 42 U.S.C. § 6601 *et seq.* (1976) (www.gpo.gov'fdsys/pkg/STATUTE-90/pdf/STATUTE-90-Pg.459.pdf).
4. See J.F. Sargent Jr. and D.A Shea, "Office of Science and Technology Policy (OSTP): History and Overview," CRS Report R43935 (Congressional Research Service, August 17, 2017). fas.org/sgp/crs/misc/R43935.pdf.
5. J. Wooley and G. Peters, *op. cit.*, "Jimmy Carter, XXXIX President of the United States: 1977–1981, Executive Order 12039–Science and Technology Policy Functions, February 24, 1978". www.presidency.ucsb.edu/ws/?pid=30416.
6. J. Wooley and G. Peters, *op cit.* "George Bush, XLI President of the United States: 1989-1993, Executive Order 12700–President's Council of Advisors on Science and Technology, January 19, 1990". www.presidency.ucsb.edu/ws/index.php?pid=23546.
7. For a report on the first PCAST meeting, see I. Goodwin, *Physics Today*, **43** (3), pp. 49–50 (1990). physicstoday.scitation.org/doi/10.1063/1.2810483.
8. Donald Trump re-chartered PSAC but had not filled any of its seats as of March 2018, more than fifteen months after he took office.
9. Donald Trump indicated he would revamp the operation of NSTC, but at the time of this book's writing the White House had not released any plans.
10. "Home > Energy Explained > Nonrenewable Resources > Oil and Petroleum Products > Imports and Exports," U.S Energy Information Administration (EIA). www.eia.gov/energyexplained/index.cfm?page=oil_imports.
11. 91 Stat. 565, Public Law 95-91, August 4, 1977, S. 826, pp. 565–613; 42 U.S.C. § 17101 *et seq.* (1977) (www.gpo.gov/fdsys/pkg/STATUTE-91/pdf/STATUTE-91-Pg565.pdf).
12. C. Marquis, "The 43rd President: Man in the News; Edmund Spencer Abraham," *The New York Times*, January 3, 2001. www.nytimes.com/2001/01/03/us/the-43rd-president-man-in-the-news-edmund-spencer-abraham.html.
13. C, Davenport and D.E. Sanger, "'Learning Curve' as Rick Perry Pursues a Job He Initially Misunderstood," *The New York Times*, January 18, 2017. www.nytimes.com/2017/01/18/us/politics/rick-perry-energy-secretary-donald-trump.html?_r=0.
14. "Department of Homeland Security: Science and Technology". www.dhs.gov/science-and-technology/about-st.
15. Even forty years later, I recall it vividly. A C-SPAN video is available at www.c-span.org/video/?153913-1/president-carters-fireside-chat-energy.
16. J. Wooley and G. Peters, *op. cit.*, "Report to the American People—Remarks from the White House Library, February 2, 1977". www.presidency.ucsb.edu/ws/index.php?pid=7455&st=&st1=.
17. J.E. Carter, private communication.
18. J. Wooley and G. Peters, *op. cit.*, "Energy Address to the Nation, April 5, 1979. http://www.presidency.ucsb.edu/ws/?pid=32159.
19. W. Sweet and S. Stencel, "Public Confidence and Energy," *Editorial Research Reports 1979, Vol. I* (CQ Press, Washington, DC). library.cqpress.com/cqresearcher/document.php?id=cqresrre1979052500.

20. Stories have long circulated—See N.A. Lewis, "New Reports Say 1980 Reagan Campaign Tried to Delay Hostage Release," *The New York Times*, April 15, 1991 (www.nytimes.com/1991/04/15/world/new-reports-say-1980-reagan-campaign-tried-to-delay-hostage-release.html). That Republican operatives might have urged Iran to delay the hostage's release until after the 1980 election in order to help Ronald Reagan, but no definitive proof has been found.
21. Hagen Research carried out the survey for ScienceCounts and Research!America.
22. J. Wooley and G. Peters, *op. cit.*, "Ronald Reagan XL President of the United States: 1981-1989, Radio Address to the Nation on the Federal Role in Scientific Research, April 2, 1988". www.presidency.ucsb.edu/ws/?pid=35637.
23. *Ibid.*, "Remarks at the Annual Convention of the National Association of Evangelicals in Orlando Florida, March 8, 1983". www.presidency.ucsb.edu/ws/index.php?pid=41023&st=evangelicals&st1=http://www.presidency.ucsb.edu/ws/index.php?pid=41023&st=evangelicals&st1=.
24. J. Gerstenzang, "Weinberger Sees End of "Mutual Suicide Pact," *Los Angeles Times*, October 10, 1985 (articles.latimes.com/1985-10-10/news/mn-15630_1_defense-strategy.
25. J. Wooley and G. Peters, *op cit.*, "Address to the Nation on Defense and National Security, March 23, 1983". www.presidency.ucsb.edu/ws/index.php?pid=41093&st=strategic&st1=defense.
26. D Kimball and K. Reif, "The Anti-Ballistic Missile (ABM) Treaty as a Glance" (Arms Control Association, Fact Sheets & Briefs, August 2012) www.armscontrol.org/factsheets/abmtreaty.
27. N. Bloembergen, C.K.N. Patel *et al.*, "Report to the American Physical Society of the Study Group on Science and Technology of Directed Energy Weapons," Rev. Mod. Phys. **59**, S1 (July 1987). doi:10.1103/RevModPhys.59.S1.
28. See "A Timeline of HIV and AIDS" (www.hiv.gov/hiv-basics/overview/history/hiv-and-aids-timeline).
29. J. Wooley and G. Peters, *op cit.*, "Remarks at the American Foundation for AIDS Research Awards Dinner, May 31, 1987". www.presidency.ucsb.edu/ws/index.php?pid=34348.
30. For a comprehensive look at White House science and technology policy during the presidency of George H.W. Bush, see D.A. Bromley, *The President's Scientists: Reminiscences of a White House Science Advisor* (Yale University Press, New Haven, 1994).
31. M. Riordan, L. Hoddeson and A.W. Kolb, *Tunnel Visions: The Rise and Fall of the Superconducting Super Collider*, The University of Chicago Press (Chicago, 2015), pp 184–189.
32. A. McDaniel, "25 Years Ago Today, George H.W. Bush Vomited on the Prime Minister of Japan," *Newsweek*, January 8, 2017 (http://www.newsweek.com/25-years-ago-today-george-h-w-bush-vomited-prime-minister-japan-538581).
33. M. Riordan *et al.*, *op. cit.*, p. 228.
34. *Ibid.*, p. 232.
35. *Ibid.*, p. 229.
36. *Ibid.*, p. 233.
37. *Ibid.*, p. 236.
38. George H.W. Bush's term began on January 20, 1989 and ended on January 20, 1993. D. Allan Bromley served as Bush's Science Advisor and Director of the Office of Science and Technology Policy (OSTP) from his swearing-in on October 13, 1989 until Bush's departure on January 20, 1993.

Chapter 9: Crossing new intersections 1992–2000

1. R.M. Solow, "A Contribution to the Theory of Economic Growth," *Q. J. Econ.* **70**, 65 (1956).
2. R.M. Solow, "Technical Change and the Aggregate Production Function," *Rev. Econ. Stat.* **39**, 312 (1957).

3. R.M. Solow, "Investment and Technical Progress," in *Mathematical Methods in the Social Sciences, 1959,* ed. by K.J. Arrow, S. Karlin, and P. Suppes (Stanford University Press, 1960), pp. 89-104.

4. For highlights of Edwin Mansfield's work, see A.M. Diamond Jr., "Edwin Mansfield's Contributions to the Economics of Technology," *Research Policy* **32**, 1607 (2003).

5. E. Mansfield, "Academic Research and Industrial Innovation," *Research Policy* **20** (1), 1 (1991).

6. D.A. Bromley, *The President's Scientists: Reminiscences of a White House Science Advisor* (Yale University Press, New Haven, 1994), p. 221.

7. E. Mansfield, "Academic Research and Industrial Innovation: A Further Note" *Research Policy* **21** (3), 295 (1992).

8. Bromley cites only the 28 percent figure in his book, and during the twelve years we worked together prior to his death in 2005, he never mentioned that he was aware of the revision upward to 40 percent.

9. E. Mansfield, "Basic Research and Productivity Increase in Manufacturing," *Amer. Econ. Rev.* **70** (5), 863 (1980).

10. Personal notes.

11. His book, *Earth in the Balance: Ecology and the Human Spirit* (Houghton Mifflin, Boston, 1992) received the Robert F. Kennedy Center for Justice and Human Rights Award in 1993.

12. M.J. Boskin and L.J. Lau, "Capital Formation and Economic Growth," *Technology and Economics* (The National Academies Press, Washington, 1991), pp. 47-54.

13. *Ibid.*, p. 52 (www.nap.edu/read/1767/chapter/4#52).

14. "Societies Call for 7% Funding Increase in Joint Statement on Scientific Research," *APS News* **6** (4), April 1997 (www.aps.org/publications/apsnews/199704/statement.cfm).

15. www.congress.gov/bill/105th-congress/senate-bill/124/text?q=%7B%22search%22%3A%5B%22S.+124+105th+Congress%22%5D%7D&r=3.

16. www.congress.gov/bill/105th-congress/senate-bill/1305/text?q=%7B%22search%22%3A%5B%22S.+1305+105th%22%5D%7D&r=1.

17. www.congress.gov/bill/105th-congress/senate-bill/2217/text?q=%7B%22search%22%3A%5B%22S.+2217%22%5D%7D&r=1.

18. www.congress.gov/bill/106th-congress/senate-bill/296/text?q=%7B%22search%22%3A%5B%22S.+296%22%5D%7D&r=1.

19. A.H. Dupree, *Science in the Federal Government*, The Johns Hopkins University Press (Baltimore, Md., 1986), pp. 215–231.

20. Public Law 115-141, March 23, 2018, H.R. 1625 (www.congress.gov/bill/115th-congress/house-bill/1625/text).

21. 88 Stat. 297, Public Law 93-344, July 12, 1974, H.R. 7130, 31 U.S.C. § 301 *et. seq.* (1974) (www.gpo.gov/fdsys/pkg/STATUTE-88/pdf/STATUTE-88-Pg297.pdf).

22. See, for example, M. Hourihan, "The Federal Budget Process 101," American Association for the Advancement of Science. www.aaas.org/news/federal-budget-process-101 (AAAS, 15 July 2014).

23. W. Churchill, "Speech, House of Commons, November 11, 1947," in *Winston S. Churchill: His Complete Speeches, 1897–1963 (In 8 Volumes),* ed. by Robert Rhodes James, (Chelsea House Publishers & R.R. Bowker, Co., New York, 1974), Vol. 7, p. 7566.

24. For accounts of the quest to sequence the human genome, see J. Shreeve, *The Genome War: How Craig Venter Tried to Capture the Code of Life and Save the World* (Alfred A. Knopf, New York, 2004) and K. Davies, *Cracking the Genome: Inside the Race to Unlock Human DNA* (Johns Hopkins University Press, Baltimore, Md., 2001).

25. 94 Stat. 96, Public Law 96-517, December 12, 1980, H.R. 6933, 35 U.S.C. § 200 *et seq.* (1980) (www.gpo.gov/fdsys/pkg/STATUTE-94/pdf/STATUTE-94-Pg3015.pdf).

26. L. Roberts, "Why Watson Quit as Project Head," *Science* **256**, 301–302 (17 April 1992).

27. In his 1994 memoire (D.A. Bromley, *op. cit.*) he never once mentioned it.

28. N.F. Lane, private communication.

29. T. Friend, *USA Today*, March 13, 2000.

30. J. Wooley and G. Peters, *The American Presidency Project*, "William J. Clinton, XLII President of the United States: 1993-2001, Remarks on Presenting the National Medals of Science and Technology, March 14, 2000". www.presidency.ucsb.edu/ws/?pid=58246.

31. N.F. Lane, private communication.

32. finance.yahoo.com/quote/%5EIXIC/history?period1=951886800&period2=1083297600&interval=1d&filter=history&frequency=1d.

33. J. Wooley and G. Peters, *op. cit.*, "Remarks on the Completion of the First Survey of the Human Genome, June 26, 2000". www.presidency.ucsb.edu/ws/index.php?pid=58701&st=&st1=.

34. *Time*, July 3, 2000. content.time.com/time/covers/0,16641,20000703,00.html.

35. International Human Genome Sequencing Consortium, "Initial Sequencing and Analysis of the Human Genome," *Nature* **409**, 860 (2001). www.nature.com/articles/35057062.

36. J. Craig Venter *et al.*, "The Sequence of the Human Genome," *Science* **291**, 1304 (2001). science.sciencemag.org/content/291/5507/1304.

37. A. Liptak, "Justices, 9-0, Bar Patenting Human Genes," *The New York Times*, June 13, 2013. www.nytimes.com/2013/06/14/us/supreme-court-rules-human-genes-may-not-be-patented. html; *Association for Molecular Pathology v. Myriad Genetics, Inc.*, No. 12-398 [569 U.S. 576 (2013)].

38. A. Jackson, "Nature's 10 People Who Mattered in 2013," Of Schemes and Memes, Community Blog, *Nature*. blogs.nature.com/ofschemesandmemes/2013/12/19/natures-10-people-who-mattered-in-2013.

39. C. Venter and D. Cohen, *National Perspectives Quarterly*, **21**, 73–77 (2004).

40. H. Varmus, "Squeeze on Science," *The Washington Post*, p. A33 (October 4, 2000).

41. J. Wooley and G. Peters, *op. cit.*, "Remarks at the California Institute of Technology in Pasadena, California, January 21, 2000". www.presidency.ucsb.edu/ws/index.php?pid=58609.

42. Beginning with his arrival in 1921, Einstein spent virtually all of his American career at the Institute for Advanced Study on the campus of Princeton University. It is often overlooked that he gave his first public address that year in the Great Hall of the City College of New York (CCNY), the "Harvard of the Proletariat," as the institution was and is still known.

43. 114 Stat. 3088, Public Law 106-580, December 29, 2000, H.R. 1795, 42 U.S.C. § 201 *et seq.* (2000) (www.gpo.gov/fdsys/pkg/PLAW-106publ580/html/PLAW-106publ580.htm).

44. Senators Connie Mack, a Republican from Florida, and Tom Harkin, a Democrat from Iowa, two of the principal supporters.

45. Ken Dill, who was a then a distinguished member of the faculty at the University of California, San Francisco, was leading the NIBIB effort.

46. Arthur Bienenstock, a renowned physicist on leave from Stanford University, was the OSTP Associate Director for Science at the time. He successfully shepherded the proposal through the Clinton Administration.

47. J. Wooley and G. Peters, *The American Presidency Project*, "Barack Obama XLIV President of the United States: 2009-2017, Remarks on Science and Technology, April 2, 2013". www.presidency.ucsb.edu/ws/?pid=103411.

48. Polling and focus groups conducted in 2016 by Edge Research, an Arlington, Va. survey firm, and studies conducted in 2017 by Civilian, a San Diego, Calif. marketing and advertising firm,

both on behalf of ScienceCounts, were very revelatory. Most people have a good feeling about science in the abstract, but as a practical matter, apart from medicine, they do not see very much of a connection between science and their daily lives. They regard innovations, such as the smart phone or tablet, as inventions created in someone's garage, rather than the adaptation of many scientific discoveries.

49. See "The Nobel Prize in Physics 1976: Burton Richter, Samuel C.C. Ting," Press Release, The Royal Swedish Academy of Sciences, 18 October 1976. www.nobelprize.org/nobel_prizes/ physics/laureates/1976/press.html.

50. See "The Nobel Prize in Physics 1995: Martin L. Perl and Frederick Reines," Press Release, The Royal Swedish Academy of Sciences, 11 October 1995. www.nobelprize.org/nobel_prizes/ physics/laureates/1995/press.html.

51. The National Synchrotron Light Source at Brookhaven National Laboratory in Upton, New York was commissioned in 1983.

52. The Advanced Light Source (ALS) at Lawrence Berkeley National Laboratory in Berkeley, Calif. was commissioned in 1993, and the Advanced Photon Source (APS) at Argonne National Laboratory in DuPage County, Ill. was commissioned in 1994.

53. K. Hodgson, "SSRL to Upgrade SPEAR Storage Ring," *Synchrotron Rad. News* **12**, 41 (1999).

54. The first step in the passage of a bill is a committee or subcommittee draft of legislative language. The process is called "marking up."

55. "Stove-piping" is sometimes called "siloing." Both terms convey the organization of activities or responsibilities into separate structures that have poor channels of communication among them.

56. Vernon J. Ehlers, a Michigan Republican was the first research physicist elected to Congress. He received his doctoral degree from the University of California-Berkeley and eventually became chairman of the Physics Department at Calvin College in Grand Rapids, Michigan. He served in the Michigan state legislature from 1983 until 1993, when he won a special election to the House of Representatives. In 1999, he was joined by Rush Holt Jr., a Democrat from New Jersey, who had received his Ph.D. from New York University and taught physics and public policy at Swarthmore College for eight years, prior to a two-year stint at the State Department and an eight-year term as assistant director of the Princeton Plasma Physics Laboratory. (Holt's father, Rush Holt, Sr., a Democrat who turned against the Coal Miner's Union, represented West Virginia in the United States Senate for one term, from 1935 until 1941.) In 2008, Bill Foster, an Illinois Democrat won a special election to replace former House Speaker Dennis Hastert, who had resigned after Democrats regained control of the House of Representatives in 2007. Foster, who had received his Ph.D. from Harvard and spent 22 years as a high-energy physicist at Fermi National Laboratory, rounded out what was called the House "Physics Caucus." He lost his bid for reelection in 2010, but won the next time around in 2012. Ehlers retired in 2011 and Holt departed in 2015, leaving Foster as the only physicist in the House or Senate.

57. John Olver, a Democrat, represented the western part of Massachusetts from 1991 until 2012. He had received his Ph.D. from the Massachusetts Institute of Technology (MIT) and taught chemistry at the University of Massachusetts Amherst before running for the House of Representatives. He retired after his district was combined with an adjacent one, following the 2010 census, which caused Massachusetts to lose a House seat.

58. Jerry McNerney, a California Democrat, won election to the House of Representatives in 2006. He had received his Ph.D. in mathematics from the University of New Mexico before working on national security issues at Sandia National Laboratory and later as an energy consultant specializing in wind power.

59. Weston Vivian, a Michigan Democrat served only one term in the House of Representatives, from 1965 to 1967. He had received a Master's Degree from MIT and a Ph.D. in electrical engineering from the University of Michigan before embarking on a consulting career.

60. Harrison Schmitt, a Republican and native of New Mexico, served one term in the United States Senate. He had received a Ph.D. in geology from Harvard University before joining NASA in 1965 as a member of the first cadre of scientist-astronauts. He flew on the 17th and last Apollo Mission, landing on the lunar surface in 1972. He ran successfully for the Senate from his home state in 1976 but losing to Jeff Bingaman in 1982.

61. See https://www.aaas.org/program/science-technology-policy-fellowships.

Chapter 10: Years of anxiety 2001–2008

1. See www.compete.org/about/about-council.

2. L. Santos, "Review of Findings of the President's Commission on Industrial Competitiveness: Memo on March 29, 1985 Committee Hearing on the Industrial Competitiveness Report," United States Senate Committee on Finance. www.finance.senate.gov/imo/media/doc/HRG99-75.pdf.

3. In 2013, The Task Force on American Innovation, "a non-partisan alliance of leading American companies and business associations, research university associations, and scientific societies" (www.innovationtaskforce.org/) produced a sequel to its 2005 study. The updated report, *American Exceptionalism, American Decline? Research, the Knowledge Economy and the 21st Century Challenge, can be found at* www.innovationtaskforce.org/task-force-on-american-innovation-american-exceptionalism-american-decline/.

4. *Rising Above the Gathering Storm: Energizing and Employing America for a Brighter Economic Future* (The National Academies Press, Washington, D.C., 2005), pp. vii–ix.

5. J. Bingaman, R.M. Simon and A.L. Rosenberg, "Needed: A Revitalized National S&T Policy," *Issues in Science and Technology* **20** (3), Spring 2004 (issues.org/20-3/p_bingaman/).

6. J. Bingaman, "Maintaining America's Competitive Edge," *APS News* **14** (6), June 2005 (www.aps.org/publications/apsnews/200506/backpage.cfm).

7. *Rising Above the Gathering Storm: Energizing and Employing America for a Brighter Economic Future, op. cit.*, pp. ES-1–ES-10.

8. N. Augustine, private communication.

9. *The National Summit on Competitiveness: Investing in U.S. Innovation* (A National Gathering of Executives Concerned About America's Future Competitiveness, Washington, D.C., December 6, 2005, unpublished).

10. R. Pear, "Physicist Said to Be Top Choice For Science Adviser to President," *The New York Times*, June 25, 2001 (www.nytimes.com/2001/06/25/us/physicist-said-to-be-top-choice-for-science-adviser-to-president.html).

11. T.J. Gay and A.F. Starace, "DAMOP 2005," in *Division of Atomic, Molecular and Optical Physics Newsletter, March 2005*, L. Cocke, editor (www.aps.org/units/damop/newsletters/upload/march05.pdf).

12. To appreciate the complexity of the issue, see for example, D.B. Audretsch *et al.*, "The Economics of Science and Technology," *J. Tech Transfer* **27**, 155 (2002). maryannfeldman.web.unc.edu/files/2011/11/Economics-of-Sci-and-Tech_2002.pdf.

13. T. Baer and F. Schlachter, "Lasers in Science and Industry: A Report to OSTP on the Contribution of Lasers to American Jobs and the American Economy" (www.laserfest.org/lasers/baer-schlachter.pdf).

14. National Research Council, "Impact of Photonics on the National Economy," *Optics and Photonics: Essential Technologies for Our Nation*, pp. 20–62 (The National Academies Press, Washington, D.C., 2013).

15. Tim Berners Lee, a computer scientist working at CERN, the European Center for Nuclear Physics in Geneva, Switzerland, was tasked with developing a communications tool with a graphics interface that would allow more than 10,000 high-energy physicists and engineers around the world to exchange information and data and to coordinate their work on the Large Hadron Collider, a multi-national project costing almost $10 billion. The hypertext transfer protocol, with familiar the letters, http, is the foundation of data communication on the Web. The precursors to the popular Web browser, Mozilla-Firefox, were Netscape and its progenitor, Mosaic, which was developed with federal funds at the National Center for Supercomputing Applications at the University of Illinois Urbana-Champagne.

16. See "Where the Future Becomes Now," Defense Advanced Research Projects Agency, About Us, History and Timeline. www.darpa.mil/about-us/darpa-history-and-timeline?PP=0.

17. See www.darpa.mil/attachments/DARPA_Directors_Sheet-web.pdf.

18. M. Boroush, "National Patterns of R&D Resources: 2014–15 Data Update—Detailed Statistical Tables, NSF 17-311, March 2017 (National Center for Science and Engineering Statistics, National Science Foundation, Alexandria, Va., 2017). https://www.nsf.gov/statistics/2017/nsf17311/pdf/nsf17311.pdf.

19. By 2015, after a small uptick during the early Obama years, it had fallen still further to 0.63 percent. See M. Boroush, *op. cit.*

20. J. Wooley and G. Peters, *op. cit.*, "George W. Bush XLII President of the United States: 2001–2009, Address Before a Joint Session of Congress on the State of the Union, January 31, 2006". www.presidency.ucsb.edu/ws/index.php?pid=65090.

21. Domestic Policy Council and Office of Science and Technology Policy, "American Competitiveness Initiative: Leading the World in Innovation," (georgewbush-whitehouse.archives.gov/stateoftheunion/2006/aci/aci06-booklet.pdf).

22. K.H. Fealing, J.L. Lane, J.H. Marburger III and S.S. Shipp, *The Science of Science Policy: A Handbook* (Stanford University Press, Stanford, Calif. 2011)

23. A on-line video of her speech can be found at www.iop.harvard.edu/forum/innovation-agenda-commitment-competitiveness.

24. "The Innovation Agenda: A Commitment to Competitiveness To Keep America #1" was developed by the House Democrats in late 2005 and was the basis for the party's science and technology agenda in 2007 following the Democratic takeover of both the House and Senate in the 2006 election. It is unpublished but is available at https://www.democraticleader.gov/wp-content/uploads/2016/02/Innovation-Agenda-110th-Congress.pdf. The following were some of its key agenda items: To create "A New Generation of Innovators—Educate 100,000 new scientists, engineers and mathematicians in the next four years... Place a highly qualified teacher in every math and science K-12 classroom... Create a special visa for the best and brightest international doctoral and post-doctoral scholars in science, technology, engineering and mathematics. Make college tuition tax deductible for students studying math, science, technology and engineering," doubling "funding for the National Science Foundation, basic research in the physical sciences across all agencies and collaborative research partnerships;" creating "regional Centers of Excellence for basic research;" and permanently extending a "globally competitive R&D [research and development] tax credit." To achieve "A Sustained Commitment to Research and Development—Double overall funding for the National Science Foundation, basic research in the physical sciences across all agencies...; restore the basic, long-term research agenda at the Defense Advanced Research Projects Agency (DARPA) to conduct long-range, high-risk, and high-reward research. Create regional Centers of Excellence for basic research that will attract the best minds and top researchers to develop far-reaching technological innovations and new industries... Modernize and permanently extend a globally

competitive R&D tax credit to increase domestic investment, create more U.S. jobs, and allow companies to pursue long-term projects with the certainty the credit will not expire." To "Bridge the Digital Divide—Implement a national broadband policy that doubles federal funding to promote broadband for all Americans, especially in rural and underserved communities; create new avenues of Internet access including wireless broadband technologies... Ensure the continued growth of Internet-based services and provide a stable regulatory framework... Enact a broadband tax credit for telecommunications companies..." To achieve "Energy Independence in 10 Years—Substantially reduce the use of petroleum based fuels by rapidly expanding production and distribution of synthetic and bio-based fuels...and by deploying new engine technologies for fuel-flexible, hybrid, plug-in hybrid and biodiesel vehicles. Create a new DARPA-like initiative within the Department of Energy...to develop high-risk, high-reward technologies and build markets for the next generation of revolutionary technologies..." To create "A Competitive Small Business Environment for Innovation—Bridge the 'valley of death' that destroys innovative ideas before they become marketable products due to lack of financing and technical support... Reward risk taking and entrepreneurship... Protect the intellectual property of American innovators..." The House Democrats updated their vision with "Innovation 2.0 in 2016 (htherivardreport.com/castro-pelosi-promote-innovation-agenda-to-keep-america-1/).

25. "Science the Endless Frontier," A Report to the President by Vannevar Bush, Director of the Office of Scientific Research and Development, July, 1945, U.S. Government Printing Office, Washington, D.C. (www.nsf.gov/od/lpa/nsf50/vbush1945.htm).
26. V. Ehlers, *Unlocking Our Future: Toward a New National Science Policy*, Committee on Science, U.S. House of Representatives, One Hundred Fifth Congress (Committee Print 105-B, September 1998). www.gpo.gov/fdsys/pkg/GPO-CPRT-105hprt105-b/pdf/GPO-CPRT-105hprt105-b.pdf.
27. V.J. Ehlers, private communication.
28. 121 Stat. 574, Public Law 110-69, August 9, 2007, H.R. 2272, 20 U.S.C. § 9801 note (2007) (www.congress.gov/110/plaws/publ69/PLAW-110publ69.pdf).
29. See www.congress.gov/bill/110th-congress/senate-bill/761/cosponsors.
30. See www.congress.gov/bill/110th-congress/house-bill/2272/cosponsors.
31. The principal agencies were the Department of Energy's Office of Science, NASA, the National Institute of Standards and Technology and the National Science Foundation.

Chapter 11: Recovery and reinvention 2009–2016

1. G. Langer, "Poll: Bush Approval Rating 92 Percent," ABC, October 10, 2001 (abcnews.go.com/Politics/story?id=120971&page=1).
2. See www.realclearpolitics.com/epolls/other/president_bush_job_approval-904.html.
3. In some cases, Administration appointees told us they were under orders not to cooperate with any Obama initiatives.
4. See, for example, www.aaas.org/page/historical-trends-federal-rd.
5. 123 Stat. 115, Public Law 111-5, February 17, 2009, H.R. 1, 26 U.S.C. § 1 note (2009) (www.congress.gov/111/plaws/publ5/PLAW-111publ5.pdf).
6. For a record of congressional actions, see www.congress.gov/bill/111th-congress/house-bill/1/actions.
7. In 2009, according to Office of Management and Budget data, exclusive of the science stimulus boost, total federal R&D spending was $145.6 billion. Of that, defense accounted for $85.3 billion and non-defense, $60.3 billion. The vast majority of defense R&D spending ($67.5 billion) was housed in the development accounts of the Department of Defense (DOD). The research accounts of DOD received $14.0 billion, and the Department of energy's (DOE) nuclear

weapons programs received $3.83 billion. Non-defense R&D spending was split among a many players: Agriculture ($2.44 billion), Education ($312 million), DOE ($6.48 billion), Environmental Protection ($563 million), Homeland Security ($1.10 billion), Interior ($702 million), Justice ($94 million), International Assistance ($152 million) NASA ($8.79 billion), National Institute of Standards and Technology ($553 million), National Oceanographic and Atmospheric Administration ($785 million), National Science Foundation ($4.77 billion), NIH ($29.8 billion), Nuclear Regulatory Commission (($101 million), Smithsonian ($216 million), State ($103 million), Transportation ($925 million), Veterans Affairs ($943 million) and others ($175 million). The total federal R&D spending represented 13.1 percent of the discretionary budget and 1.01 percent of the nation's gross domestic product (GDP). In 2017, those percentages had declined to 9.6 percent and 0.7 percent, respectively.

8. 124 Stat. 3982, Public Law 111-358, January 4, 2011, H.R. 5116, 42 U.S.C. § 11861 note (2011) (www.congress.gov/111/plaws/publ358/PLAW-111publ358.pdf)
9. For a record of congressional actions, see www.congress.gov/bill/111th-congress/house-bill/5116/actions.
10. J.P. Holdren and M. Smith, "Exit Memo: Office of Science and Technology Policy" (The White House, President Barack Obama, Cabinet Exit Memos, Washington, D.C. 2017). obamawhitehouse.archives.gov/administration/cabinet/exit-memos/office-science-and-technology-policy.
11. See "New START: Treaty Text," Diplomacy In Action, U.S. Department of State. www.state.gov/t/avc/newstart/c44126.htm.
12. See, for example, apps.washingtonpost.com/g/documents/world/full-text-of-the-iran-nuclear-deal/1651/.
13. For the complete text of the United Nations Framework Convention on Climate Change (known familiarly as the 2015 Paris Climate Accord), see unfccc.int/sites/default/files/english_paris_agreement.pdf. A concise summary appears at www.nrdc.org/sites/default/files/paris-climate-agreement-IB.pdf.
14. D.K Shipler, "The Summit; Reagan and Gorbachev Sign Missile Treaty and Vow to Work for Greater Reductions," *The New York Times*, December 9, 1987
15. J. Bresolin and B. Gautam, "Fact Sheet: The Nunn-Lugar Cooperative Threat Reduction Program," The Center for Arms Control and Non-Proliferation, Washington D.C., June, 2014 (armscontrolcenter.org/fact-sheet-the-nunn-lugar-cooperative-threat-reduction-program/).
16. P.I. Bernstein and J.D. Wood, "The Origins of Nunn-Lugar and Cooperative Threat Reduction," Case Studies Series ed. by J.A. Larsen and E.R. Mahan, Center for the Study of Weapons of Mass Destruction, National Defense University, Washington, D.C., April 2010 (ndupress.ndu.edu/portals/68/documents/casestudies/cswmd_casestudy-3.pdf).
17. For a record of Senate actions on START I, see www.congress.gov/treaty-document/102nd-congress/20.
18. For a record of Senate actions on START II, see www.congress.gov/treaty-document/103rd-congress/1/.
19. See "Treaty Between The United States of America and The Union of Soviet Socialist Republics on The Limitation of Anti-Ballistic Missile Systems (ABM Treaty), Diplomacy in Action, U.S. Department of State. www.state.gov/t/avc/trty/101888.htm.
20. See "Comprehensive Nuclear Test-Ban Treaty (CTBT)," Diplomacy in Action, U.S. Department of State. www.congress.gov/treaty-document/111th-congress/5.
21. For a record of Senate actions on New START, see www.congress.gov/treaty-document/111th-congress/5.

22. The full text of the Galvin Report can be found at "The Galvin Report," Secretary of Energy Advisory Board Task Force on Alternative Futures for the Department of Energy National Laboratories, R. Galvin, chair, February 1995. scipp.ucsc.edu/~haber/UC_CORP/galvin.htm.

23. E. Klein, "Letter from Los Alamos: The Hunting of Wen Ho Lee," *Vanity Fair*, December 2000, pp. 142 ff. (edwardklein.com/pdfs/december_2000_VF_the_hunting_of_wen_ho_lee.pdf).

24. E. Schmitt, "Spying Furor Brings Vote In Senate For New Unit," *The New York Times*, July 22, 1999 (www.nytimes.com/1999/07/22/world/spying-furor-brings-vote-in-senate-for-new-unit.html).

25. 113 Stat. 512, Public Law 106-65, October 5, 1999, S. 1059, U.S.C. § 3211 *et seq.* (1999) (www.gpo.gov/fdsys/pkg/PLAW-106publ65/pdf/PLAW-106publ65.pdf).

26. "Department of Energy: National Laboratories Need Clearer Missions and Batter Management," United States General Accounting Office Report to the Secretary of Energy (GAO/RECD-95-10), Washington, D.C., January, 1995 (www.gao.gov/assets/160/154864.pdf).

27. It was not the first time GAO had weighed in so strongly. In 1993, it had issued a report with a similar warning in its title, "Management Problems Require a Long-Term Commitment to Change" United States General Accounting Office Report to the Secretary of Energy (GAO/RECD-93-72), Washington, D.C., August, 1993 (www.gao.gov/assets/220/218381.pdf).

28. V. Ehlers, *Unlocking Our Future: Toward a New National Science Policy*, Committee on Science, U.S. House of Representatives, One Hundred Fifth Congress (Committee Print 105-B, September 1998). www.gpo.gov/fdsys/pkg/GPO-CPRT-105hprt105-b/pdf/GPO-CPRT-105hprt105-b.pdf.

29. 100 Stat. 1758, Public Law 99-502, October 20, 1986, H.R. 3773, 15 U.S.C. § 3701 *et seq.* (1986) (www.congress.gov/bill/99th-congress/house-bill/3773/text?q=%7B%22search%22%3A%5B%22Public+Law+99+502%22%5D%7D&r=2).

30. Section 319 of the "Consolidated Appropriations Act of 2014," 128 Stat. 6, Public Law 113-76, January 14, 2014, H.R. 3547, 1 U.S.C. § 1 note (2014) (www.gpo.gov/fdsys/pkg/PLAW-113publ76/pdf/PLAW-113publ76.pdf).

31. T.J. Glothier (co-chair), J.L Cohon (co-chair) et al., "Securing America's Future: Realizing the Potential of the Department of Energy's National Laboratories—Final Report of the Commission to Review the Effectiveness of the National Energy Laboratories," U.S. Department of Energy, October 28, 2015 (www.energy.gov/sites/prod/files/2015/10/f27/Final%20Report%20Volume%201.pdf; www.energy.gov/sites/prod/files/2015/10/f27/Final%20Report%20Volume%202.pdf).

32. There was scant evidence in the files of CIA analysts to back up the Administration's assertion that WMDs existed. (Anonymous private communication.)

33. R. Pielke, Jr., "Ernest Moniz and the Physics of Diplomacy," *The Guardian*, April 8, 2015 (https://www.theguardian.com/science/political-science/2015/apr/08/ernest-moniz-and-the-physics-of-diplomacy).

34. www.state.gov/documents/organization/245317.pdf.

35. www.state.gov/e/eb/tfs/spi/iran/jcpoa/.

36. M. Lander, "Trump Abandons Iran Nuclear Deal He Long Scorned," *The New York Times*, May 8, 2018 (www.nytimes.com/2018/05/08/world/middleeast/trump-iran-nuclear-deal.html).

37. For temperature trends, see "Climate Change: Vital Signs of the Planet, Global Temperature," Goddard Institute for Space Studies, NASA. climate.nasa.gov/vital-signs/global-temperature/.

38. For polar icecap trends, see "Arctic Sea Ice Reaches Another Record Low," Goddard Space Flight Center, NASA, March 7, 2017. https://www.nasa.gov/feature/goddard/2017/sea-ice-extent-sinks-to-record-lows-at-both-poles.

39. For sea level trends, see "Climate Change: Vital Signs of the Planet, Sea Level," Goddard Institute for Space Studies, NASA. climate.nasa.gov/vital-signs/sea-level/.

40. "Climate Change Indicators: Weather and Climate," U.S. Environmental Protection Agency. https://www.epa.gov/climate-indicators/weather-climate.
41. "Trends in Atmospheric Carbon Dioxide: Full Mauna Loa CO_2 Record," Global Greenhouse Gas Reference Network, Earth System Research Laboratory, Global Monitoring Division, National Oceanographic and Atmospheric Administration, U.S. Department of Commerce. www.esrl.noaa.gov/gmd/ccgg/trends/full.html.
42. See J. Warrick, "Reagan, Bush 41 Memos Reveal Sharp Contrast With Today's GOP On Climate and the Environment," *The Washington Post*, December 3, 2015, especially the embedded Bernthal memorandum (www.washingtonpost.com/news/energy-environment/wp/2015/12/03/reagan-bush-41-memos-reveal-how-republicans-used-to-think-about-climate-change-and-the-environment/?utm_term=.f4ce4961c10d).
43. N. Popovich and L. Albeck-Ripka, "How Republicans Think About Climate Change—In Maps," *The New York Times*, December 14, 2017 (www.nytimes.com/interactive/2017/12/14/climate/republicans-global-warming-maps.html).
44. M. Mildenberger, J. R. Marlon, P.D. Howe and A. Leiserowitz, "The Spatial Distribution of Republican and Democratic Climate Opinions at State and Local Scales," Climate Change **145** (3), 539–548 (2017).
45. G. Thrush and C.B. Brown, "Obama's Health Care Conversion," *Politico*, September 22–23, 2013 (www.politico.com/story/2013/09/obama-health-care-conversion-obamacare-097185).
46. See www.congress.gov/bill/111th-congress/house-bill/3590.
47. See https://www.congress.gov/bill/111th-congress/house-bill/2454.
48. For a technical but reasonably accessible book on energy and climate change, see B. Richter, *Beyond Smoke and Mirrors: Climate Change and Energy in the 21st Century* (Cambridge University Press, New York, 2010).
49. Wien's Law is usually stated as $\lambda_{max} = 2.9 \times 10^6/T$, where λ_{max} is the wavelength in nanometers (nm) where the emitted radiation is most intense, and T is the temperature in Kelvin of the body emitting the radiation. The relation applies precisely only to "black bodies," which are perfect emitters or absorbers. The Sun and the Earth are approximately so.
50. B. Richter, *op. cit.*, p. 196.
51. See "Greenhouse Gas Emissions in the United States 2009," U.S. Energy Information Administration, U.S. Department of Energy, Washington, D.C., March 2011 (www.eia.gov/environment/emissions/ghg_report/pdf/0573%282009%29.pdf).
52. B. Richter, *op. cit.*, pp. 197–200.
53. 121 Stat. 1492, Public Law 110-140, December 19, 2007, H.R. 6, 42 U.S.C. § 1401 note (2007) (www.congress.gov/110/plaws/publ140/PLAW-110publ140.pdf).
54. For a discussion of energy efficiency, see B. Richter (co-chair), D. Goldston (co-chair) *et al.*, "Energy Future: Think Efficiency—How America Can Look Within to Achieve Energy Security and Reduce Global Warming," *Rev. Mod. Phys.* **80**, S1 (2008) (www.aps.org/energyefficiencyreport/report/aps-energyreport.pdf) and M.S. Lubell and B. Richter, "Energy Efficiency: Transportation and Buildings," *AIP Conf. Proc.* **1401**, 107 (2011).
55. C. Curtis, "President Obama Announces New Fuel Economy Standards," The White House, President Barack Obama, Blog, July 29, 2011 (obamawhitehouse.archives.gov/blog/2011/07/29/president-obama-announces-new-fuel-economy-standards).
56. "Driving Efficiency: Cutting Costs for Families at the Pump and Slashing Dependence on Oil," Obama Administration Fuel Economy Standards in the Year 2025 (obamawhitehouse.archives.gov/sites/default/files/fuel_economy_report.pdf).
57. Two Independents, Bernie Sanders of Vermont and Angus King of Maine, caucused with the 51 Democrats.

58. "Paris Agreement," United Nations, 2015 (unfccc.int/sites/default/files/english_paris_agreement.pdf).
59. J. Kirkland, "How Moniz Pushed Energy Innovation Onto Paris' Main Stage," *E&E News Energywire*, Thurs., December 17, 2015 (www.eenews.net/stories/1060029626).
60. "FACT SHEET: President's Budget Proposal to Advance Mission Innovation," Briefing Room, The White House, Office of the Press Secretary, February 6, 2016 (obamawhitehouse.archives.gov/the-press-office/2016/02/06/fact-sheet-presidents-budget-proposal-advance-mission-innovation).
61. "Testimony of Secretary Ernest Moniz, U.S. Department of Energy, Before the Subcommittee on Energy and Power, Committee on Energy and Commerce, U.S. House of Representatives, March 2, 2016" (U.S. Department of Energy—www.energy.gov/sites/prod/files/2016/03/f30/3.2.16%20Final%20FY%202017%20Sec%20Moniz%20HEC%20Budget%20Hearing%20Testimony.pdf).
62. J. Taylor, "Obama's Energy Secretary Champions Nuclear Power To Fight Global Warming," *Forbes*, September 20, 2016 (www.forbes.com/sites/jamestaylor/2016/09/20/obamas-energy-secretary-champions-nuclear-power-to-fight-global-warming/#71fce0e47517).
63. "Federal Science Budget Tracker—FYI Science Policy News from AIP, Fiscal Year 2017," American Institute of Physics, College Park, Md. (www.aip.org/fyi/federal-science-budget-tracker/FY2017).

Chapter 12: Loose change

1. T. Friedman, *Thank You for Being Late: An Optimist's Guide to Thriving in the Age of Accelerations* (Farrar, Strauss and Giroux, New York, 2016), especially pp. 28–35.
2. Charles Townes, Arthur Schawlow and Theodore Maiman are generally credited with inventing the laser. Townes, a Columbia University professor of physics, and Schawlow, a Bell Labs staff member, developed the theory, and Maiman, a Hughes Aircraft Company scientist, produced the first working model. Gordon Gould, a Columbia University doctoral student at the time and later a staff scientist at the private research company TRG, fought Maiman over the patent rights but never succeeded in the United States.
3. J. Gertner, *The Idea Factory: Bell Labs and the Great Age of American Innovation* (The Penguin Press, New York, 2012).
4. W. Isaacson, "Inventing the Future: 'The Idea Factory,' by Jon Gartner," *The New York Times*, April 6, 2012 (www.nytimes.com/2012/04/08/books/review/the-idea-factory-by-jon-gertner.html).
5. D.E. Stokes, *Pasteur's Quadrant: Basic Science and Technological Innovation* (Brookings Institution Press, Washington, D.C., 1997).
6. Sixty-seventh Congress, Sess. I, Public Law 67-15, June 10, 1921, H.R. 6567, 42 U.S.C. § 27–28 (1921) (uscode.house.gov/statviewer.htm?volume=42&page=27#).
7. J. Gertner, *op. cit.*, pp. 300-301.
8. For the complete hearing transcript and video recording see www.c-span.org/video/?283308-1/energy-secretary-nomination-hearing.
9. Reuters, "A Third Solar Company Files for Bankruptcy," *The New York Times*, September 6, 2011 (archive.nytimes.com/www.nytimes.com/gwire/2011/09/06/06greenwire-solyndra-bankruptcy-reveals-dark-clouds-in-sol-45598.html).
10. K. Fehrenbacher, "Why the Solyndra Mistake Is Still Important to Remember," *Fortune*, August 27, 2015 (fortune.com/2015/08/27/remember-solyndra-mistake/).
11. G. Avalos, "With $535M Federal Loan, Solyndra Begins Work On Fremont Solar-Panel Plant," *The Mercury News*, September 4, 2009 (www.mercurynews.com/2009/09/04/with-535m-federal-loan-solyndra-begins-work-on-fremont-solar-panel-plant-2/).

12. 50 U.S.C. § 161—"An Act Authorizing the Conservation, Production, and Exploitation of Helium Gas, a Mineral Resource Pertaining to the National Defense, and to the Development of Commercial Aeronautics, and for Other Purposes," Sixty-Eighth Congress, Sess. II, Ch. 426. March 3, 1925, pp. 1110, 1111 (uscode.house.gov/statviewer.htm?volume=43&page=1110#).
13. 74 Stat. 918, Public Law 86-777, September 13, 1960, H.R. 10548, 50 U.S.C. § 161 *et seq.* (1960) (www.gpo.gov/fdsys/pkg/STATUTE-74/pdf/STATUTE-74-Pg918.pdf).
14. 110 Stat. 3315, Public Law 104-273, October 9, 1996, H.R. 4168, 50 U.S.C. 167 Note (1960) (www.gpo.gov/fdsys/pkg/PLAW-104publ273/pdf/PLAW-104publ273.pdf).
15. See R. Jaffe, J. Price, *et al.*, "Energy Critical Elements: Securing Materials for emerging Technologies" (American Physical Society Panel on Public Affairs and the Materials Research Society, Washington, D.C., 2011) (www.aps.org/policy/reports/popa-reports/upload/elementsreport.pdf) and S.R. Bare, M. Lilly *et al.*, "The U.S. Research Community's Liquid Helium Crisis," (American Physical Society, Materials Research Society and American Chemical Society, Washington, D.C., 2016) (www.aps.org/policy/reports/popa-reports/upload/HeliumReport.pdf).
16. 127 Stat. 534, Public Law 113-40, October 2, 2013, H.R. 527, 50 U.S.C 167 Note (2013) (www.gpo.gov/fdsys/pkg/PLAW-113publ40/pdf/PLAW-113publ40.pdf).
17. G. Collins, "Ode to Helium," *The New York Times*, May 3, 2013 (www.nytimes.com/2013/05/04/opinion/collins-an-ode-to-helium.html).
18. "Last Minute Legislation Averts Helium Supply Crisis," *APS News* **22** (10), November 2013, ed. by A. Chodos, (American Physical Society, College Park, Md.).
19. M. Elsesser, private communication.
20. P. Gwynne, "Physicists Seek to Cut Helium Costs," *Physics World*, June 24, 2014.
21. See, for example, "The Rise and Fall of Corporate R&D: Out of the Dusty Labs," Briefing, *The Economist*, March 1, 2007 (www.economist.com/briefing/2007/03/01/out-of-the-dusty-labs).
22. 94 Stat. 96, Public Law 96-517, December 12, 1980, H.R. 6933, 35 U.S.C. § 200 *et seq.* (1980) (www.gpo.gov/fdsys/pkg/STATUTE-94/pdf/STATUTE-94-Pg3015.pdf).
23. See files.taxfoundation.org/legacy/docs/fed_individual_rate_history_nominal.pdf.
24. See www.taxpolicycenter.org/briefing-book/how-are-capital-gains-taxed.
25. 121 Stat. 574, Public Law 110-69, August 9, 2007, H.R. 2272, 20 U.S.C. § 9801 note (2007) (www.congress.gov/110/plaws/publ69/PLAW-110publ69.pdf).
26. See http://www.sciencephilanthropyalliance.org/.
27. "U.S. Research Institutions Received Over $2.3 Billion in Private Funding for Basic Science in 2017," *Alliance News*, Science Philanthropy Alliance, June 7, 2018.
28. "Trends in Basic Research by Agency," *AAAS Report: Research and Development Series* (American Association for the Advancement of Science, Washington, D.C., 2018) (www.aaas.org/sites/default/files/BasicRes;.jpg).
29. "Federal R&D as a Percentage of GDP," *AAAS Report: Research and Development Series* (American Association for the Advancement of Science, Washington, D.C., 2018) (www.aaas.org/sites/default/files/RDGDP;.jpg).
30. "Global Innovation Index 2018: Energizing the World with Innovation, 11th Edition," Ed. by S. Dutta, B. Lanvin and S. Wunsch-Vincent [Cornell University SC Johnson College of Business, INSEAD (The Business School of the World) and WIPO (World Intellectual Property Organization), Ithaca, Fontainebleau and Geneva, 2018] (www.globalinnovationindex.org/Home).
31. See, for example, B.P. Bosworth, "Sources of Real Wage Stagnation," Op-Ed, Brookings, December 22, 2014 (www.brookings.edu/opinions/sources-of-real-wage-stagnation/); J. Shambaugh, R. Nunn, P. Liu and G. Nantz, "Thirteen Facts About Wage Growth," Economic

Facts, The Hamilton Project, Brookings, Washington, D.C., September 2017 (www.hamiltonproject.org/assets/files/thirteen_facts_wage_growth.pdf).

32. T. Piketty, *Capital in the Twenty-First Century*, translated by A. Goldhammer, (Belknap Press of Harvard University Press, Cambridge, Mass., 2014).

33. See, for example, J. Hawksworth and R. Berriman, "Will Robots Really Steal Our Jobs: An International Analysis of the Potential Long Term Impact of Automation," Pricewaterhouse-Coopers, LLC, London, United Kingdom, 2018 (www.pwc.com/hu/hu/kiadvanyok/assets/pdf/impact_of_automation_on_jobs.pdf).

34. T. Friedman, *The World Is Flat* (Farrar, Straus and Giroux, New York, 2005).

35. J. Hawksworth and R. Berriman *op. cit.*, p. 2.

36. T. Friedman, *Thank You for Being Late, op. cit.*, pp. 33.

37. See National Center for Education Statistics (NCES), Program for International Student Assessment (PISA) 2015 Results (nces.ed.gov/surveys/pisa/pisa2015/index.asp)

38. NCES, *op. cit.*, Science Literacy: School Poverty Indicator (nces.ed.gov/surveys/pisa/pisa2015/pisa2015highlights_3e.asp).

39. NCES, *op. cit.*, Mathematics Literacy: School Poverty Indicator (nces.ed.gov/surveys/pisa/pisa2015/pisa2015highlights_5d.asp).

40. C. DeNavas-Walt and B.D. Proctor, "Income and Poverty in the United States: 2014," Current Population Reports, United States Census Bureau, U.S. Department of Commerce, Economic and Statistics Administration, September 2015 (www.census.gov/content/dam/Census/library/publications/2015/demo/p60-252.pdf).

41. OECD Income Distribution Database (IDD): Gini, Poverty, Methods and Concepts: www.oecd.org/social/income-distribution-database.htm.

42. R.D. Putnam, *Our Kids: The American Dream in Crisis* (Simon and Schuster, New York, 2015).

43. J. DeParle, "'Our Kids,' by Robert D. Putnam," *The New York Times*, March 4, 2015 (www.nytimes.com/2015/03/08/books/review/our-kids-by-robert-d-putnam.html).

44. T. Friedman, *op. cit.*

45. For a layman's description of CRISPR, see B. Plumer, E. Barclay, J. Bulluz and U. Irfan, "A Simple Guide to CRISPR, One of the Biggest Science Stories of the Decade," *Vox*, July 23, 2018 (www.vox.com/2018/7/23/17594864/crispr-cas9-gene-editing).

46. "The Precision Medicine Initiative," Obama White House Archives, January 30, 2015 (obamawhitehouse.archives.gov/precision-medicine).

Index

Note: Page numbers followed by *f* indicate figures.